AQA Chemistry
Third edition

Lawrie Ryan
Ray Peacock
Editor: Lawrie Ryan

Message from AQA

This textbook has been approved by AQA for use with our qualification. This means that we have checked that it broadly covers the specification and we are satisfied with the overall quality. Full details of our approval process can be found on our website.

We approve textbooks because we know how important it is for teachers and students to have the right resources to support their teaching and learning. However, the publisher is ultimately responsible for the editorial control and quality of this book.

Please note that when teaching the *AQA GCSE Chemistry* or *AQA GCSE Combined Science: Trilogy* course, you must refer to AQA's specification as your definitive source of information. While this book has been written to match the specification, it cannot provide complete coverage of every aspect of the course.

A wide range of other useful resources can be found on the relevant subject pages of our website: www.aqa.org.uk.

OXFORD
UNIVERSITY PRESS

OXFORD
UNIVERSITY PRESS

Great Clarendon Street, Oxford, OX2 6DP, United Kingdom

Oxford University Press is a department of the University of Oxford. It furthers the University's objective of excellence in research, scholarship, and education by publishing worldwide. Oxford is a registered trade mark of Oxford University Press in the UK and in certain other countries

© Lawrie Ryan 2016

British Library Cataloguing in Publication Data
Data available

978 0 19 835938 8

10 9 8 7 6 5 4 3 2 1

Paper used in the production of this book is a natural, recyclable product made from wood grown in sustainable forests. The manufacturing process conforms to the environmental regulations of the country of origin.

Printed in Great Britain by Bell and Bain Ltd. Glasgow

Acknowledgements

Many thanks to the following people for their help and support in producing this textbook. Each one has added value to my initial efforts: John Scottow, Annie Hamblin, Sadie Garratt, Emma-Leigh Craig, Amie Hewish, Sue Orwin.

AQA examination questions are reproduced by permission of AQA.

Index compiled by INDEXING SPECIALISTS (UK) Ltd., Indexing house, 306A Portland Road, Hove, East Sussex, BN3 5LP United Kingdom.

COVER: PAUL D STEWART/SCIENCE PHOTO LIBRARY
p4: Brostock/Shutterstock; p8: Pixelspieler/Shutterstock; p22(R): Science Photo Library; p22(L): Sheila Terry/Science Photo Library; p23(B): Olga Popova/Shutterstock; p23(T): Science Photo Library; p26: Martyn F. Chillmaid/Science Photo Library; p27(T): Trevor Clifford Photography/Science Photo Library; p27(C): Trevor Clifford Photography/Science Photo Library; p27(B): Trevor Clifford Photography/Science Photo Library; p33: Dirk Wiersma/Science Photo Library; p44: Oleg Znamenskiy/Shutterstock; p45: Joel Arem/Science Photo Library; p46: Michael J Thompson/Shutterstock; p48: Africa Studio/Shutterstock; p51: Ambelrip/Shutterstock; p52: Author Provided; p55(T): John Guillemin/Bloomberg/Getty Images; p55(C): Dan Kosmayer/123RF; p55(B): Jon Summers/123RF; p56: Steve Gschmeissner/Science Photo Library; p58(T): William West/AFP/Getty Images; p58(B): Blend Images/Shutterstock; p59: PaulFleet/iStockphoto; p72: MARTYN F. CHILLMAID/SCIENCE PHOTO LIBRARY; p78: TRL Ltd. /Science Photo Library; p84(T): Tr3gin/Shutterstock; p84(C): IM_photo/Shutterstock; p84(B): Xxlphoto/123RF; p88: Terence Walsh/Shutterstock; p90: MARTYN F. CHILLMAID/SCIENCE PHOTO LIBRARY; p94: Nigel Cattlin/Alamy Stock Photo; p95: Eco Images/Getty Images; p96(T): SCIENCE PHOTO LIBRARY; p96(B): SCIENCE PHOTO LIBRARY; p106: Attl Tibor/Shutterstock; p109(T): Studioshots/Alamy Stock Photo; p109(C): Elena Elisseeva/Shutterstock; p109(B): Anocha Tumsuk/123RF; p112(L): Author Provided; p112(R): Christopher Wood/Shutterstock; p113: C.byatt-norman/Shutterstock; p114: MARTYN F. CHILLMAID/SCIENCE PHOTO LIBRARY; p115: Praisaeng/Shutterstock; p120: Huguette Roe/Shutterstock; p122(R): Martin Bond/Science Photo Library; p122(L): Toby Melville/Reuters/Corbis; p130: Emzet70/Shutterstock; p132(T): Lightpoet/Shutterstock; p132(B): EcoPrint/Shutterstock; p133: MaraZe/Shutterstock; p136(T): Chris Sargent/Shutterstock; p136(B): Adam88x/123RF; p138(T): Andrew Lambert Photography/Science Photo Library; p138(B): Andrew Lambert Photography/Science Photo Library; p139: MARTYN F. CHILLMAID/SCIENCE PHOTO LIBRARY; p140: GiphotoStock/Science Photo Library; p141: SCIENCE PHOTO LIBRARY; p148: Xxlphoto/123RF; p151: Ttstudio/Shutterstock; p152: Yocamon/Shutterstock; p153: TonLammerts/Shutterstock; p154: Jaochainoi/Shutterstock; p158: Andrew Lambert Photography/Science Photo Library; p161: Joerg Beuge/Shutterstock; p162: Piotr Marcinski/Shutterstock; p163: Trevor Clifford Photography/Science Photo Library; p164: MARTYN F. CHILLMAID/SCIENCE PHOTO LIBRARY; p168(BL): Robbi/Shutterstock; p168(BR): Yeko Photo Studio/Shutterstock; p170: David Pereiras/Shutterstock; p171: Coburn77/123RF; p172: Rido/Shutterstock; p174(T): Leah-Anne Thompson/Shutterstock; p174(B): Leonid Andronov/Shutterstock; p180: Anne Gilbert/Alamy Stock Photo; p181: Alexander Raths/Shutterstock; p182: Andrew Lambert Photography/Science Photo Library; p186(T): Martyn F. Chillmaid/Science Photo Library; p186(B): Andrew Lambert Photography/Science Photo Library; p188: Andrew Lambert Photography/Science Photo Library; p189: Andrew Lambert Photography/Science Photo Library; p190(T): Archive Holdings Inc./Getty Images; p190(B): Andrew Brookes, National Physical Laboratory/Science Photo Library; p191: Dept. of Physics, Imperial College/Science Photo Library; p194(T): Rainer Albiez/Shutterstock; p194(B): NASA/Science Photo Library; p195(T): David Seymour/123RF; p195(B): Gunnar Assmy/123RF; p196: Martin Kunzel/123RF; p200: Zacarias Pereira da Mata/Shutterstock; p201: Rhonda Roth/Shutterstock; p202(T): Martin_33/iStockphoto; p202(B): Gemphoto/Shutterstock; p203: Kzenon/Shutterstock; p206: Art Directors & TRIP/Alamy Stock Photo; p208: Worker/Shutterstock; p210: Antikainen/iStockphoto; p211: Sigur/Shutterstock; p212: Alice Nerr/Shutterstock; p214: Ruud Morijn Photographer/Shutterstock; p215(L): Valentyn Volkov/Shutterstock; p215(R): Anaken2012/Shutterstock; p216: Ktd/Shutterstock; p217(T): John Birdsall/REX Shutterstock; p217(B): Kodda/Shutterstock; p218: Anthony Jay D. Villalon/Shutterstock; p220: HTU/Shutterstock; p221: MARTYN F. CHILLMAID/SCIENCE PHOTO LIBRARY; p222(TL): Ng Wei Keong/Shutterstock; p222(TC): Nikitabuida/Shutterstock; p222(TR): Pareto/iStockphoto; p222(B): Pio3/Shutterstock; p223: RainerPlendl/iStockphoto; p224: Anaken2012/Shutterstock; p225: Rtimages/Shutterstock; p227: Chuyu/123RF; p226: Didecs/Shutterstock; p228: Fotokostic/Shutterstock; p230: Peter Bowater/Science Photo Library; p232: Robert Brook/Science Photo Library; p234(T): Martyn F. Chillmaid/Science Photo Library; p234(B): Sputnik/Science Photo Library; p246(TL): David Mccarthy/Science Photo Library; p246(TR): MARTYN F. CHILLMAID/SCIENCE PHOTO LIBRARY; p246(B): Photong/Shutterstock; p247(TL): Efired/Shutterstock; p247(TC): Africa Studio/Shutterstock; p247(TR): Shahril KHMD/Shutterstock; p247(B): Just Keep Drawing/Shutterstock; p248(T): Everett Historical/Shutterstock; p248(B): Andrew Lambert Photography/Science Photo Library; p249(T): Omphoto/Shutterstock; p249(B): Antb/Shutterstock; p250(T): Andrew Fletcher/Shutterstock; p250(B): Andrew Lambert Photography/Science Photo Library; p252: Kagai19927/Shutterstock; p253: Massawfoto/Shutterstock; p254(L): SpeedKingz/Shutterstock; p254(R): Pete Niesen/Shutterstock; p256: Stuart Jenner/Shutterstock; p257: YanLev/Shutterstock; p262: Runi/Shutterstock; p264: Annto/Shutterstock; p265: Kletr/Shutterstock; p266: Alexei Novikov/Shutterstock; p267: Hxdbzxy/Shutterstock; p268: Author Provided;

Section Opener 1: Zebrah/Shutterstock;

Section Opener 2: Albert Russ/Shutterstock;

Section Opener 3: Gary L Jones/Shutterstock;

Section Opener 4: Paul D Stewart/Science Photo Library;

Contents

This book has been written for the *AQA GCSE Chemistry* and *AQA GCSE Combined Science: Trilogy* courses, making them completely co-teachable. Chemistry only lessons are easily identifiable with their own black-bordered design, and are also formatted in italics in the below contents list for quick access.

Practical work is a vital part of chemistry, helping to support and apply your scientific knowledge, and develop your investigative and practical skills. As part of your GCSE Chemistry course, there are 8 required practicals that you must carry out. Questions in your exams could draw on any of the knowledge and skills you have developed in carrying out these practicals.

A Required practical feature box has been included in this student book for each of your required practicals. Further support is available on Kerboodle.

Required practicals		Topic
1	**Prepare a salt from an insoluble metal carbonate or oxide.**	C5.5
	Prepare with the appropriate apparatus and techniques, a pure, dry sample of a soluble salt from an insoluble carbonate or oxide.	C5.6
2	**Use titration to investigate reacting volumes.**	C4.7
	Use titration to find out how much of an acid is needed to completely react with an alkali.	
3	**Investigate the electrolysis of a solution**	C6.4
	Investigate the electrolysis of different aqueous solutions using inert electrodes.	
4	**Investigating temperature changes.**	C7.1
	Use appropriate apparatus to investigate the variables that affect energy changes in reactions involving at least one solution.	
5	**Investigating the effect of concentration on rate of reaction.**	C8.4
	Investigate how changes in concentration affect rates of reactions using a method involving measuring the volume of a gas produced and a method involving a change in colour or turbidity.	
6	**Calculate R_f values.**	C12.2
	Use paper chromatography to find out the R_f values of the dyes found in different food colourings.	
7	**Use chemical tests to identify unknown compounds.**	C12.5
	Use a range of chemical tests to identify negative and positive ions in ionic compounds.	
8	**Purify and test water.**	C14.2
	Analyse and purify water from different sources, including pH, dissolved solids and distillation.	

Learning objectives

- Learning objectives at the start of each spread tell you the content that you will cover.
- Any objectives marked with the higher tier icon **H** are only relevant to those who are sitting the higher tier exams.

This book has been written by subject experts to match the new 2016 specifications. It is packed full of features to help you understand your course and achieve the very best you can.

Key words are highlighted in the text. You can look them up in the glossary at the back of the book if you are not sure what they mean.

The diagrams in this book are as important for your understanding as the text, so make sure you revise them carefully.

Synoptic link

Synoptic links show how the content of a topic links to other parts of the course. This will support you with the synoptic element of your assessment.

There are also links to the Mathematical skills for chemistry chapter, so you can develop your maths skills whilst you study.

Practical

Practicals are a great way for you to see science in action for yourself. These boxes may be a simple introduction or reminder, or they may be the basis for a practical in the classroom. They will help your understanding of the course.

Required practical

These practicals have important skills that you will need to be confident with for part of your assessment. Your teacher will give you additional information about tackling these practicals.

Study tip

Hints giving you advice on things you need to know and remember, and what to watch out for.

Anything in the Higher Tier spreads and boxes must be learnt by those sitting the higher tier exam. If you will be sitting foundation tier, you will not be assessed on this content.

Higher

Go further

Go further feature boxes encourage you to think about science you have learnt in a different context and introduce you to science beyond the specification. You do not need to learn any of the content in the Go further boxes.

Using maths

This feature highlights and explains the key maths skills you need. There are also clear step-by-step worked examples.

Summary questions

Each topic has summary questions. These questions give you the chance to test whether you have learnt and understood everything in the topic. The questions start off easier and get harder, so that you can stretch yourself.

The Literacy pen ✏ shows activities or questions that help you develop literacy skills.

Any questions marked with the higher tier icon **H** are for students sitting the higher tier exams.

Key points

Linking to the Learning objectives, the Key points boxes summarise what you should be able to do at the end of the topic. They can be used to help you with revision.

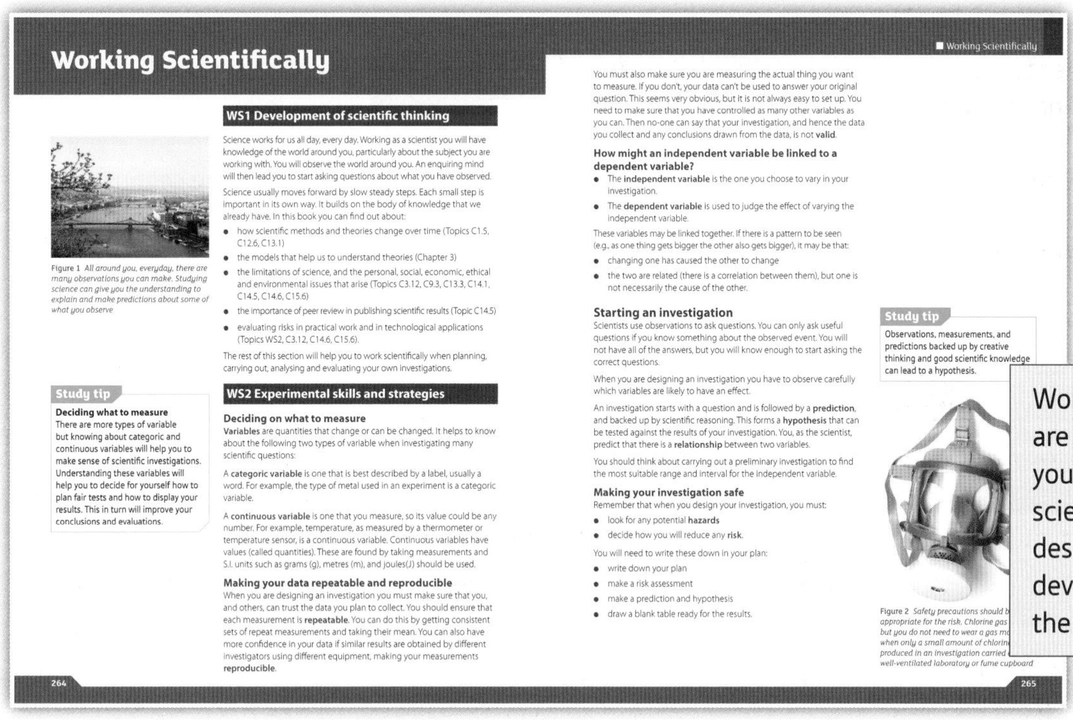

Working Scientifically

WS1 Development of scientific thinking

Science works for us all day, every day. Working as a scientist you will have knowledge of the world around you, particularly about the subject you are working with. You will observe the world around you. An enquiring mind will then lead you to start asking questions about what you have observed.

Science usually moves forward by slow steady steps. Each small step is important in its own way. It builds on the body of knowledge that we already have. In this book you can find out about:

- how scientific methods and theories change over time (Topics C1.5, C12.6, C13.1)
- the models that help us to understand theories (Chapter 3)
- the limitations of science, and the personal, social, economic, ethical and environmental issues that arise (Topics C3.12, C9.3, C13.3, C14.1, C14.5, C14.6, C15.6)
- the importance of peer review in publishing scientific results (Topic C14.5)
- evaluating risks in practical work and in technological applications (Topics WS2, C3.12, C14.6, C15.6).

The rest of this section will help you to work scientifically when planning, carrying out, analysing and evaluating your own investigations.

Figure 1 *All around you, everyday, there are many observations you can make. Studying science can give you the understanding to explain and make predictions about some of what you observe*

Study tip

Deciding what to measure

There are more types of variable but knowing about categoric and continuous variables will help you to make sense of scientific investigations. Understanding these variables will help you to decide for yourself how to plan fair tests and how to display your results. This in turn will improve your conclusions and evaluations.

WS2 Experimental skills and strategies

Deciding on what to measure

Variables are quantities that change or can be changed. It helps to know about the following two types of variable when investigating many scientific questions:

A **categoric variable** is one that is best described by a label, usually a word. For example, the type of metal used in an experiment is a categoric variable.

A **continuous variable** is one that you measure, so its value could be any number. For example, temperature, as measured by a thermometer or temperature sensor, is a continuous variable. Continuous variables have values (called quantities). These are found by taking measurements and S.I. units such as grams (g), metres (m), and joules (J) should be used.

Making your data repeatable and reproducible

When you are designing an investigation you must make sure that you, and others, can trust the data you plan to collect. You should ensure that each measurement is **repeatable**. You can do this by getting consistent sets of repeat measurements and taking their mean. You can also have more confidence in your data if similar results are obtained by different investigators using different equipment, making your measurements **reproducible**.

■ Working Scientifically

You must also make sure you are measuring the actual thing you want to measure. If you don't, your data can't be used to answer your original question. This seems very obvious, but it is not always easy to set up. You need to make sure that you have controlled as many other variables as you can. Then no-one can say that your investigation, and hence the data you collect and any conclusions drawn from the data, is not **valid**.

How might an independent variable be linked to a dependent variable?

- The **independent variable** is the one you choose to vary in your investigation.
- The **dependent variable** is used to judge the effect of varying the independent variable.

These variables may be linked together. If there is a pattern to be seen (e.g., as one thing gets bigger the other also gets bigger), it may be that:

- changing one has caused the other to change
- the two are related (there is a correlation between them), but one is not necessarily the cause of the other.

Starting an investigation

Scientists use observations to ask questions. You can only ask useful questions if you know something about the observed event. You will not have all of the answers, but you will know enough to start asking the correct questions.

When you are designing an investigation you have to observe carefully which variables are likely to have an effect.

An investigation starts with a question and is followed by a **prediction**, and backed up by scientific reasoning. This forms a **hypothesis** that can be tested against the results of your investigation. You, as the scientist, predict that there is a **relationship** between two variables.

You should think about carrying out a preliminary investigation to find the most suitable range and interval for the independent variable.

Making your investigation safe

Remember that when you design your investigation, you must:

- look for any potential **hazards**
- decide how you will reduce any **risk**.

You will need to write these down in your plan:

- write down your plan
- make a risk assessment
- make a prediction and hypothesis
- draw a blank table ready for the results.

Study tip

Observations, measurements, and predictions backed up by creative thinking and good scientific knowledge can lead to a hypothesis.

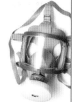

Figure 2 *Safety precautions should be appropriate for the risk. Chlorine gas ... but you do not need to wear a gas ma... when only a small amount of chlorine ... produced in an investigation carried ... well-ventilated laboratory or fume cupboard*

264 265

Working scientifically skills are an important part of your course. The Working scientifically section describes and supports the development of some of the key skills you will need.

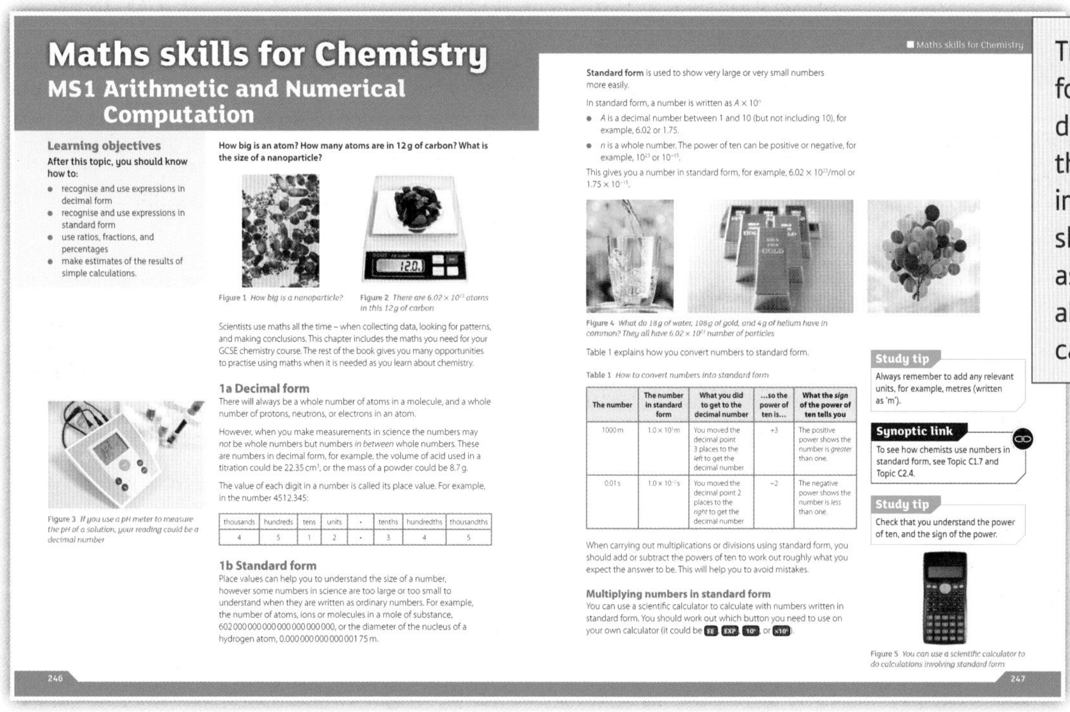

Maths skills for Chemistry

MS1 Arithmetic and Numerical Computation

Learning objectives

After this topic, you should know how to:

- recognise and use expressions in decimal form
- recognise and use expressions in standard form
- use ratios, fractions, and percentages
- make estimates of the results of simple calculations.

How big is an atom? How many atoms are in 12 g of carbon? What is the size of a nanoparticle?

Figure 1 *How big is a nanoparticle?* Figure 2 *There are 6.02 × 10²³ atoms in this 12 g of carbon*

Scientists use maths all the time – when collecting data, looking for patterns, and making conclusions. This chapter includes the maths you need for your GCSE chemistry course. The rest of the book gives you many opportunities to practise using maths when it is needed as you learn about chemistry.

1a Decimal form

There will always be a whole number of atoms in a molecule, and a whole number of protons, neutrons, or electrons in an atom.

However, when you make measurements in science the numbers may *not* be whole numbers but numbers *in between* whole numbers. These are numbers in decimal form, for example, the volume of acid used in a titration could be 22.35 cm³, or the mass of a powder could be 8.7 g.

The value of each digit in a number is called its place value. For example, in the number 4512.345:

thousands	hundreds	tens	units	·	tenths	hundredths	thousandths
4	5	1	2	·	3	4	5

Figure 3 *If you use a pH meter to measure the pH of a solution, your reading could be a decimal number*

1b Standard form

Place values can help you to understand the size of a number, however some numbers in science are too large or too small to understand when they are written as ordinary numbers. For example, the number of atoms, ions or molecules in a mole of substance, 602 000 000 000 000 000 000 000, or the diameter of the nucleus of a hydrogen atom, 0.000 000 000 000 001 75 m.

■ Maths skills for Chemistry

Standard form is used to show very large or very small numbers more easily.

In standard form, a number is written as $A \times 10^n$

- *A* is a decimal number between 1 and 10 (but not including 10), for example, 6.02 or 1.75.
- *n* is a whole number. The power of ten can be positive or negative, for example, 10^{23} or 10^{-15}.

This gives you a number in standard form, for example, 6.02×10^{23}/mol or 1.75×10^{-15}.

Figure 4 *What do 18 g of water, 108 g of gold, and 4 g of helium have in common? They all have 6.02 × 10²³ number of particles*

Table 1 explains how you convert numbers to standard form.

Table 1 *How to convert numbers into standard form*

The number	The number in standard form	What you did to get to the decimal number	...so the power of ten is...	What the sign of the power of ten tells you
1000 m	1.0×10^3 m	You moved the decimal point 3 places to the left to get the decimal number	+3	The positive power shows the number is *greater* than one.
0.01 s	1.0×10^{-2} s	You moved the decimal point 2 places to the right to get the decimal number	−2	The negative power shows the number is *less* than one.

When carrying out multiplications or divisions using standard form, you should add or subtract the powers of ten to work out roughly what you expect the answer to be. This will help you to avoid mistakes.

Multiplying numbers in standard form

You can use a scientific calculator to calculate with numbers written in standard form. You should work out which button you need to use on your own calculator (it could be EE, EXP, 10ˣ, or ×10ˣ).

Study tip

Always remember to add any relevant units, for example, metres (written as 'm').

Synoptic link

To see how chemists use numbers in standard form, see Topic C1.7 and Topic C2.4.

Study tip

Check that you understand the power of ten, and the sign of the power.

Figure 5 *You can use a scientific calculator to do calculations involving standard form*

246 247

The Mathematical skills for chemistry chapter describes and supports the development of the important mathematical skills you will need for all aspects of your course. It also has questions so you can test your skills.

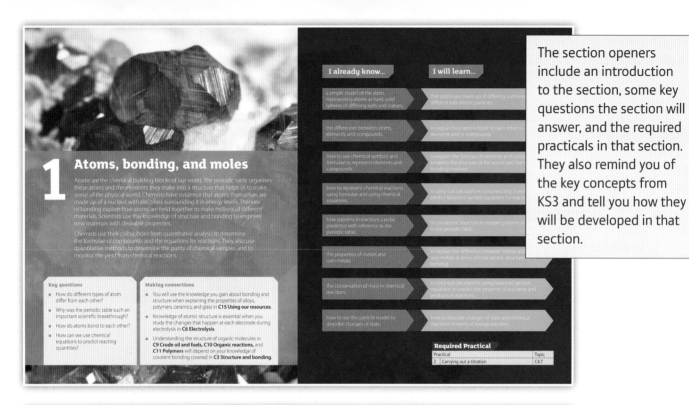

The section openers include an introduction to the section, some key questions the section will answer, and the required practicals in that section. They also remind you of the key concepts from KS3 and tell you how they will be developed in that section.

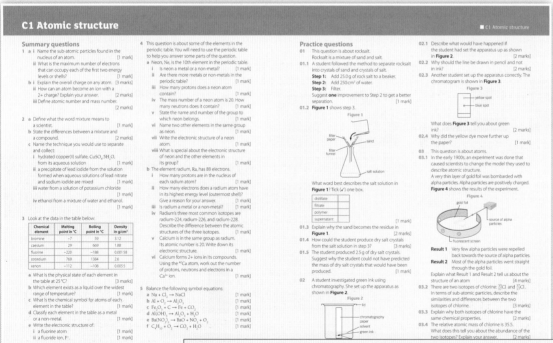

At the end of every chapter there are summary questions and practice questions. The questions test your literacy, maths, and working scientifically skills, as well as your knowledge of the concepts in that chapter. The practice questions can also call on your knowledge from any of the previous chapters to help support the synoptic element of your assessment.

There are also further practice questions at the end of the book to cover all of the content from your course.

This book is also supported by Kerboodle, offering digital support for building your practical, maths, and literacy skills.

If your school subscribes to Kerboodle, you will find a wealth of additional resources to help you with your studies and revision:

- animations, videos, and revision podcasts
- webquests
- maths and literacy skills activities and worksheets
- on your marks activities to help you achieve your best
- practicals and follow-up activities
- interactive quizzes that give question-by-question feedback
- self-assessment checklists

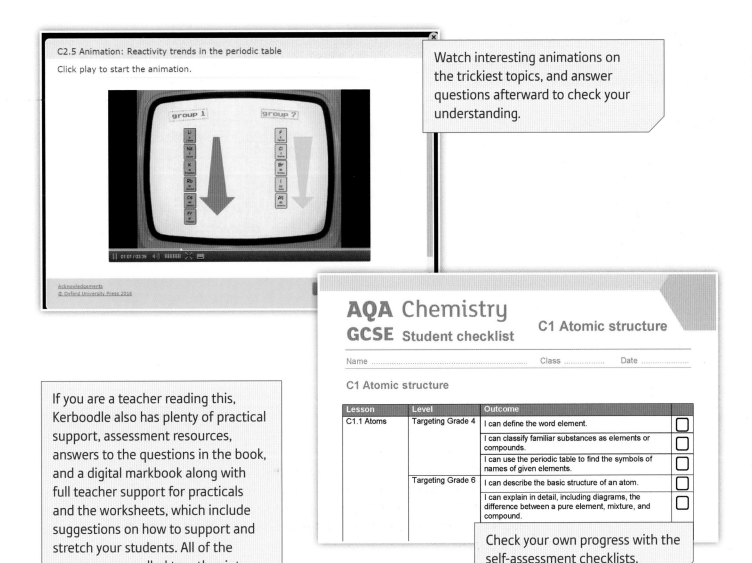

C2.5 Animation: Reactivity trends in the periodic table

Click play to start the animation.

group 1   group 7

01:01 / 03:39

Acknowledgements
© Oxford University Press 2016

Watch interesting animations on the trickiest topics, and answer questions afterward to check your understanding.

AQA Chemistry
GCSE Student checklist   **C1 Atomic structure**

Name ....   Class ....   Date ....

C1 Atomic structure

Lesson	Level	Outcome	
C1.1 Atoms	Targeting Grade 4	I can define the word element.	☐
		I can classify familiar substances as elements or compounds.	☐
		I can use the periodic table to find the symbols of names of given elements.	☐
	Targeting Grade 6	I can describe the basic structure of an atom.	☐
		I can explain in detail, including diagrams, the difference between a pure element, mixture, and compound.	☐

If you are a teacher reading this, Kerboodle also has plenty of practical support, assessment resources, answers to the questions in the book, and a digital markbook along with full teacher support for practicals and the worksheets, which include suggestions on how to support and stretch your students. All of the resources are pulled together into ready-to-use lesson presentations.

Check your own progress with the self-assessment checklists.

1 Atoms, bonding, and moles

Atoms are the chemical building blocks of our world. The periodic table organises these atoms and the elements they make into a structure that helps us to make sense of the physical world. Chemists have evidence that atoms themselves are made up of a nucleus with electrons surrounding it in energy levels. Theories of bonding explain how atoms are held together to make millions of different materials. Scientists use this knowledge of structure and bonding to engineer new materials with desirable properties.

Chemists use their calculations from quantitative analysis to determine the formulae of compounds and the equations for reactions. They also use quantitative methods to determine the purity of chemical samples and to monitor the yield from chemical reactions.

Key questions

- How do different types of atom differ from each other?

- Why was the periodic table such an important scientific breakthrough?

- How do atoms bond to each other?

- How can we use chemical equations to predict reacting quantities?

Making connections

- You will use the knowledge you gain about bonding and structure when explaining the properties of alloys, polymers, ceramics, and glass in **C15 Using our resources**.

- Knowledge of atomic structure is essential when you study the changes that happen at each electrode during electrolysis in **C6 Electrolysis**.

- Understanding the structure of organic molecules in **C9 Crude oil and fuels, C10 Organic reactions,** and **C11 Polymers** will depend on your knowledge of covalent bonding covered in **C3 Structure and bonding**.

I already know...

I will learn...

I already know...	I will learn...
a simple model of the atom, representing atoms as hard, solid spheres of differing sizes and masses.	that atoms are made up of differing numbers of three different sub-atomic particles.
the differences between atoms, elements, and compounds.	to explain how atoms bond to each other in elements and in compounds.
how to use chemical symbols and formulae to represent elements and compounds.	to explain the formula of elements and compounds, knowing the structure of the atoms and the type of bonding involved.
how to represent chemical reactions using formulae and using chemical equations.	to carry out calculations using reacting masses to predict balanced symbol equations for reactions.
how patterns in reactions can be predicted with reference to the periodic table.	to use atomic structure to explain patterns in reactivity in the periodic table.
the properties of metals and non-metals.	to explain the difference between metals and non-metals in terms of their atomic structures and bonding.
the conservation of mass in chemical reactions.	to carry out calculations using balanced symbol equations to predict the amounts of reactants and products in reactions.
how to use the particle model to describe changes of state.	how to describe changes of state and chemical reactions in terms of energy transfers.

Required Practical

Practical		Topic
2	Carrying out a titration	C4.7

Learning objectives

After this topic, you should know:

- the definition of an element
- that each type of atom has a chemical symbol
- the basic structure of the periodic table
- the basic structure of an atom.

Look at the things around you and the substances that they are made from. You will find wood, metal, plastic, glass – the list is almost endless. There are millions of different substances catalogued by scientists.

All substances are made of tiny particles called **atoms**. There are about 100 different types of atom found naturally on Earth. These can combine in a huge variety of ways, giving all those different substances.

A relatively small number of substances are made up of only one type of atom. These substances are called **elements**. An atom is the smallest part of an element that can exist. As there are only about 100 different types of atom, it follows that there are only about 100 different elements.

Elements can have very different properties. Elements such as silver, chromium, copper, and gold are shiny, solid metals (Figure 1). Other elements such as oxygen, nitrogen, argon, and chlorine are non-metals, and are gases at room temperature.

Figure 1 *An element contains only one type of atom – in this case gold*

Chemical symbols

The name used for an element depends on the language being spoken. For example, the element sulfur is called *schwefel* in German and *azufre* in Spanish. However, the world of science forms a global community, and scientists from many nations communicate with each other and publish their findings. So it is important that there are symbols for elements that all nationalities can understand. These symbols are shown in the **periodic table** (Figure 2).

Group numbers

																	0
1	2			H 1 Hydrogen								3	4	5	6	7	He 2
Li 3	Be 4											B 5	C 6	N 7	O 8	F 9	Ne 10
Na 11	Mg 12											Al 13	Si 14	P 15	S 16	Cl 17	Ar 18
K 19	Ca 20	Sc 21	Ti 22	V 23	Cr 24	Mn 25	Fe 26	Co 27	Ni 28	Cu 29	Zn 30	Ga 31	Ge 32	As 33	Se 34	Br 35	Kr 36
Rb 37	Sr 38	Y 39	Zr 40	Nb 41	Mo 42	Tc 43	Ru 44	Rh 45	Pd 46	Ag 47	Cd 48	In 49	Sn 50	Sb 51	Te 52	I 53	Xe 54
Cs 55	Ba 56	Lanthanum see below	Hf 72	Ta 73	W 74	Re 75	Os 76	Ir 77	Pt 78	Au 79	Hg 80	Tl 81	Pb 82	Bi 83	Po 84	At 85	Rn 86
Fr 87	Ra 88	Actinium see below															

The alkali metals

The alkaline earth metals

The transition metals

The halogens

The noble gases

La 57	Ce 58	Pr 59	Nd 60	Pm 61	Sm 62	Eu 63	Gd 64	Tb 65	Dy 66	Ho 67	Er 68	Tm 69	Yb 70	Lu 71
Ac 89	Th 90	Pa 91	U 92	Np 93	Pu 94	Am 95	Cm 96	Bk 97	Cf 98	Es 99	Fm 100	Md 101	No 102	Lr 103

Lanthanides

Actinides

Figure 2 *The periodic table shows the symbols for each type of atom*

- The symbols in the periodic table represent atoms. For example, O represents an atom of oxygen and Na represents an atom of sodium.
- The elements in the table are arranged in columns, called **groups**. Each group contains elements with similar chemical properties.
- The 'staircase' drawn on the right of the periodic table in bold black is the dividing line between metals and non-metals. The elements to the left of the line are the metals. Those to the right of the line are the non-metals. However, a few elements lying next to the dividing line are called metalloids or semi-metals, as they have some metallic and some non-metallic properties. Examples include silicon, Si, and germanium, Ge, from Group 4.

Atoms, elements, and compounds

The vast majority of substances you come across are not elements. They are made up of different types of atom bonded together and are called **compounds**. Look at the diagram of a water molecule in Figure 3. A sample of pure water will always have twice as many hydrogen atoms as oxygen atoms. So its chemical formula is written as H_2O. If there is no subscript after an atom's symbol in a chemical formula, it is read as '1', that is, the ratio of H atoms : O atoms is 2 : 1.

Chemical bonds hold the atoms tightly together in compounds. Some compounds are made from just two types of atom (e.g., water or carbon dioxide, CO_2). However, most compounds consist of more than two different types of atom.

All atoms are made up of a tiny central **nucleus** with **electrons** orbiting around it (Figure 4).

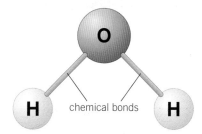

Figure 3 *A grouping of two or more atoms bonded together is called a **molecule**. Chemical bonds hold the hydrogen and oxygen atoms together in this water molecule. Water is an example of a compound*

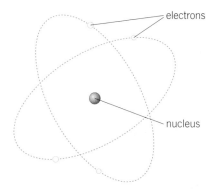

Figure 4 *Each atom consists of a tiny nucleus surrounded by electrons*

Synoptic link

For more information on the periodic table of elements, see Topic C2.1.

1. **a** Arrange these elements into a table showing metals and non-metals: phosphorus, P, barium, Ba, vanadium, V, mercury, Hg, krypton, Kr, potassium, K, and uranium, U. [2 marks]
   b Would you classify hydrogen as a metal or a non-metal? Explain why. [1 mark]
2. Explain why when you mix two elements together you can often separate them quite easily by physical means, yet when two elements are chemically combined in a compound, they are usually difficult to separate. [2 marks]
3. Draw diagrams to explain the difference between an element and a compound. [2 marks]
4. Describe the basic structure of an atom. [2 marks]
5. Find out the Latin words from which the symbols of the following metallic elements are derived:
   a sodium, Na [1 mark]
   b gold, Au [1 mark]
   c lead, Pb [1 mark]
   d potassium, K. [1 mark]
6. Explain what information can be deduced from the chemical formula of carbon dioxide, CO_2. [2 marks]

Key points

- All substances are made up of atoms.
- The periodic table lists all the chemical elements, with eight main groups each containing elements with similar chemical properties.
- Elements contain only one type of atom.
- Compounds contain more than one type of atom.
- An atom has a tiny nucleus at its centre, surrounded by electrons.

C1.2 Chemical equations

Learning objectives

After this topic, you should know:

- what happens to the atoms in a chemical reaction
- how the mass of reactants compares with the mass of products in a chemical reaction
- why there can be an apparent loss or gain in mass during reactions involving gases in open containers
- how to write balanced symbol equations, including state symbols, to represent reactions.

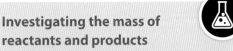

Investigating the mass of reactants and products

You are given solutions of lead nitrate and potassium iodide.

Add a small volume of each solution together in a test tube.

- **What do you see happen?**

The formula of lead nitrate is $Pb(NO_3)_2$ and potassium iodide is KI.

The **precipitate** (a solid formed in the reaction between two solutions) is lead iodide, PbI_2.

- Write a word equation and a balanced symbol equation, including state symbols, for the reaction.
- How do you think that the mass of reactants compares with the mass of the products?

Now plan an experiment to test your answer to this question. Your teacher must check your plan before you start the practical work.

Safety: Wear chemical splash-proof eye protection. Lead nitrate and lead iodide are toxic. Wash your hands after the experiment.

Chemical equations show the **reactants** (the substances you start with) and the **products** (the new substances made) in a reaction.

You can represent the test for hydrogen gas using a **word equation**:

$$\text{hydrogen} + \text{oxygen} \rightarrow \text{water}$$
$$\text{(reactants)} \qquad \text{(product)}$$

In chemical reactions the atoms get rearranged. You can investigate what happens to the mass of reactants compared with mass of products in a reaction in the practical box.

Using **symbol equations** helps you to see how much of each substance is involved in a reaction.

For example, calcium carbonate decomposes (breaks down) on heating. You can show the reaction using a symbol equation like this:

$$CaCO_3 \rightarrow CaO + CO_2$$

This equation is **balanced** – there is the same number of each type of atom on both sides of the equation. This is very important, because atoms cannot be created or destroyed in a chemical reaction. This also means that:

The total mass of the products formed in a reaction is equal to the total mass of the reactants.

This is called the **Law of conservation of mass**. In reactions involving gases, this law can appear to be broken when the reactions are carried out in open containers, such as test tubes and conical flasks. For example, if you weigh a sample of calcium carbonate before you heat it, then weigh it again after heating, it will appear to lose mass. This is because the carbon dioxide gas formed in the reaction has escaped into the air. Similarly, a piece of copper increases in mass when heated in air. However, the apparent extra mass comes from the oxygen gas that the copper reacts with to make copper oxide.

You can check if an equation is balanced by counting the number of each type of atom on either side of the equation. If the numbers are equal, then the equation is balanced.

Adding state symbols

You can also add **state symbols** to a balanced symbol equation to give extra information. The state symbols used are (s) for solids, (l) for liquids, (g) for gases, and (aq) for substances dissolved in water, called **aqueous solutions**.

So the balanced symbol equation, including state symbols, for the decomposition of calcium carbonate is:

$$CaCO_3(s) \rightarrow CaO(s) + CO_2(g)$$

Making an equation balance

In the case of hydrogen reacting with oxygen, it is not so easy to balance the equation. First of all, you write the formula of each reactant and product:

$$H_2 + O_2 \rightarrow H_2O$$

Counting the atoms on either side of the equation you see that there are:

Reactants	Products
2 H atoms, 2 O atoms	2 H atoms, 1 O atom

So you need another oxygen atom on the product side of the equation. You cannot simply change the formula of H_2O to H_2O_2. (H_2O_2 – hydrogen peroxide – is a bleaching agent, which is certainly not suitable to drink.) But you can have two water molecules in the reaction – this is shown in a symbol equation as:

$$H_2 + O_2 \rightarrow \mathbf{2}H_2O$$

Counting the atoms on either side of the equation again, you get:

Reactants	Products
2 H atoms, 2 O atoms	4 H atoms, 2 O atoms

Although the oxygen atoms are balanced, you now need two more hydrogen atoms on the reactant side. You do this by putting '2' in front of H_2:

$$2H_2 + O_2 \rightarrow 2H_2O$$

Now you have:

Reactants	Products
4 H atoms, 2 O atoms	4 H atoms, 2 O atoms

The equation is balanced (Figure 1).

Study tip

When balancing a chemical equation, you can NEVER change a chemical formula.

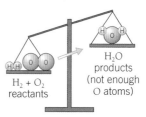

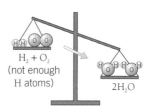

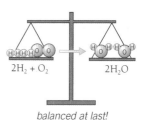

Figure 1 *Balancing an equation*

1 a Explain why all symbol equations must be balanced. [2 marks]
  b Balance the equation: $H_2 + Cl_2 \rightarrow HCl$ [1 mark]
2 a A mass of 33.6 g of magnesium carbonate, $MgCO_3$, completely decomposed when it was heated. It made 16.0 g of magnesium oxide, MgO. Calculate the mass of carbon dioxide, CO_2, produced in this reaction. [1 mark]
  b Write a word equation and a balanced symbol equation, including state symbols, to show the reaction in part **a**. [1 mark]
3 Balance these symbol equations:
  a $KNO_3 \rightarrow KNO_2 + O_2$ [1 mark]
  b $Li + O_2 \rightarrow Li_2O$ [1 mark]
  c $Fe + O_2 \rightarrow Fe_2O_3$ [1 mark]
  d $Fe_2O_3 + CO \rightarrow Fe + CO_2$ [1 mark]
4 Sodium metal, Na, reacts with water to form a solution of sodium hydroxide, NaOH, and gives off hydrogen gas, H_2. Write a balanced symbol equation, including state symbols, for this reaction. [2 marks]

Key points

- No new atoms are ever created or destroyed in a chemical reaction: the total mass of reactants = the total mass of products.
- There is the same number of each type of atom on each side of a balanced symbol equation.
- You can include state symbols to give extra information in balanced symbol equations. These are (s) for solids, (l) for liquids, (g) for gases, and (aq) for aqueous solutions.

C1.3 Separating mixtures

Learning objectives

After this topic, you should know:

- what a mixture is
- how to separate the components in a range of mixtures by:
  - filtration
  - crystallisation
  - simple distillation.

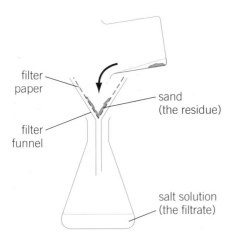

Figure 1 *Filtering a mixture of sand, salt, and water in the lab*

Figure 2 *Filtering in the home – some people like filter coffee made from ground-up coffee beans. The solid bits that are insoluble in water get left on the filter paper as a residue*

When analytical chemists working on forensic or medical investigations are given an unknown sample to identify, it is often a mixture of different substances.

A mixture is made up of two or more substances (elements or compounds) that are not chemically combined together.

Mixtures are different to chemical compounds. Look at Table 1.

Table 1 *The differences between compounds and mixtures*

Compounds	Mixtures
Compounds have a fixed composition (the ratio of elements present is always the same in any particular compound).	Mixtures have no fixed composition (the proportions vary depending on the amount of each substance mixed together).
Chemical reactions must be used to separate the elements in a compound.	The different elements or compounds in a mixture can be separated again more easily (by physical means using the differences in properties of each substance in the mixture).
There are chemical bonds between atoms of the different elements in the compound.	There are no chemical bonds between atoms of the different substances in a mixture.

Before the substances in a mixture are identified, they are separated from each other. As it states in Table 1, you can use physical means to achieve the separation. The techniques available include:

- filtration
- crystallisation
- distillation
- chromatography (see Topic C1.4).

These separation techniques all rely on differences in the physical properties of the substances in the mixture, such as different solubilities in a solvent or different boiling points.

Filtration

The technique of filtration is used to separate substances that are insoluble in a particular solvent from those that are soluble in the solvent. For example, you have probably tried to separate a mixture of sand, salt (sodium chloride, and water) before in science lessons (Figure 1).

The sand that you collect on the filter paper can then be washed with distilled water to remove any salt solution left on it. The wet sand is finally dried in a warm oven to evaporate any water off and leave the pure, dry sand.

Crystallisation

To obtain a sample of pure salt (sodium chloride, NaCl) from the salt solution following filtration, you would need to separate the sodium chloride in the solution (called the filtrate) from the water. You can do this by evaporating the water from the sodium chloride solution.

The best way to do this is by heating it in an evaporating dish on a water bath (Figure 3). Using a water bath is a gentler way of heating than heating the evaporating dish directly on a tripod and gauze.

Heating should be stopped when the solution is at the point of crystallisation. This is when small crystals first appear around the edge of the solution or when crystals appear in a drop of solution extracted from the dish with a glass rod. The rest of the water is then left to evaporate off the saturated solution at room temperature to get a good sample of sodium chloride crystals. A flat-bottomed crystallisation dish or Petri dish can be used for this final step, to give a large surface area for the water to evaporate from.

Distillation

Crystallisation separates a soluble solid from a solvent but sometimes you need to collect the solvent itself instead of just letting it evaporate off into the air. For example, some countries with a lack of fresh water sources purify seawater to obtain usable water. Distillation allows us to do this.

In simple distillation, a solution is heated and boiled to evaporate the solvent. The vapour given off then enters a condenser. This is an outer glass tube with water flowing through it that acts as a cooling 'jacket' around the inner glass tube from the flask. Here the hot vapour is cooled and condensed back into a liquid for collection in a receiving vessel (Figure 4). Any dissolved solids will remain in the heated flask.

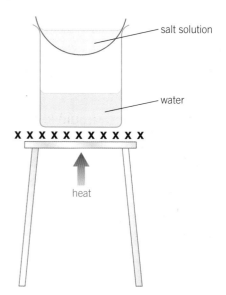

Figure 3 *Crystallising sodium chloride from its solution in water*

Synoptic link

To find out more about crystallisation, see Topic C5.5.

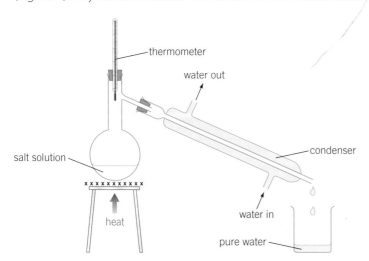

Figure 4 *Distilling pure water from salt solution*

1 Define what a mixture is. [2 marks]

2 'A mixture has no fixed composition, whereas a compound has.' Explain what this means, using hydrogen, oxygen, and water to illustrate your answer. [3 marks]

3 Explain how the process of distillation can be used to remove dissolved impurities from a sample of water. [4 marks]

4 Sulfur is soluble in the flammable liquid xylene but not in water. Sodium nitrate is soluble in water but not in xylene. Describe and explain TWO ways to separate a mixture of sulfur powder and sodium nitrate to collect pure samples of each solid. ✔ [6 marks]

Key points

- A mixture is made up of two or more substances that are not chemically combined together.
- Mixtures can be separated by physical means, such as filtration, crystallisation, and simple distillation. (The physical techniques of separation, fractional distillation, and chromatography, are discussed in Topic C1.4).

C1.4 Fractional distillation and paper chromatography

Learning objectives

After this topic, you should know:

- why fractional distillation is needed to separate some liquids
- how fractional distillation works
- how paper chromatography works.

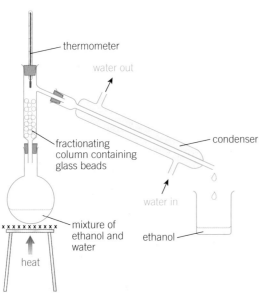

Figure 1 *The fractional distillation of the miscible liquids, ethanol, and water*

Synoptic link

One application of fractional distillation is in oil refineries – see Topic C9.2.

Synoptic link

You can find out how you can use chromatography to identify substances in Topic C12.2.

Fractional distillation

You saw how simple distillation works in Topic C1.3. Distillation can also be used to separate mixtures of miscible liquids, such as ethanol and water. The word miscible describes liquids that dissolve in each other, mixing completely. They do not form the separate layers seen in mixtures of immiscible liquids that have been allowed to settle, such as the oil and water layers in a salad dressing.

The miscible liquids will have different boiling points, so you can use this to distil off and collect the liquid with the lowest boiling point first.

However, it is difficult to get pure liquids from mixtures of liquids with similar boiling points by simple distillation, as vapour is given off from each liquid before they actually reach their boiling point. So to aid separation you can add a fractionating column to the apparatus for distillation (Figure 4 in Topic C1.3). This is usually a tall glass column filled with glass beads, fitted vertically on top of the flask being heated (Figure 1).

The vapours must pass over and between the glass beads in the fractionating column before they reach the condenser. The temperature in the fractionating column is highest at the bottom of the column, getting lower as the vapours rise up. The substance with the higher boiling point will condense more readily on the cooler glass beads nearer the bottom of the column and drip back down into the flask beneath. The substance with the lower boiling point will continue rising and pass over into the condenser, where it is cool enough to turn back into the liquid state and be collected.

In Figure 1, the boiling point of ethanol is 78 °C and that of water is 100 °C. So if the temperature reading on the thermometer can be kept at around 80 °C, the liquid collected will be mainly ethanol. You can test the difference between the starting mixture of ethanol and water and the distillate collected by applying a lighted splint to a small volume of each in an evaporating dish. Ethanol is a flammable liquid but is not flammable when mixed with an excess of water. The distillate will ignite when a flame is applied, as the ethanol collected should only have a small amount of water present. It burns with a clear blue flame – you might have seen the flame when the brandy on a Christmas pudding is lit.

Fractional distillation is used to separate ethanol from a fermented mixture in the alcoholic spirits industry and in the use of ethanol as a **biofuel**.

Carrying out paper chromatography

One technique that is used to separate (and identify) substances from mixtures in solution is paper **chromatography**. It works because some compounds in a mixture will dissolve better than others in the solvent chosen (Figure 3).

A capillary tube is used to dab a spot of the solution on a pencil line near the bottom of a sheet of absorbent chromatography paper. The paper is then placed standing in a solvent at the bottom of a beaker or tank. The solvent is allowed to soak up the paper, running through the spot of mixture. The relative solubility of the components making up the mixture in the solvent determines how far they travel up the paper. The more soluble a substance is in the solvent, the further up the paper it is carried. Different solvents can be used to maximise separation.

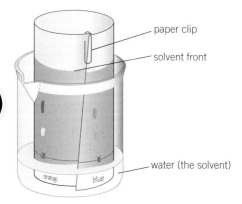

Figure 3 *A chromatogram is the paper record of the separation. If the components of the mixture are not coloured, you can sometimes spray the chromatogram with a detecting agent that colours them*

Detecting dyes in food colourings

In this experiment you can make a chromatogram to analyse various food colourings.

Set up the experiment as shown in Figure 2 and Figure 3.

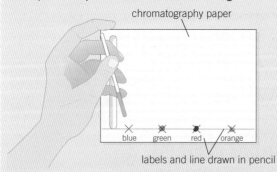

chromatography paper

blue   green   red   orange

labels and line drawn in pencil

Figure 2 *Setting up a chromatogram*

● What can you deduce from your chromatography experiment?

1 a Draw and label the apparatus you could use to separate a mixture of ethanol and water. [2 marks]
  b What is this method of separation called? [1 mark]

2 Explain why you would be able to collect a more concentrated sample of ethanol from a mixture of water and ethanol using the apparatus drawn in question **1** than by using simple distillation. [4 marks]

3 a Describe a method to separate the dyes in coloured inks. [4 marks]
  b A paper chromatogram from a mixture of two substances, A and B, was obtained using a solvent of propanone. Substance B was found to travel further up the paper than substance A. What does this tell you about substances A and B. [1 mark]

4 Look at the boiling points of the three liquids in the table:

Liquid	Boiling point in °C
water	100
ethanol	78
propanol	97

A mixture was made by stirring together equal volumes of these three miscible liquids.
Evaluate the effectiveness of fractional distillation as a way of separating this mixture into the three pure liquids. [3 marks]

Go further

There are other ways to analyse mixtures by chromatography. If you study A Level Chemistry, you will meet thin-layer chromatography (TLC) – where the solvent runs up a solid pasted onto a microscope slide. You will also find out about gas chromatography (GC) – where the substances in the mixture are vaporised and carried by a gas through a long coiled tube packed with a solid.

Key points

● Fractional distillation is an effective way of separating miscible liquids, using a fractionating column. The separation is possible because of the different boiling points of the liquids in the mixture.
● Paper chromatography separates mixtures of substances dissolved in a solvent as they move up a piece of chromatography paper. The different substances are separated because of their different solubilities in the solvent used.

C1.5 History of the atom

Learning objectives

After this topic, you should know:

- how and why the atomic model has changed over time
- that scientific theories are revised or replaced by new ones in the light of new evidence.

Early ideas about atoms

The ancient Greeks were the first to have ideas about particles and atoms. However, it was not until the early 1800s that these ideas became linked to strong experimental evidence when John Dalton put forward his ideas about atoms. From his experiments, he suggested that substances were made up of atoms that were like tiny, hard spheres. He also suggested that each chemical element had its own atoms that differed from others in their mass. Dalton believed that these atoms could not be divided or split. They were the fundamental building blocks of nature.

In chemical reactions, he suggested that the atoms re-arranged themselves and combined with other atoms in new ways. In many ways, Dalton's ideas are still useful today. For example, they help to visualise elements, compounds, and molecules, as well as the models still used to describe the different arrangement and movement of particles in solids, liquids, and gases.

Evidence for electrons in atoms

At the end of the 1800s, a scientist called J.J. Thomson discovered the **electron**. This is a tiny, negatively charged particle that was found to have a mass about 2000 times smaller than the lightest atom. Thomson was experimenting by applying high voltages to gases at low pressure (Figure 1).

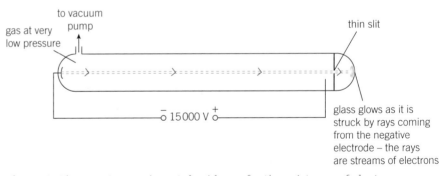

Figure 1 *Thomson's experimental evidence for the existence of electrons*

Thomson did experiments on the beams of particles. They were attracted to a positive charge, showing they must be negatively charged themselves. He called the tiny, negatively charged particles electrons. These electrons must have come from inside atoms in the tube. So Dalton's idea that atoms could not be divided or split had to be revised. Thomson proposed a different model for the atom. He said that the tiny negatively charged electrons must be embedded in a cloud of positive charge. He knew that atoms themselves carry no overall charge, so any charges in an atom must balance out. He imagined the electrons as the bits of plum in a plum pudding (Figure 2).

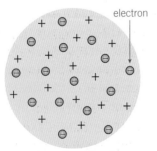

Figure 2 *Thomson's 'plum pudding' model of the atom*

Evidence for the nucleus

The next breakthrough in understanding the atom came about 10 years later. Geiger and Marsden were doing an experiment with radioactive particles. They were firing dense, positively charged particles (called alpha particles) at the thinnest piece of gold foil they could make (Figure 3). They expected the particles to pass straight through the gold atoms with their diffuse cloud of positive charge (as in Thompson's plum-pudding model). However, their results shocked them (Figure 3).

Their results were used to suggest a new model for the atom (Figure 4). Rutherford suggested that Thomson's atomic model was not possible. The positive charge must be concentrated at a tiny spot in the centre of the atom. Otherwise the large, positive particles fired at the foil could never be repelled back towards their source. It was proposed that the electrons must be orbiting around this **nucleus** (centre of the atom), which contains very dense positively charged **protons** (Figure 4).

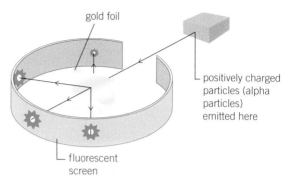

Figure 3 *The alpha particle scattering experiment, carried out by Geiger and Marsden, that changed the 'plum pudding' theory*

Evidence for electrons in shells (energy levels)

The next important development came in 1914, when Niels Bohr revised the atomic model again. He noticed that the light given out when atoms were heated only had specific amounts of energy. He suggested that the electrons must be orbiting the nucleus at set distances, in certain fixed energy levels (or shells). The energy must be given out when excited electrons fall from a high to a low energy level. Bohr matched his model to the energy values observed (Figure 5).

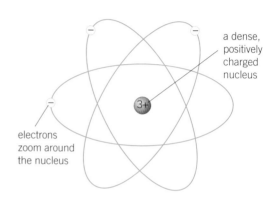

Figure 4 *Rutherford's nuclear model of the atom*

Evidence for neutrons in the nucleus

Scientists at the time speculated that there were two types of sub-atomic particles inside the nucleus. They had evidence of protons but a second sub-atomic particle in the nucleus was also proposed to explain the missing mass that had been noticed in atoms. These **neutrons** must have no charge and have the same mass as a proton.

Because neutrons have no charge, it was very difficult to detect them in experiments. It was not until 1932 that James Chadwick did an experiment that could only be explained by the existence of neutrons.

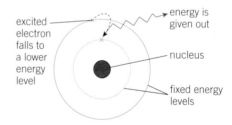

Figure 5 *Bohr's model of the atom*

1 a Which one of Dalton's ideas listed below about atoms do scientists no longer believe?
       A Elements contain only one type of atom.
       B Atoms get re-arranged in chemical reactions.
       C Atoms are solid spheres that cannot be split into simpler particles. [1 mark]
   b Which two of the following substances from Dalton's list of elements are not actually chemical elements?
                soda   oxygen   carbon   gold   lime   [2 marks]

2 a Which sub-atomic particle did J.J. Thomson discover?   [1 mark]
   b Describe J.J. Thomson's 'plum pudding' model of the atom. [2 marks]

3 State two ways in which Rutherford changed Thomson's model of the atom. [2 marks]

4 Explain why Bohr revised Rutherford's model of the atom. [2 marks]

Key points

- The ideas about atoms have changed over time.
- New evidence has been gathered from the experiments of scientists who have used their model of the atom to explain their observations and calculations.
- Key ideas were proposed successively by Dalton, Thomson, Rutherford, and Bohr, before arriving at the model of the atom you use at GCSE level today.

C1.6 Structure of the atom

After this topic, you should know:

- the location, relative charge, and relative mass of the protons, neutrons, and electrons in an atom
- what the atomic number and mass number of an atom represent
- why atoms have no overall charge
- that atoms of a particular element have the same number of protons.

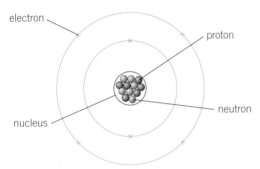

Figure 1 *Understanding the structure of an atom gives important clues to the way chemicals react together. This atom of carbon has six protons and six neutrons in its nucleus, and six electrons orbiting the nucleus. Although the number of protons and electrons are always equal in an atom, the number of neutrons can differ (but in this carbon atom they are the same)*

Synoptic link

You can find out how to calculate the relative atomic mass of an element in Topic C4.1.

You have now seen how ideas about atoms have developed over time. At GCSE level, you use the model in which a very small nucleus is in the centre of every atom. This nucleus contains two types of sub-atomic particle, called **protons** and **neutrons**. A third type of sub-atomic particle orbits the nucleus. These really tiny particles are called **electrons**.

Protons have a positive charge. Neutrons have no charge, that is, they are neutral. So the nucleus itself has an overall positive charge. The electrons orbiting the nucleus are negatively charged. The relative charge on a proton is +1 and the relative charge on an electron is −1.

The mass of an atom is concentrated in its nucleus. A proton and a neutron each have the same mass. The electrons are so light that their mass can be ignored when working out the relative mass of atoms (Table 1).

Table 1 *The relative charge and mass of sub-atomic particles*

Type of sub-atomic particle	Relative charge	Relative mass
proton	+1	1
neutron	0	1
electron	−1	very small (it would take almost 2000 electrons to have the same mass as one proton or neutron)

Because every atom contains equal numbers of protons and electrons, the positive and negative charges cancel out. So there is no overall charge on any atom. For example, a carbon atom is neutral. It has six protons, so you know it must have six electrons (Figure 1).

Atomic number

All the atoms of a particular element have the same number of protons. For example, hydrogen atoms all have one proton in their nucleus, carbon atoms have six protons in their nucleus and sodium atoms have 11 protons in their nucleus.

The number of protons in each atom of an element is called its **atomic number**.

The elements in the periodic table are arranged in order of their atomic number (number of protons). If you are told that the atomic number of an element is eight, you can identify it using the periodic table. It will be the eighth element listed – oxygen. Knowing the atomic number of an element, you also know its number of electrons (as this will equal its number of protons). So oxygen atoms have eight protons and eight electrons.

Mass number

As you know, the protons and neutrons in an atom's nucleus make up the vast majority of the mass of the atom.

The number of protons plus neutrons in the nucleus of an atom is called its **mass number**. A beryllium atom, Be has four protons and five neutrons, so its mass number will be $4 + 5 = 9$.

Given the atomic number and mass number, you can work out how many protons, electrons, and neutrons are in an atom. For example, an argon atom has an atomic number of 18 and a mass number of 40.

● Its atomic number is 18, so it has 18 protons. Remember that atoms have an equal number of protons and electrons. So argon also has 18 electrons.

● Argon's mass number is 40, so you know that:
18 (the number of protons) + the number of neutrons = 40

● Therefore, argon must have 22 neutrons (as 18 + 22 = 40).

You can summarise the last part of the calculation as:

number of neutrons = mass number – atomic number

Worked example

Lead has an atomic number of 82 and a mass number of 207.

How many protons, neutrons, and electrons does it contain?

Solution

atomic number = number of protons p = number of electrons e
$= 82$

mass number = number of protons p + number of neutrons n
$= 207$

So substituting in the value of p, you get $82 + n = 207$

You can rearrange the equation so that its subject (i.e., the quantity you want to find out) is n, by subtracting 82 from both sides of the equation:

$n = 207 - 82$

$= 125$

So the lead atom has 82 protons, 125 neutrons, and 82 electrons.

1 Draw a table showing the location, relative charge, and relative mass of the three sub-atomic particles. [3 marks]

2 An atom has 27 protons and 32 neutrons. Use the periodic table in Topic C1.1 to name this element and give its symbol, atomic number, and mass number. [2 marks]

3 Explain why all atoms are neutral. [2 marks]

4 How many protons, electrons, and neutrons do the following atoms contain?
   a A nitrogen atom, with atomic number 7 and mass number 14. [1 mark]
   b A chlorine atom, with atomic number 17 and mass number 35. [1 mark]
   c A silver atom, with atomic number 47 and mass number 108. [1 mark]
   d A uranium atom, with atomic number 92 and mass number 235. [1 mark]

Study tip

In an atom, the number of protons is always equal to the number of electrons. You can find out the number of protons and electrons in an atom by looking up its atomic number in the periodic table.

Key points

● Atoms are made of protons, neutrons, and electrons.
● Protons have a relative charge of +1, and electrons have a relative charge of –1. Neutrons have no electric charge. They are neutral.
● The relative masses of a proton and a neutron are both 1.
● Atoms contain an equal number of protons and electrons, so carry no overall charge.
● Atomic number = number of protons (= number of electrons).
● Mass number = number of protons + neutrons
● Atoms of the same element have the same number of protons (and hence electrons) in their atoms.

C1.7 Ions, atoms, and isotopes

Learning objectives

After this topic, you should know:

- how to work out the number of protons, neutrons, and electrons in an ion
- how to represent an atom's atomic number and mass number
- how to estimate the size and scale of atoms, using SI units and the prefix 'nano'
- the definition of isotopes.

Synoptic link

For more information about the formation of ions from atoms, see Topic C3.2.

Standard form and nanometres

When dealing with very large or very small numbers, scientists use 'powers of ten' to express a number. For example, a distance of one million metres (1 000 000 m) is written as 1×10^6 m. One millionth of a metre is written as 1×10^{-6} m (see Maths skills MS1b).

Chemists use units called nanometres (nm) to quote distances on an atomic level, where 1 nm is 1×10^{-9} m. The radius of an atom is about 0.1 nm (i.e., 1×10^{-10} m). Compare this to the approximate radius of its nucleus, which is about 1×10^{-14} m. This shows what a small space is occupied by the nucleus of an atom, as its radius is less than $\frac{1}{10\,000}$ the radius of a single atom. Therefore the vast majority of an atom is space, occupied by an atom's electrons.

What is an ion?

You have seen that atoms are neutral because they have an equal number of protons (each carrying a positive charge) and electrons (carrying a negative charge). However, sometimes atoms can lose or gain electrons, for example, when metals react with non-metals. If an atom gains one or more electrons, it gains an overall negative charge because it has more electrons than protons. You say that the atom has become a negative **ion**. If it loses one or more electrons, it becomes a positive ion because it has more protons than electrons.

An ion is a charged atom (or group of atoms).

Think about an oxygen atom (atomic number 8 and mass number 16). What happens when it gains two electrons? The atomic number of oxygen is 8, so the original oxygen atom has eight protons (8+) and eight electrons (8–). When it gains two electrons, there are still eight protons (8+) but now it has 10 electrons (10–). So the overall charge on the negative ion formed is 2–. You write the formula of the ion as O^{2-}.

The O^{2-} ion has eight protons, eight neutrons, and 10 electrons.

If a lithium atom (atomic number 3 and mass number 7) loses one electron, it has 2 electrons (2–) and 3 protons (3+). Therefore it forms a positive ion with a single positive charge, Li^+.

This Li^+ ion will have three protons, four neutrons, and two electrons.

Representing the atomic number and mass number

You can show the atomic number and mass number of an atom like this:

$$\text{mass number} \quad {}^{12}_{6}\text{C (carbon)} \quad {}^{23}_{11}\text{Na (sodium)}$$
$$\text{atomic number}$$

Given this information, you can work out the numbers of protons, neutrons, and electrons in an atom. The bottom number is its atomic number, giving you the number of protons (which equals the number of electrons). Then you can calculate the number of neutrons by subtracting its atomic number from its mass number (see Topic C1.6).

Sodium, ${}^{23}_{11}$Na has an atomic number of 11 and its mass number is 23.

So a sodium atom has 11 protons and 11 electrons, as well as $(23 - 11) = 12$ neutrons.

The size of atoms

It has been estimated that a person has about 7 billion, billion, billion atoms in their body. That huge number is written as 7 followed by 27 zeros:

7 000 000 000 000 000 000 000 000 000

You cannot see the atoms because each individual atom is incredibly small. An atom is about a tenth of a billionth of a metre across.

Isotopes

Atoms of the same element always have the same number of protons. However, they can have different numbers of neutrons.

Atoms of the same element with different numbers of neutrons are called **isotopes**.

Isotopes always have the same atomic number but different mass numbers. For example, two isotopes of carbon are $^{12}_{6}C$ (also written as carbon-12) and $^{13}_{6}C$ (carbon-13). The carbon-12 isotope has six protons and six neutrons in the nucleus. The carbon-13 isotope has six protons and seven neutrons, that is, one more neutron than carbon-12.

Sometimes extra neutrons make the nucleus unstable, so it is radioactive. However, not all isotopes are radioactive – they are simply atoms of the same element that have different masses.

Samples of different isotopes of an element have different *physical* properties. For example, they will have a different density and they may or may not be radioactive. However, they always have the same *chemical* properties, because their reactions depend on their electronic structures. As their atoms will have the same number of protons, and therefore electrons, the electronic structure will be same for all isotopes of an element.

For example, look at the three isotopes of hydrogen in Figure 2. The three isotopes are called hydrogen (hydrogen-1), deuterium (or hydrogen-2), and tritium (or hydrogen-3). Each has a different mass and tritium is radioactive. However, all have identical chemical properties, for example, they all react with oxygen to make water:

$$2H_2(g) + O_2(g) \rightarrow 2H_2O(l)$$

1 State how many protons, neutrons, and electrons there are in each of the following atoms or ions:

   a $^{11}_{5}B$ [1 mark]   b $^{14}_{7}N$ [1 mark]   c $^{24}_{12}Mg$ [1 mark]

   d $^{37}_{17}Cl$ [1 mark]   e $^{127}_{53}I$ [1 mark]   f $^{19}_{9}F^-$ [1 mark]

   g $^{31}_{15}P^{3-}$ [1 mark] h $^{39}_{19}K^+$ [1 mark]   i $^{27}_{13}Al^{3+}$ [1 mark]

2 a Define the word isotopes.   [1 mark]
   b Look at Figure 1. Which isotope of carbon is shown?   [1 mark]

3 The atomic radius of a boron atom is 9×10^{-11} m.
   a Give its atomic radius in nanometres.   [1 mark]
   b Calculate the approximate radius of its nucleus (in nm), given
     that it will be about one ten thousandth the radius of the
     boron atom. Give your answer in standard form.   [1 mark]

4 a Which physical property will always differ in pure samples
     of each isotope of the same element?   [1 mark]
   b Explain why the isotopes of the same element have
     identical chemical properties.   [2 marks]

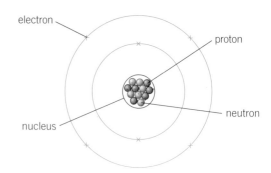

Figure 1 *An atom of carbon*

$^{1}_{1}H$ hydrogen

$^{2}_{1}H$ deuterium

$^{3}_{1}H$ tritium

Figure 2 *The isotopes of hydrogen – they have identical chemical properties but different physical properties, such as density*

Key points

- Atoms that gain electrons form negative ions. If atoms lose electrons they form positive ions.
- You can represent the atomic number and mass number of an atom using the notation: $^{24}_{12}Mg$, where magnesium's atomic number is 12 and its mass number is 24.
- Isotopes are atoms of the same element with different numbers of neutrons. They have identical chemical properties, but their physical properties, such as density, can differ.

C1.8 Electronic structures

Learning objectives

After this topic, you should know:

- how the electrons are arranged in an atom
- the electronic structures of the first 20 elements in the periodic table
- how to represent electronic structures in diagrams and by using numbers.

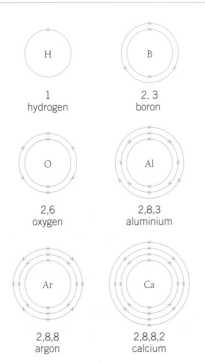

Figure 2 *Once you know the pattern, you should be able to draw the energy levels (shells) and electrons in the atoms of any of the first 20 elements, when given their atomic number*

The model of the atom that you use at GCSE level has electrons arranged around the nucleus in **shells**, rather like the layers of an onion. Each shell represents a different energy level.

The lowest energy level is shown by the shell that is nearest to the nucleus. The electrons in an atom occupy the lowest available energy level (the available shell closest to the nucleus).

Electron shell diagrams

An energy level (or shell) can only hold a certain number of electrons.

- The first, and lowest, energy level (nearest the nucleus) can hold up to two electrons.
- The second energy level can hold up to eight electrons.
- Once there are eight electrons in the third energy level, the fourth begins to fill up.

Beyond the first 20 elements in the periodic table the situation gets more complex. You only need to know the full arrangement of electrons in atoms of the first 20 elements.

You can draw diagrams to show the arrangement of electrons in an atom. For example, a sodium atom has an atomic number of 11 so it has 11 protons, which means it also has 11 electrons. Figure 1 shows how you can represent an atom of sodium.

Figure 1 *A simple way of representing the arrangement of electrons in the energy levels (shells) of a sodium atom*

To save drawing atoms all the time, you can write down the numbers of electrons in each energy level. This is called an **electronic structure**. For example, the sodium atom in Figure 1 has an electronic structure of 2,8,1. You start at the lowest energy level (innermost or first shell), recording the numbers in each successive energy level or shell. The numbers of electrons in each shell are separated from each other by a comma.

Silicon, whose atoms have 14 electrons, is in Group 4 of the periodic table. It has the electronic structure 2,8,4. This represents two electrons in the lowest energy level (first shell), then eight in the next energy level, and four in its highest energy level (its outermost shell).

The best way to understand these arrangements is to look at the examples of standard notation in Figure 2.

Electronic structure and the periodic table

Look at the elements in any one of the main groups of the periodic table. Their atoms will all have the same number of electrons in their highest energy level. These electrons are often called the outer electrons, because they are in the outermost shell. Therefore, all the elements in Group 1 have one electron in their highest energy level.

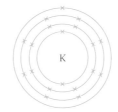

The chemical properties of an element depend on how many electrons it has. The way an element reacts is determined by the number of electrons in its highest energy level (or outermost shell). So because the elements in a particular group all have the same number of electrons in their highest energy level, they all react in a similar way.

For example, when the Group 1 elements are added to water:

lithium + water → lithium hydroxide + hydrogen
sodium + water → sodium hydroxide + hydrogen
potassium + water → potassium hydroxide + hydrogen

The elements in Group 0 of the periodic table are called the **noble gases**. They are very unreactive elements. Their atoms all have a very stable arrangement of electrons with eight electrons in the outer shell, except for helium which has two electrons in the outer shell.

1   a Which shell represents the lowest energy level in an atom? [1 mark]
    b How many electrons can each of the lowest two energy
       levels hold?   [1 mark]

2 Using the periodic table, draw the arrangement of electrons in the
   following atoms and label each one with its electronic structure.
    a He   [1 mark]   b Be   [1 mark]
    c Cl   [1 mark]   d Ar   [1 mark]
3   a Write the electronic structure of potassium (atomic number 19).
                                                                [1 mark]
    b How many electrons does a potassium atom have in its
       highest energy level (outermost shell)?   [1 mark]
4 Give the name and symbol of the atom shown in Figure 3.   [1 mark]
5   a Why do the Group 1 metals all react in a similar way with
       oxygen?   [1 mark]
    b Write word equations for the reactions of lithium, sodium,
       and potassium with oxygen to form their oxides.   [3 marks]
    c The Group 1 metals also react with chlorine gas, Cl_2, to form
       chlorides, such as lithium chloride, LiCl, and sodium chloride, NaCl.
       Write a balanced symbol equation for the reactions of lithium, Li,
       and sodium, Na, with chlorine gas (see Topic C1.2).   [2 marks]

Synoptic link

For more information on the reactions of elements and their electronic structures, see Chapter C2.

Go further

If you study A Level Chemistry you will find out that after the first shell, each shell is actually made up of sub-shells. These sub-shells are labelled as s, p, d, or f sub-shells.

Figure 3 *See question 4*

Key points

- The electrons in an atom are arranged in energy levels or shells.
- The lowest energy level (1st shell) can hold up to 2 electrons and the next energy level (2nd shell) can hold up 8 electrons.
- The 4th shell starts to fill after 8 electrons occupy the 3rd shell.
- The number of electrons in the outermost shell of an element's atoms determines the way in which that element reacts.

C1 Atomic structure

Summary questions

1 a i Name the sub-atomic particles found in the nucleus of an atom. [1 mark]
       ii What is the maximum number of electrons that can occupy each of the first two energy levels or shells? [1 mark]
    b i Explain the overall charge on any atom. [3 marks]
       ii How can an atom become an ion with a 2+ charge? Explain your answer. [2 marks]
       iii Define atomic number and mass number. [2 marks]

2 a Define what the word mixture means to a scientist. [1 mark]
    b State the differences between a mixture and a compound. [2 marks]
    c Name the technique you would use to separate and collect:
       i hydrated copper(II) sulfate, $CuSO_4.5H_2O$, from its aqueous solution [1 mark]
       ii a precipitate of lead iodide from the solution formed when aqueous solutions of lead nitrate and sodium iodide are mixed [1 mark]
       iii water from a solution of potassium chloride [1 mark]
       iv ethanol from a mixture of water and ethanol. [1 mark]

3 Look at the data in the table below:

Chemical element	Melting point in °C	Boiling point in °C	Density in g/cm³
bromine	−7	59	3.12
caesium	29	669	1.88
fluorine	−220	−188	0.001 58
strontium	769	1384	2.6
xenon	−112	−108	0.005 5

    a What is the physical state of each element in the table at 25 °C? [3 marks]
    b Which element exists as a liquid over the widest range of temperature? [1 mark]
    c What is the chemical symbol for atoms of each element in the table? [1 mark]
    d Classify each element in the table as a metal or a non-metal. [1 mark]
    e Write the electronic structure of:
       i a fluorine atom [1 mark]
       ii a fluoride ion, F⁻. [1 mark]

4 This question is about some of the elements in the periodic table. You will need to use the periodic table to help you answer some parts of the question.
    a Neon, Ne, is the 10th element in the periodic table.
       i Is neon a metal or a non-metal? [1 mark]
       ii Are there more metals or non-metals in the periodic table? [1 mark]
       iii How many protons does a neon atom contain? [1 mark]
       iv The mass number of a neon atom is 20. How many neutrons does it contain? [1 mark]
       v State the name and number of the group to which neon belongs. [1 mark]
       vi Name two other elements in the same group as neon. [1 mark]
       vii Write the electronic structure of a neon atom. [1 mark]
       viii What is special about the electronic structure of neon and the other elements in its group? [1 mark]
    b The element radium, Ra, has 88 electrons.
       i How many protons are in the nucleus of each radium atom? [1 mark]
       ii How many electrons does a radium atom have in its highest energy level (outermost shell)? Give a reason for your answer. [1 mark]
       iii Is radium a metal or a non-metal? [1 mark]
       iv Radium's three most common isotopes are radium-224, radium-226, and radium-228. Describe the difference between the atomic structures of the three isotopes. [1 mark]
       v Calcium is in the same group as radium. Its atomic number is 20. Write down its electronic structure. [1 mark]
       vi Calcium forms 2+ ions in its compounds. Using the ⁴⁰Ca atom, work out the number of protons, neutrons and electrons in a Ca^{2+} ion. [1 mark]

5 Balance the following symbol equations:
    a $Na + Cl_2 \rightarrow NaCl$ [1 mark]
    b $Al + O_2 \rightarrow Al_2O_3$ [1 mark]
    c $Fe_2O_3 + C \rightarrow Fe + CO_2$ [1 mark]
    d $Al(OH)_3 \rightarrow Al_2O_3 + H_2O$ [1 mark]
    e $Ba(NO_3)_2 \rightarrow BaO + NO_2 + O_2$ [1 mark]
    f $C_4H_{10} + O_2 \rightarrow CO_2 + H_2O$ [1 mark]

Practice questions

01 This question is about rocksalt.
Rocksalt is a mixture of sand and salt.

01.1 A student followed the method to separate rocksalt into crystals of sand and crystals of salt.
Step 1: Add 25.0 g of rock salt to a beaker.
Step 2: Add 250 cm³ of water.
Step 3: Filter.
Suggest **one** improvement to Step 2 to get a better separation. [1 mark]

01.2 **Figure 1** shows step 3.

Figure 1

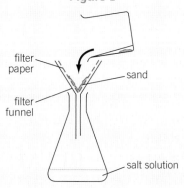

filter paper
sand
filter funnel
salt solution

What word best describes the salt solution in **Figure 1**? Tick (✓) one box.

distillate	
filtrate	
polymer	
supernatant	

[1 mark]

01.3 Explain why the sand becomes the residue in **Figure 1**. [2 marks]

01.4 How could the student produce dry salt crystals from the salt solution in step 3? [3 marks]

01.5 The student produced 2.5 g of dry salt crystals. Suggest why the student could not have predicted the mass of dry salt crystals that would have been produced. [1 mark]

02 A student investigated green ink using chromatography. She set up the apparatus as shown in **Figure 2**.

Figure 2

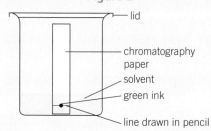

lid
chromatography paper
solvent
green ink
line drawn in pencil

02.1 Describe what would have happened if the student had set the apparatus up as shown in **Figure 2**. [2 marks]

02.2 Why should the line be drawn in pencil and not in ink? [2 marks]

02.3 Another student set up the apparatus correctly. The chromatogram is shown in **Figure 3**.

Figure 3

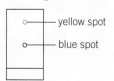

yellow spot
blue spot

What does **Figure 3** tell you about green ink? [2 marks]

02.4 Why did the yellow dye move further up the paper? [1 mark]

03 This question is about atoms.

03.1 In the early 1900s, an experiment was done that caused scientists to change the model they used to describe atomic structure.
A very thin layer of gold foil was bombarded with alpha particles. Alpha particles are positively charged. **Figure 4** shows the results of the experiment.

Figure 4

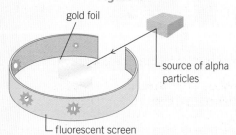

gold foil
source of alpha particles
fluorescent screen

Result 1 Very few alpha particles were repelled back towards the source of alpha particles.

Result 2 Most of the alpha particles went straight through the gold foil.

Explain what Result 1 and Result 2 tell us about the structure of an atom [4 marks]

03.2 There are two isotopes of chlorine: $^{35}_{17}Cl$ and $^{37}_{17}Cl$. In terms of sub-atomic particles, describe the similarities and differences between the two isotopes of chlorine. [3 marks]

03.3 Explain why both isotopes of chlorine have the same chemical properties. [2 marks]

03.4 The relative atomic mass of chlorine is 35.5. What does this tell you about the abundance of the two isotopes? Explain your answer. [2 marks]

2.1 Development of the periodic table

Learning objectives

After this topic, you should know:

- how the periodic table was developed over time
- how testing a prediction can support or refute a new scientific idea.

Imagine trying to understand the chemical elements:

- without knowing much about atoms
- with some chemical compounds mistakenly thought to be elements
- without knowing a complete list of the elements.

This is the task that faced scientists at the start of the 1800s.

During the 19th century, chemists were finding new elements almost every year. They were also trying very hard to find patterns in the behaviour of the elements. This would allow them to organise the elements and understand more about chemistry.

One of the first suggestions came from John Dalton. He arranged the elements in order of their atomic weights, which had been measured in various chemical reactions. In 1808 he published a table of elements in his book *A New System of Chemical Philosophy*.

Figure 1 *Looking for patterns in the chemical elements in the early part of the 19th century was a bit like solving a crossword puzzle. Some answers were clear, since scientists did have some correctly identified elements. However, they only had a vague idea about some, as some compounds were wrongly thought to be elements. They did not even know the clues for other answers, as there were still undiscovered elements*

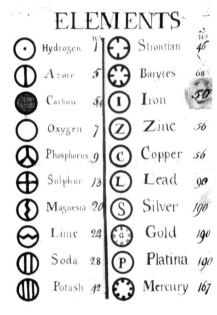

Figure 2 *Dalton and his table of elements*

In 1864, John Newlands built on Dalton's ideas. Newlands also arranged the known elements in order of mass but noticed that the properties of every *eighth* element seemed similar.

He produced a table showing his 'law of octaves' (Figure 3). However, he assumed that all the elements had been found. He did not take into account that chemists were still discovering new ones. So he filled in his octaves, even though some of his elements were not similar at all. His table only really worked for the known elements up to calcium, before the pattern broke down.

Other scientists ridiculed his ideas and refused to accept them.

H 1	F 8	Cl 15	Co and Ni 22	Br 29	Pd 36	I 42	Pt and Ir 50
Li 2	Na 9	K 16	Cu 23	Rb 30	Ag 37	Cs 44	Os 51
Be 3	Mg 10	Ca 17	Zn 24	Sr 31	Cd 38	Ba and V 45	Hg 52
B 4	Al 11	Cr 19	Y 25	Ce and La 33	U 40	Ta 46	Tl 53
C 5	Si 12	Ti 18	In 26	Zr 32	Sn 39	W 47	Pb 54
N 6	P 13	Mn 20	As 27	Bi and Mo 34	Sb 41	Nb 48	Bi 55
O 7	S 14	Fe 21	Se 28	Rh and Ru 35	Te 43	Au 49	Th 56

Figure 3 *Newlands and his table of octaves. Looking at Newlands' octaves, a fellow chemist commented that putting the elements in alphabetical order would probably produce just as many groups of elements with similar properties*

Mendeleev's breakthrough

In 1869, the Russian chemist Dmitri Mendeleev cracked the problem. At this time around 50 elements had been identified. Mendeleev arranged all of these in a table. He placed them in the order of their atomic weights. Then he arranged them so that a periodic (regularly occuring) pattern in their properties could be seen.

He left gaps for elements that had not yet been discovered. Then he used his table to predict what their properties should be. A few years later, new elements were discovered with properties that closely matched Mendeleev's predictions. Then there were not many doubts left that his table was a breakthrough in scientific understanding.

However, not all elements fit in with Mendeleev's pattern. One example is argon. Argon atoms have a greater average relative mass than potassium atoms. Ordering by atomic weights would result in argon (a noble gas) being in the same group as reactive metals such as sodium and lithium, and would group potassium (an extremely reactive metal) with the unreactive noble gases. So argon's position in the periodic table must be before potassium to maintain the periodic pattern, even though the average mass of its atoms is heavier than that of potassium's atoms. When Mendeleev was working the noble gases had not been discovered, but he met the same problem with other elements. Mendeleev simply changed their order where necessary to keep elements with similar properties in the same group. This problem was a mystery for decades.

It was not until the start of the 20th century that scientists began to find out more about the structure of the atom. Only then could they solve this issue of certain elements breaking the periodic pattern. The elements in the periodic table are in order of their number of protons (their atomic number). The existence of isotopes accounted for the oddly heavy atomic weights of some elements.

1 Why did Newlands' fellow scientists refuse to accept his law of octaves? [1 mark]

2 a Look at the periodic table in Appendix 1. Name two elements, other than argon and potassium, that do not appear in order of their relative atomic masses. [2 marks]

b Explain why this was a problem for Mendeleev and how it was eventually solved. [5 marks]

3 Explain how Mendeleev persuaded any doubters that his periodic table really was a useful tool for understanding the chemical elements. [3 marks]

Figure 4 *Dmitri Mendeleev on a Russian stamp issued in his honour in 1969. He is remembered as the father of the modern periodic table. Using the table, chemists could now make sense of the chemical elements*

Key points

- The periodic table of elements developed as chemists tried to classify the elements. It arranges them in an order in which similar elements are grouped together.
- The periodic table is so named because of the regularly repeating patterns in the properties of elements.
- Mendeleev's periodic table left gaps for the unknown elements, which when discovered matched his predictions, and so his table was accepted by the scientific community.

C2.2 Electronic structures and the periodic table

Learning objectives

After this topic, you should know:

- how atomic structure is linked to the periodic table
- how metals and non-metals differ, including the electronic structures of their atoms and their positions in the periodic table
- why the noble gases are so unreactive.

Synoptic link

You will see in Chapter C3 how an atom's outer electrons are transferred or shared when atoms react and combine with each other.

Synoptic links

You first looked at the positions of metals and non-metals in Topic C1.1.

You also saw in Topic C1.7 how atoms can gain or lose electrons to form ions.

The chemical elements are placed in order of their atomic (proton) number in the periodic table. This arranges the elements so that they line up in groups (vertical columns) with similar properties. There are eight main groups in the periodic table (Figure 1).

The periodic table also gives an important summary of the electronic structures of all the elements. Elements in the same group of the periodic table react in similar ways because their atoms have the same number of electrons in the highest occupied energy level (outer shell).

The group number in the periodic table tells you the number of electrons in the outermost shell (highest occupied energy level) of an atom.

For example, all the atoms of Group 2 elements have two electrons in their outermost shell (highest energy level) and those in Group 6 have six electrons in their outermost shell.

Metals, non-metals, and electronic structures

You will have studied the different properties of metals and non-metals before. The main difference is that metals conduct electricity but non-metals generally are electrical insulators. Notable exceptions are some forms of carbon. In general, metals also have much higher melting and boiling points. Comparing solid examples, you find metals are ductile (can be drawn out into wires) and malleable (can be hammered into shapes without smashing), whereas non-metal solids are brittle.

The non-metal elements are found in the top right-hand corner of the periodic table (above the 'staircase' in Figure 1). The atoms of elements in Group 5, Group 6, and Group 7 can gain electrons to form negative ions. The atoms of Group 5 elements tend to gain three electrons, those in Group 6 gain two electrons, and those in Group 7 gain one electron. These then have the electronic structure of the noble gas at the end of their row (called a period).

The metal elements are found on the left-hand side and centre of the periodic table. Elements in Group 1, Group 2, or Group 3 tend to lose electrons and form positive ions. They attain the electronic structure of the noble gas at the end of the period one row (period) above them.

Group 0 – the noble gases

Look at the Group 0 noble gases, shaded purple in Figure 1. The atoms of noble gases have eight electrons in their outermost shell, making the atoms very stable. The exception is the first of the noble gases, helium, which has just two electrons but this complete first shell is also a very stable electronic structure.

The stable electronic structure of the noble gases explains why they exist as single atoms. They are monatomic (single-atom) gases. They have no tendency to react and modify their electronic structures by forming molecules. However, chemists have managed to make a few compounds of the larger noble gases. These contain the most reactive non-metallic elements, fluorine and oxygen, for instance in the compounds XeF_6 and XeO_4. As with many groups in the periodic table, there are trends in properties. For example, the boiling points of the noble gases get higher going down Group 0. Helium, at the top of the group, boils at −269 °C, whereas radon, at the bottom, boils at −62 °C.

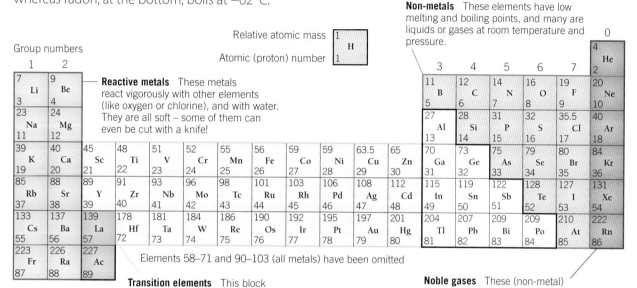

Figure 1 *The modern periodic table. The elements are arranged in order of atomic number*

1 a What does periodic mean in the term periodic table? [1 mark]
  b In the periodic table, what is:
    i a group [1 mark]   ii a period? [1 mark]

2 a How do the numbers of metal and non-metal elements compare with each other? Estimate an approximate percentage difference. 🖩 [1 mark]
  b How do the electronic structures of the atoms of a metallic element change when they react? [1 mark]

3 How many electrons do atoms of the following elements have in their highest energy level (outermost shell)?
  a beryllium, Be [1 mark]   b boron, B [1 mark]
  c potassium, K [1 mark]   d helium, He [1 mark]
  e argon, Ar [1 mark]   f radium, Ra [1 mark]
  g radon, Rn [1 mark]   h iodine, I [1 mark]

4 Explain why elements in many groups of the periodic table have similar chemical properties. [2 marks]

5 Explain why the noble gases are so unreactive. [2 marks]

Key points

- The atomic (proton) number of an element determines its position in the periodic table.
- The number of electrons in the outermost shell (highest energy level) of an atom determines its chemical properties.
- The group number in the periodic table equals the number of electrons in the outermost shell.
- The atoms of metals tend to lose electrons, whereas those of non-metals tend to gain electrons.
- The noble gases in Group 0 are unreactive because of their very stable electron arrangements.

C2.3 Group 1 – the alkali metals

Learning objectives

After this topic, you should know:

- how the Group 1 elements behave
- how the properties of the Group 1 elements change going down the group.

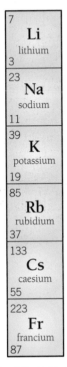

Figure 1 *The Alkali Metals (Group 1)*

Figure 2 *The Alkali Metals have to be stored in oil*

The first group (Group 1), on the left-hand side of the periodic table, is called the **alkali metals** (Figure 1). This group consists of the metals lithium, Li, sodium, Na, potassium, K, rubidium, Rb, caesium, Cs, and francium, Fr. You will probably only see the first three of these as rubidium and caesium are too reactive to use in schools and francium is an extremely unstable radioactive element.

Properties of the alkali metals

All the alkali metals are very reactive. They have to be stored in oil (Figure 2). This stops them reacting with oxygen in the air. Their reactivity increases as you go down the group. So lithium is the least reactive alkali metal and francium is the most reactive.

All the alkali metals have a very low density compared with other metals. In fact, the densities of lithium, sodium and potassium are all less than $1\,g/cm^3$, so they float on water. The alkali metals are also all very soft and can be cut with a knife. They have a silvery, shiny surface when you first cut them. However, this quickly goes dull as the metals react with oxygen in the air. This forms a layer of oxide on the shiny surface, for example:

$$sodium\ +\ oxygen\ \rightarrow\ sodium\ oxide$$
$$4Na(s)\ +\ O_2(g)\ \rightarrow\ 2Na_2O(s)$$

In a jar of oxygen gas, hot alkali metals burn vigorously, forming white smoke of their oxides.

The properties of this unusual group of metals are due to their electronic structure. The atoms of alkali metals all have one electron in their outermost shell (highest energy level). This gives them similar properties. It also makes them very reactive because they only need to lose one electron to get the stable electronic structure of a noble gas. They react with non-metals, losing their single outer electron. They form a metal ion carrying a 1+ charge, for example, Na^+ and K^+. They always form ionic compounds.

Melting points and boiling points

The Group 1 metals melt and boil at relatively low temperatures for metals. Going down the group, the melting points and boiling points get lower and lower. In fact, caesium turns into a liquid at just 29 °C.

Reaction with water

When you add lithium, sodium, or potassium to water, the metal floats on the water, moving around and fizzing. The fizzing happens because the metal reacts with the water to form hydrogen gas. Potassium reacts so vigorously with the water that the hydrogen produced ignites. It burns with a lilac flame, coloured by the potassium ions formed in the reaction. The reaction between an alkali metal and water also produces a metal hydroxide. This is why they are called alkali metals. The hydroxides of the alkali metals are all soluble in water. The solution is colourless with a high pH. (**Universal indicator** turns purple.)

sodium   +   water   →   sodium hydroxide   +   hydrogen

$2Na(s) + 2H_2O(l) \rightarrow 2NaOH(aq) + H_2(g)$

potassium   +   water   →   potassium hydroxide   +   hydrogen

$2K(s) + 2H_2O(l) \rightarrow 2KOH(aq) + H_2(g)$

Reactions of alkali metals with water

The reaction of the alkali metals with water can be demonstrated by dropping a small piece of the metal into a trough of water (Figure 3). This must be done with great care. The reactions are vigorous, releasing a lot of energy. Hydrogen gas is also given off (Figure 4).

- Describe your observations in detail, including a word equation and balanced symbol equation for each alkali metal used.

Safety: This is a demonstration and must be carried out by a teacher. Wear eye protection when watching this demonstration.

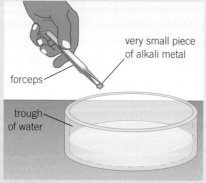

Figure 3 *Reacting alkali metals with water*

Other reactions

The alkali metals also react vigorously with non-metals such as chlorine gas. They produce metal chlorides, which are white solids. The metal chlorides all dissolve readily in water to form colourless solutions.

The reactions get more and more vigorous as you go down the group. That is because it becomes easier to lose the single electron in the outer shell to form ions with a 1+ charge.

sodium   +   chlorine   →   sodium chloride

$2Na(s) + Cl_2(g) \rightarrow 2NaCl(s)$

They react in a similar way with fluorine, bromine, and iodine. All of these ionic compounds of the alkali metals and non-metals are also white and dissolve easily in water. The solutions formed are all colourless.

1. Why are the alkali metals stored under oil? [1 mark]
2. Describe the trend in the melting points of the alkali metals as their atomic number increases. [1 mark]
3. Explain why the alkali metals form ions with a 1+ charge. [3 marks]
4. Write a balanced symbol equation (including state symbols) and a description of the product formed for the reaction of caesium, Cs, with:
   a iodine [3 marks]   b bromine. [3 marks]
5. Caesium is near the bottom of Group 1 in the periodic table. What do you think would happen if it was dropped into water containing a little universal indicator solution? Include an explanation of your expected observations and a balanced symbol equation with state symbols in your answer. [5 marks]

lithium

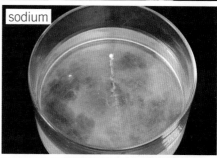
sodium

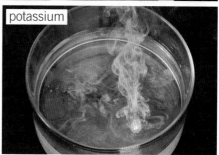
potassium

Figure 4 *Lithium, sodium, and potassium reacting with water. The bottom photo shows the hydrogen produced burning with a lilac flame*

Key points

- The elements in Group 1 of the periodic table are called the alkali metals.
- Their melting points and boiling points decrease going down the group.
- The metals all react with water to produce hydrogen and an alkaline solution containing the metal hydroxide.
- They form 1+ ions in reactions to make ionic compounds. These are generally white and dissolve in water, giving colourless solutions.
- The reactivity of the alkali metals increases going down the group.

C2.4 Group 7 – the halogens

Learning objectives

After this topic, you should know:

- how the Group 7 elements behave
- how the properties of the Group 7 elements change going down the group.

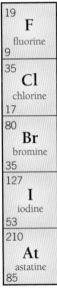

Figure 1 *The Group 7 elements*

Synoptic links

For further information on the bonding between non-metal atoms in substances, see Topic C3.5.

Synoptic links

For further information on the bonding between non-metal and metal ions in substances, see Topic C3.2.

Properties of the halogens

The Group 7 elements are called the **halogens**. They are a group of toxic non-metals that have coloured vapours. They have fairly typical properties of non-metals.

- They have low melting points and boiling points. Their melting points and boiling points increase going down the group (Table 1).
- They are poor conductors of heat and electricity.

As elements, the halogens all exist as molecules made up of pairs of atoms. These are called diatomic molecules. The atoms in each pair are joined to each other by a covalent bond.

Table 1 *The melting and boiling points of Group 7 halogens*

Group 7 halogen	F—F F_2	Cl—Cl Cl_2	Br—Br Br_2	I—I I_2
Melting point in °C	−220	−101	−7	114
Boiling point in °C	−188	−35	59	184

Reactions of the halogens

The electronic structure of the halogens determines the way they react with other elements. They all have seven electrons in their outermost shell (highest energy level). So they need to gain just one more electron to achieve the stable electronic structure of a noble gas. When they react with non-metals they gain an extra electron by sharing a pair of electrons with another atom, for example, with hydrogen (Table 2).

Table 2 *The reactions of the halogens with hydrogen*

	How the halogens react with hydrogen
$F_2(g) + H_2(g) \rightarrow 2HF(g)$	Explosive, even at −200 °C and in the dark.
$Cl_2(g) + H_2(g) \rightarrow 2HCl(g)$	Explosive in sunlight but slow in the dark.
$Br_2(g) + H_2(g) \rightarrow 2HBr(g)$	Only at over 300 °C in the presence of a platinum catalyst.
$I_2(g) + H_2(g) \rightarrow 2HI(g)$	Only at over 300 °C in the presence of a platinum catalyst (very slow, reversible).

Table 1 shows a general trend that you find in Group 7 – the elements get less reactive going down the group.

The halogens also all react with metals. In this case, the halogen atoms gain a single electron to give them a stable arrangement of electrons. They form ions with a 1– charge, for example, F^-, Cl^-, Br^-. Examples include sodium chloride, NaCl, and iron(III) bromide, $FeBr_3$ (which are called ionic compounds).

Displacement reactions between halogens

You can carry out test-tube reactions to check the order of reactivity of the halogens, using the following rule.

A more reactive halogen will displace a less reactive halogen from solutions of its salts.

You use solutions of the halogens and their salts in water, for example, chlorine dissolved in water mixed with potassium bromide solution. The colour of the solution after mixing will be due to the less reactive of the pair of halogens, which is left in solution as the aqueous molecule. For example, $Cl_2(aq)$ will be very pale green in solution, $Br_2(aq)$ will be yellow, whereas $I_2(aq)$ is a darker red/brown colour.

Displacement reactions

Add bromine water to potassium iodide solution in a test tube. Then try some other combinations of solutions of halogens and potassium halides.

- Record your results in a table.
- Explain your observations.

Safety: Wear chemical splash-proof eye protection. Chlorine and bromine are toxic.

Bromine displaces iodide ions from solution because it is more reactive than iodine. Chlorine will displace both iodide ions and bromide ions.

For example, chlorine will displace bromide ions, which form bromine molecules:

chlorine + potassium bromide → potassium chloride + bromine
$$Cl_2(aq) + 2KBr(aq) \rightarrow 2KCl(aq) + Br_2(aq)$$

Obviously fluorine, the most reactive of the halogens, will displace all of the others. However, it reacts so violently with water that you cannot carry out reactions in aqueous solutions.

1 In the Group 7 elements, state the trend going down the group in:
   a their melting points [1 mark]   b their reactivity? [1 mark]
2 a Using the data in Table 1, give the state of each Group 7 element at 20°C. [2 marks]
   b What is the general name given to the Group 7 elements? [1 mark]
3 a Write the electronic structures of the ions in lithium fluoride, Li^+F^-. [2 marks]
   b Describe how both the iodine and hydrogen atoms in hydrogen iodide, HI, manage to gain an electron. [1 mark]
4 Write a balanced symbol equation, including state symbols, for the reaction of:
   a sodium metal with iodine vapour [3 marks]
   b chlorine water with sodium iodide solution. [3 marks]
5 Explain in detail what happens when a solution of bromine is mixed with sodium iodide solution. Include a balanced symbol equation with state symbols in your answer. [6 marks]

Synoptic link

For further information on the state symbols (s), (l), (g), and (aq) that are used in equations, see Topic C1.2.

Study tip

In Group 7, reactivity *decreases* as you go down the group. However, in Group 1, reactivity *increases* going down the group.

Key points

- The halogens all form ions with a single negative charge in their ionic compounds with metals.
- The halogens form covalent compounds by sharing electrons with other non-metals.
- A more reactive halogen can displace a less reactive halogen from a solution of one of its salts.
- The reactivity of the halogens decreases going down the group.

C2.5 Explaining trends

After this topic, you should know:

- the trends in reactivity in Group 1 and Group 7
- how electronic structure can explain trends in reactivity in these groups.

As you saw in Topic C2.3, the Group 1 elements get more reactive going down the group:

Li
Na
K   getting *more* reactive
Rb
Cs

The opposite trend is observed in the Group 7 elements (see Topic C2.4):

F
Cl
Br   getting *less* reactive
I
At

You can explain these trends by looking at the electronic structures and how the atoms tend to lose or gain electrons in their reactions.

Reactivity within groups

As you go down a group in the periodic table, the number of shells occupied by electrons increases, by one extra electron shell per period. This means that the atoms become larger going down any group.

This has two effects:

- larger atoms lose electrons more easily going down a group
- larger atoms gain electrons less easily going down a group.

This happens because the outer electrons (which are negatively charged) are further away from the attractive force of the nucleus (which is positively charged because of its protons). Also, the inner shells of electrons 'screen' or 'shield' the outer electrons from the positive charge of the nucleus. You can see this effect with the alkali metals and the halogens. Remember that the atoms of alkali metals tend to lose electrons when they form chemical bonds. On the other hand, the atoms of the halogens tend to gain electrons.

Explaining the trend in Group 1

Reactivity increases going down Group 1 because the atoms get larger so the single electron in the outermost shell (highest energy level) is attracted less strongly to the positive nucleus. The electrostatic attraction with the nucleus gets weaker because the distance between the outer electron and the nucleus increases.

Also the outer electron experiences a shielding effect from inner shells of electrons. This reduces the attraction between the oppositely charged outer electron and the nucleus.

The size of the positive charge on the nucleus does become larger as you go down a group, as more protons are present inside the nucleus. This suggests that the attraction for the outer electron should get stronger. However, the greater distance and the shielding effect of inner electrons outweigh the increasing nuclear charge. So the change from Li to Li$^+$ takes more energy than Na changing to Na$^+$ (Figure 1).

Figure 1 *Sodium's outer electron is further from the nuclear charge and is shielded by more inner shells of electrons than lithium's outer electron*

Therefore, in Group 1, the outer electron gets easier to remove going down the group and the elements get more and more reactive.

Explaining the trend in Group 7

Reactivity decreases going down Group 7. To explain this, you consider the same factors you looked at with the alkali metals:

- the size of the atom
- the shielding effect of inner electrons, and
- the nuclear charge.

When Group 7 elements react, their atoms gain an electron in their outermost shell (highest energy level). Going down the group, the outermost shell's electrons get further away from the attractive force of the nucleus, so it is harder to attract and gain an extra incoming electron.

The outer shell will also be shielded by more inner shells of electrons, again reducing the electrostatic attraction of the nucleus for an incoming electron.

The effect of the increased nuclear charge going down the group (which helps atoms gain an incoming electron) is outweighed by the effect of increased distance and shielding by more inner electrons. So chlorine is less reactive than fluorine. The attraction for the incoming electron when F changes to F$^-$ is much greater than when Cl changes to Cl$^-$ (Figure 2).

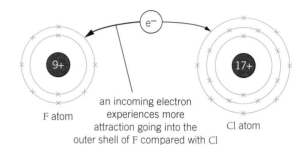

Figure 2 *F forms F$^-$ more readily than Cl forms Cl$^-$*

1 Which one of these atoms will have electrons in its outermost shell (highest energy level) that experience the greatest attractive force between themselves and the nucleus?

   helium   lithium   fluorine   potassium   argon
   Explain your answer.   [4 marks]

2 Explain why potassium is more reactive than lithium. 🪱 [4 marks]

3 Explain why fluorine is more reactive than bromine. 🪱 [4 marks]

4 Predict the difference in reactivity, giving reasons, seen in:
   a Group 2, between magnesium and calcium   [5 marks]
   b Group 6, between oxygen and sulfur. 🪱   [5 marks]

Key points

- You can explain trends in reactivity as you go down a group in terms of the attraction between electrons in the outermost shell and the nucleus.
- This electrostatic attraction depends on:
  - the distance between the outermost electrons and the nucleus
  - the number of occupied inner shells (energy levels) of electrons, which provide a shielding effect
  - the size of the positive charge on the nucleus (called the nuclear charge).
- In deciding how easy it is for atoms to lose or gain electrons from their outermost shell, these three factors must be taken into account. The increased nuclear charge, due to extra protons in the nucleus, going down a group is outweighed by the other two factors.
- Therefore electrons are easier for the larger atoms to lose going down a group, and harder for them to gain going down a group.

C2.6 The transition elements

Learning objectives

After this topic, you should know:

- the properties of the transition elements
- how the transition elements compare with the alkali metals.

Go further

The block of elements shown in Figure 1 arise as successive outer electrons fill an electron sub-shell called the d-sub-shell – which can hold 10 electrons in total. You will find at A Level that the block of elements are more correctly called the d-block elements. The alkali metals are in the s-block and the halogens are in the p-block. Looking at the blocks in the periodic table, can you guess how many electrons an s-sub-shell and a p-sub-shell can hold?

When most people are asked to name a few metals, it is likely they will name more than one transition element, such as iron, copper, or nickel. They might name a few alloys (mixtures of metals) that also contain at least one transition element, such as steel (containing iron) or brass (containing copper and zinc).

The transition elements, which are all metals, are positioned in the large central block of the periodic table, between Group 2 and Group 3 (Figure 1). This block contains the **transition elements,** also called the transition metals.

45 Sc 21	48 Ti 22	51 V 23	52 Cr 24	55 Mn 25	56 Fe 26	59 Co 27	59 Ni 28	63.5 Cu 29	65 Zn 30
89 Y 39	91 Zr 40	93 Nb 41	96 Mo 42	98 Tc 43	101 Ru 44	103 Rh 45	106 Pd 46	108 Ag 47	112 Cd 48
	178 Hf 72	181 Ta 73	184 W 74	186 Re 75	190 Os 76	192 Ir 77	195 Pt 78	197 Au 79	201 Hg 80

Figure 1 *The transition elements are found in the central block of the periodic table. The more well-known transition metals are circled*

Physical properties

The transition elements have the properties of typical metals.

- They are good conductors of electricity and thermal energy.
- They are hard and strong.
- They have high densities.
- They have high melting points (with the exception of mercury, Hg, which is a liquid at room temperature).

The transition elements have very high melting points compared with those of the alkali metals in Group 1. They are also harder, stronger, and much more dense.

Chemical properties

The transition elements are much less reactive than the metals in Group 1. This means they do not react as readily with oxygen, chlorine, or water as the alkali metals do. For example, copper foil has to be heated strongly before it reacts with oxygen gas in the air. A black coating of copper(II) oxide forms slowly on its suface:

$$\text{copper} + \text{oxygen} \rightarrow \text{copper(II) oxide}$$
$$2Cu(s) + O_2(g) \rightarrow 2CuO(s)$$

Iron reacts much slower with chlorine gas than sodium does. For a reaction to take place, the chlorine gas has to be passed over iron and heated very strongly inside a horizontal pyrex tube. The reaction forms red iron(III) chloride:

$$\text{iron} + \text{chlorine} \rightarrow \text{iron(III) chloride}$$
$$2Fe(s) + 3Cl_2(g) \rightarrow 2FeCl_3(s).$$

You will recall the vigorous reactions of the alkali metal with water. Compare these with the the reactions of copper with water (no reaction) and iron (which rusts slowly over time).

Synoptic links

To fully explain the physical properties of metals, you will need to understand metallic bonding and the giant structures of metals – see Topic C3.9 and Topic C3.10.

So if transition elements corrode, they do so very slowly. Together with their physical properties, this makes the transition elements very useful as structural materials. However, when the main metal used in the construction industry, iron (usually in steel) rusts, it weakens structures.

Compounds of transition elements

Many of the transition elements form coloured compounds. These include some very common compounds that you use in the laboratory. For example,

- copper(II) sulfate is blue (due to its Cu^{2+} ions)
- nickel(II) carbonate is pale green (due to its Ni^{2+} ions)
- chromium(III) oxide is dark green (due to its Cr^{3+} ions)
- manganese(II) chloride is pale pink (due to its Mn^{2+} ions)

Figure 2 *Compounds of transition elements are coloured (as opposed to the mainly white compounds of the alkali metals). The colours of many minerals, rocks and gemstones are due to transition element ions. The reddish-brown colour in a rock is often due to iron(III) ions, Fe^{3+}. The blue colour of sapphires and the green of emeralds are both due to transition element ions in the structures of their crystals*

The colours of vanadium ions

Your teacher can show you the range of colours that different ions of vanadium can have.

- Find out the names and formulae of the ions responsible for each colour.

The name of a compound containing a transition element usually includes a Roman number. For example, you will have seen copper(II) sulfate or iron(III) oxide.

This is because transition elements can form more than one ion. For example, iron may exist as Fe^{2+} or Fe^{3+}. Copper can form Cu^+ and Cu^{2+}, and chromium Cr^{2+} and Cr^{3+}. Compounds of these ions are different colours. For example, iron(II) ions, Fe^{2+}, give compounds a green colour, but iron(III) ions, Fe^{3+}, give a reddish-brown colour.

Transition elements and their compounds are also very important in the chemical industry as catalysts. For example, nickel is used as a catalyst in the manufacture of margarine and iron is used as a catalyst in the Haber process to make ammonia.

1   a   List the properties of a typical transition element.   [5 marks]
    b   State how mercury is unlike a typical transition metal.   [1 mark]

2   Write down the name of the following compounds of transition elements:
    a $FeCl_2$ [1 mark]   b Cr_2O_3 [1 mark]
    c $MnBr_2$ [1 mark]   d $NiCO_3$ [1 mark]

3   Copper, Cu, can form ions that carry a 1+ or a 2+ charge. Write down the name and formula of each compound copper can form with oxide, O^{2-}, ions.   [2 marks]

4   Carry out some research to find out the name and formula of the transition metal compound used as a catalyst in the manufacture of sulfuric acid and explain which property of the transition elements enables it to work.   [5 marks]

> ### Study tip
>
> The charge on the metal ion is given in the name of many transition metal compounds, for example, copper(II) sulfate, $CuSO_4$, contains Cu^{2+} ions, whereas copper(I) oxide, Cu_2O, contains Cu^+ ions.

> ### Synoptic link
>
> You will learn more about the Haber process in Chapter C15.

> ### Key points
>
> - Compared with the alkali metals, transition elements have much higher melting points and densities. They are also stronger and harder but are much less reactive.
> - The transition elements do not react vigorously with oxygen or water.
> - A transition element can form ions with different charges, in compounds that are often coloured.
> - Transition elements and their compounds are important industrial catalysts.

C2 The periodic table

Summary questions

1 a Where in the periodic table do you find:
   i the halogens [1 mark]
   ii the noble gases [1 mark]
   iii the alkali metals [1 mark]
   iv the transition elements? [1 mark]

   b In which groups would you find the four elements described below?
   i This is a dense metal with a high melting point. It reacts only very slowly with water but will react when heated with steam. [1 mark]
   ii This is a metal that can be cut with a knife and is stored under oil. It reacts violently with water and forms ions with a 1+ charge. [1 mark]
   iii This is a very unreactive, monatomic gas. [1 mark]
   iv This toxic gas is the most reactive of the all the non-metallic elements. It forms ions with a 1− charge and will also form covalent compounds. [1 mark]

2 Astatine, At, is a halogen whose atomic number is 85. It lies at the bottom of Group 7, beneath iodine.
   a How many electrons occupy its outermost shell (highest energy level)? Explain how you worked out your answer. [1 mark]
   b Predict the state of astatine at 20°C. [1 mark]
   c For the compound sodium astatide, predict:
   i its type of bonding [1 mark]
   ii its colour [1 mark]
   iii its chemical formula [1 mark]
   iv the word equation and the balanced symbol equation for its formation from its elements [2 marks]
   v whether or not you would see signs of a reaction if a solution of sodium astatide was mixed with chlorine water. Explain how you arrived at your answer. [2 marks]
   d Place the halogens, including astatine, in order of reactivity, with the most reactive element first. Explain your answer in detail (include the trend in reactivity and the how the trend can be explained, referring to the formation of halide ions). [6 marks]

3 Rubidium, Rb, is in Group 1 of the periodic table, lying directly beneath potassium.
   a Predict the following physical properties of rubidium, including its hardness, electrical conductivity, and melting point. [3 marks]
   b i What will be the charge on a rubidium ion? [1 mark]
   ii Copy and complete the table below:

Rubidium compound	Chemical formula
rubidium iodide	
rubidium fluoride	
rubidium hydroxide	

[3 marks]

   c Write down word equations and balanced symbol equations, including state symbols, for the following reactions:
   i rubidium and water [3 marks]
   ii rubidium and chlorine. [2 marks]
   d Will the reactions of rubidium in part c be more or less vigorous than the same reactions using potassium? Explain your answer in detail by comparing the atoms of the two elements. [5 marks]

4 Copper is a typical transition element. Sodium is a Group 1 element.
   a i How would you be able to tell which compound was which if you were given unlabelled samples of copper(II) sulfate and sodium sulfate? [1 mark]
   ii Why does copper in copper(II) sulfate have a Roman numeral after its name, but sodium in sodium sulfate does not? [1 mark]
   b State how the physical properties of copper differ from those of sodium. [3 marks]
   c How does the reactivity of copper compare with that of sodium? Use a test with water to illustrate your answer. [4 marks]

5 a Explain how Dmitri Mendeleev used atomic weights to construct his periodic table. [2 marks]
   b Explain how the scientific community were influenced to accept his periodic table. [6 marks]
   c Why, in 1869, could scientists not explain why some pairs of elements in his periodic table appeared to be in the wrong order according to their atomic weights? [3 marks]

6 a Explain how the atoms of metals and non-metals change when they react to form their ions, using magnesium as an example of a metallic element and oxygen as an example of a non-metallic element. [6 marks]
   b Suggest how the radius of a metal ion compares with the radius of the metal atom it was made from. Give a reason for your suggestion. [2 marks]

Practice questions

01 Elements **A–G** are shown on the periodic table in **Figure 1. A–G** are not the symbols for the elements.

Figure 1

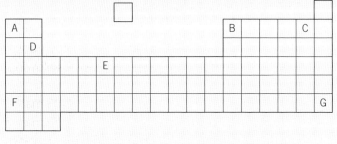

Choose one letter **A–G** to answer questions **01.1** to **01.5**.

01.1 Which element is a very unreactive gas? [1 mark]

01.2 Which element is a metal that only forms 2+ ions? [1 mark]

01.3 Which element reacts explosively with water forming an alkaline solution? [1 mark]

01.4 Which element forms coloured compounds some of which are useful as catalysts? [1 mark]

01.5 Which element forms an oxide with the formula X_2O_3? [1 mark]

01.6 Element **C** is in Group 7.
A solution of a Group 7 element will displace a less reactive Group 7 element from a solution of its salt.
Table 1 shows the colours of the solutions of some Group 7 elements and the solutions of sodium salts of some Group 7 elements.

Table 1

	Colour of solution		Colour of solution
Cl_2	colourless	NaCl	colourless
Br_2	orange	NaBr	colourless
I_2	brown	NaI	colourless

Predict the colour changes, if any, which would be seen when the following solutions are mixed together. Explain you answers.
- chlorine solution and sodium bromide solution
- iodine solution and sodium chloride solution

[5 marks]

02 This question is about Group 1 metals.
A teacher dropped a small piece of potassium into some water in a glass trough.
The equation for the reaction is shown below.
...... $K(s)$ + ... $H_2O(l) \rightarrow$ .... $KOH(aq)$ + ... $H_2(g)$

02.1 Balance the equation. [2 marks]

02.2 Using the state symbols in the equation, give **two** observations that you would expect to see when the potassium reacts with the water. Explain your answer. [4 marks]

02.3 The teacher used a safety screen. Give one other safety precaution the teacher should take. [1 mark]

02.4 A few drops of universal indicator were put in the glass trough after the reaction. The universal indicator turned blue.
Why did the universal indicator turned blue? [2 marks]

02.5 The list shows properties of metals.
Tick (✓) two properties that are typical of Group 1 metals.

high density	
low density	
form compounds with a 1+ charge that dissolve in water forming colourless solutions	
form compounds that dissolve in water forming coloured solutions	
form ions with different charges	

[2 marks]

02.6 The electronic structures of lithium and potassium are shown in **Figure 2**.

Figure 2

lithium

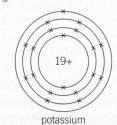

potassium

Why are lithium and potassium both in the same group of the periodic table? [1 mark]

02.7 Explain why lithium is less reactive than potassium. You should refer to **Figure 2** in your answer. [3 marks]

Learning objectives

After this topic, you should know:

- that the melting and boiling points of a substance depend on the nature of its particles and the forces between particles
- how to recognise that atoms themselves do not have the bulk properties of materials
- how to predict the states of substances at different temperatures, given appropriate data
- **H** the limitations of the particle theory.

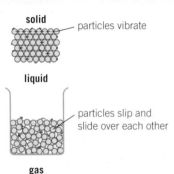

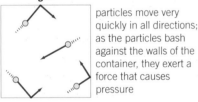

solid — particles vibrate

liquid — particles slip and slide over each other

gas — particles move very quickly in all directions; as the particles bash against the walls of the container, they exert a force that causes pressure

Figure 1 *The three states of matter.*

Cooling curve

Heat a test tube of stearic acid clamped in a water bath until its temperature reaches about 75 °C. Then remove the test tube from the hot water and monitor the temperature as it falls. Plot or print off a graph of the results.

- What is the melting point of stearic acid?
- Explain the shape of the line on your graph.

Safety: Wear eye protection.

From an early age, you can tell the differences between solids, liquids, and gases by using your senses. Later you learn that the majority of substances are classified as solids, liquids, or gases, and that these are called the three **states of matter**.

Solids have a fixed shape and volume. They cannot be compressed. **Liquids** have a fixed volume but they can flow and change their shape. Liquids occupy just slightly more space than when solid (water and ice are exceptions). **Gases** have no fixed shape or volume. They can be compressed easily.

To explain the properties of solids, liquids, and gases, the **particle theory** is used. It is based on the fact that all matter is made up of tiny particles and describes:

- the movement of the particles,
- the average distance between particles.

Look at the diagrams in Figure 1 that represent the three states of matter.

Each particle in a solid is touching its nearest neighbours and they remain in this fixed arrangement. They cannot move around but they do vibrate constantly.

The particles in a liquid are also very close together but they can move past each other. This results in a constantly changing, random arrangement of particles.

The particles in a gas have on average, much more space, between them. They can move around at high speeds in any direction. This means the particles have a random arrangement. The hotter the gas is, the faster the particles move. The pressure of a gas is caused by the particles colliding with the sides of the container. The more frequent and energetic the collisions, the higher the pressure of the gas. So, in a sealed container, the pressure of the gas increases with temperature.

Changing state

A solid turns into a liquid at its melting point. This is the same temperature at which the liquid freezes or solidifies back into the solid. The hotter a solid is, the faster its particles vibrate. Eventually, the vibrations will be so strong that the particles begin to break free from their neighbours. At this point, the solid starts to melt and become a liquid.

A liquid turns into a gas at its boiling point. The gas condenses back into the liquid at the same temperature. The hotter a liquid is, the faster its particles move around. As the temperature rises, more and more energy is transferred from the surroundings to the particles and the particles escape from the surface of the liquid. Its rate of evaporation increases. Eventually, the liquid boils and bubbles of gas rise and escape from within the liquid.

Each change of state is reversible. They are examples of physical changes. No new substances are formed in changes of state. For example, water molecules, H_2O, are the same in ice as they are in liquid water or in water vapour. It is just the movement and arrangement of the particles that differ and affect the properties of the substances at different temperatures, not any change in the particles themselves. Substances with higher melting points and boiling points have stronger forces operating between their particles.

Energy transfers during changes of state

When you monitor the temperature of a solid as you heat it to beyond its melting point, the results are surprising. The temperature stops rising at the solid's melting point. It remains constant until all the solid has melted, and only then starts to rise again (Figure 2).

At its melting point, enough energy is transferred from the surroundings to the solid for the forces between the particles in the solid to break. This enables the particles to break away from their fixed positions in the solid and start moving around. Once all the solid has melted, the transfer of energy from the surroundings to the substance causes the temperature of the liquid to continue to rise as expected.

On the other hand, changes of state in which you cool down a substance involve particles becoming closer together, such as condensing and freezing (solidifying). The cooling ceases at the boiling point and melting point as energy is transferred from the substance to the surroundings as stronger forces form between particles.

Limitations of the particle model

Higher

This simple particle model assumes that particles are made up of solid spheres with no forces operating between them. This is useful when comparing the properties of solids, liquids, and gases. However, the particles that make up substances are atoms, molecules, or ions. They can vary in size from the small He atoms in helium gas to the polymer molecules in plastics, which can contain many thousands of atoms and are not spherical. The interactions between neighbouring atoms, molecules, and ions can also distort their shapes. Atoms are mostly empty space, so real particles are not solid at all.

1 Draw a table to summarise the general properties of solids, liquids, and gases, as well as the average distance, arrangement, and movement of their particles. [6 marks]

2 Describe the changes that occur to the particles as a gas is cooled down to a temperature below its freezing point. ⊘ [6 marks]

3 Using particle theory, predict how temperature and pressure affect the density of a fixed mass of gas. ⊘ [6 marks]

4 Explain why substances have different melting points. [2 marks]

5 Evaporation is the change of state that occurs when a liquid changes to a gas below its boiling point. You can investigate the factors that affect the rate of evaporation using a wet paper towel on a high-resolution electric balance.
Plan an investigation into one factor that might affect the rate of evaporation of water from the paper towel. ⊘ [5 marks]

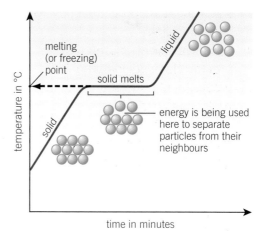

Figure 2 *The heating curve of a solid.*

Key points

- The three states of matter are solids, liquids, and gases.
- The particles in a solid are packed closely together and vibrate around fixed positions. The particles in a liquid are also close together but can slip and slide over each other in random motion. The particles in a gas have, on average, lots of space between them and zoom around randomly.
- In melting and boiling, energy is transferred from the surroundings to the substance. In freezing and condensing, energy is transferred from the substance to the surroundings.
- Ⓗ The simple particle model of solids, liquids, and gases is useful but has its limitations because the atoms, molecules, and ions that make up all substances are not solid spheres with no forces between them.

C3.2 Atoms into ions

Learning objectives

After this topic, you should know:

- what a chemical compound is
- how elements form compounds
- how atoms can form either positive or negative ions
- how the elements in Group 1 bond with the elements in Group 7.

You already know that you can mix two substances together without either of them changing. For example, you can mix sand and copper sulfate together and then separate them again. No change will have taken place to the sand or the copper sulfate.

However, in chemical reactions the situation is very different (Figure 1). When the atoms of two or more elements react chemically, they make a compound.

A compound contains two or more elements, which are chemically combined.

The compound formed is different from the elements that it is made from and you cannot get the elements back again easily. You can also react compounds together to form other compounds. However, the reaction of elements is easier to understand as a starting point.

The atoms of the noble gases, in Group 0 of the periodic table, have an arrangement of electrons that make them stable and unreactive. However, most atoms do not have this electronic structure. When atoms react, they take part in changes which give them a stable arrangement of electrons. They may do this by either:

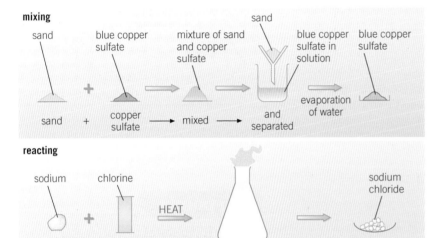

Figure 1 *The difference between mixing and reacting. Separating mixtures is usually quite easy but separating the elements from compounds once they have reacted can be difficult. Notice the state symbols used in the chemical equation – the sodium is shown as $Na(l)$, because it is heated on a combustion spoon (melting at 98 °C) before plunging it into the gas jar of chlorine – this should only be attempted as a teacher demonstration*

- sharing electrons, which is called **covalent bonding**, or
- transferring electrons, which is called **ionic bonding**.

Losing electrons to form positive ions

In ionic bonding, the atoms involved lose or gain electrons to form charged particles called ions. The ions have the electronic structure of a noble gas. So, for example, if sodium (2,8,1), from Group 1 in the periodic table, loses one electron, it is left with the stable electronic structure of neon (2,8).

However, it is also left with one more proton in its nucleus than there are electrons around the nucleus. The proton has a positive charge, so the sodium atom has now become a positively charged ion. The sodium ion has a single positive charge. The formula of a sodium ion is written as Na^+. The electronic structure of the Na^+ ion is 2,8 (Figure 2).

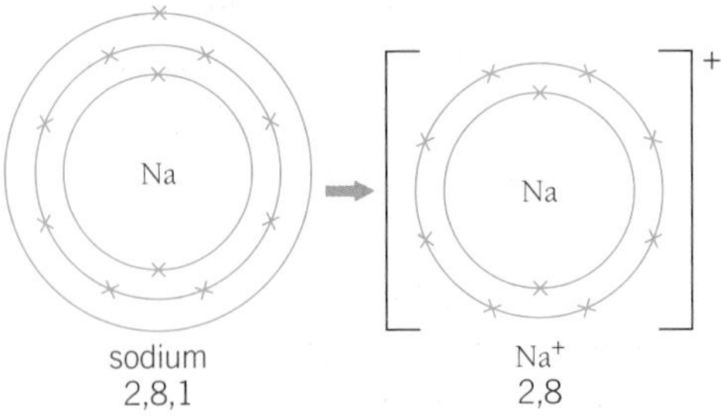

sodium
2,8,1

Na^+
2,8

Figure 2 *A positive sodium ion, Na^+, is formed when a sodium atom loses an electron during ionic bonding*

Gaining electrons to form negative ions

When non-metals react with metals, the non-metal atoms gain electrons to achieve the stable electronic structure of a noble gas. Chlorine, for example, has the electronic structure 2,8,7. It is in Group 7 of the periodic table. By gaining a single electron, it gets the stable electronic structure of argon (2,8,8). In this case, there is now one more electron than there are positive protons in the nucleus. So the chlorine atom becomes a negatively charged ion. This chloride ion carries a single negative charge. The formula of the chloride ion is written as Cl^-. Its electronic structure is 2,8,8 (Figure 3).

Representing ionic bonding

Metal atoms, which tend to lose electrons, react with non-metal atoms, which tend to gain electrons. So when sodium reacts with chlorine, each sodium atom loses an electron and each chlorine atom gains an electron. They both form stable ions. The electrostatic attraction between the oppositely charged Na^+ ions and Cl^- ions is called ionic bonding. You can show what happens in a diagram. The electrons of one atom are represented by dots, and the electrons of the other atom are represented by crosses (Figure 4).

Synoptic link

You first encountered the formation of ions in Topic C1.7.

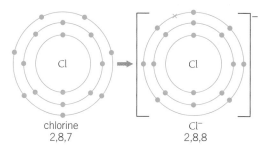

chlorine
2,8,7

Cl^-
2,8,8

Figure 3 *A negative chloride ion, Cl^-, is formed when a chlorine atom gains an electron during ionic bonding*

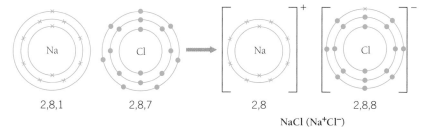

2,8,1   2,8,7   2,8   2,8,8

NaCl (Na^+Cl^-)

Figure 4 *The formation of sodium chloride, NaCl – an example of ion formation by transferring an electron*

(2,8,1)   (2,8,7)   (2,8)   (2,8,8)

Figure 5 *The dot and cross diagram to show electron transfer in sodium chloride, NaCl*

This can also be shown more simply by just showing the electrons in the outermost shell of the atoms and ions in a **dot and cross** diagram (Figure 5).

1 a When atoms bond together by *sharing* electrons, what type of bond is formed? [1 mark]

b When ions bond together as a result of *gaining* or *losing* electrons, what type of bond is this? [1 mark]

2 Write electron structures to show the ions that would be formed when the following atoms are involved in ionic bonding. For each one, state how many electrons have been lost or gained and show the charge on the ions formed.

a aluminium, Al [2 marks]
b fluorine, F [2 marks]
c potassium, K [2 marks]
d oxygen, O [2 marks]

3 Explain how and why atoms of Group 1 and Group 7 elements react with each other, in terms of their electronic structures. [4 marks]

Key points

- Elements react together to form compounds by gaining or losing electrons or by sharing electrons.
- The elements in Group 1 react with the elements in Group 7. As they react, atoms of Group 1 elements can each lose one electron to gain the stable electronic structure of a noble gas. This electron can be given to an atom from Group 7, which then also achieves the stable electronic structure of a noble gas.

C3.3 Ionic bonding

Learning objectives

After this topic, you should know:

- how ionic compounds are held together
- which elements, as well as those in Group 1 and Group 7, form ions
- how the charges on ions are related to group numbers in the periodic table.

When you eat salty food, you probably do not think about the charged particles, called sodium ions and chloride ions, that enter your body. You have seen how positive and negative ions form during some reactions. Ionic compounds are usually formed when metals react with non-metals. It is the metals that form positive ions and the non-metals that form negative ions.

The ions formed are held next to each other by very strong forces of attraction between the oppositely charged ions. This electrostatic force of attraction, which acts in all directions, is called ionic bonding.

The ionic bonds between the charged particles result in an arrangement of ions called a **giant structure** or **giant lattice** (Figure 1 in Topic C3.4).

Ions and the periodic table

You have seen how atoms in Group 1 of the periodic table have one electron in their outermost shell (or highest energy level) and form 1+ ions. Group 7 atoms have seven electrons in their outermost shell and form 1– ions. The group number gives the number of electrons in the outermost shell. So how does the group number relate to the charges on the ions formed from atoms?

Sometimes the atoms reacting need to gain or lose two electrons to gain the stable electronic structure of a noble gas. An example is when magnesium (2,8,2) from Group 2 reacts with oxygen (2,6) from Group 6. When these two elements react they form magnesium oxide, MgO. This is made up of magnesium ions with a double positive charge, Mg^{2+}, and oxide ions with a double negative charge, O^{2-}.

So you can say that when atoms form ionic bonds, atoms from:

- Group 1 form 1+ ions
- Group 2 form 2+ ions
- Group 3 form 3+ ions, when they form ions as opposed to sharing electrons
- Group 4 do not form ions (apart from tin, Sn, and lead, Pb, at the bottom of the group)
- Group 5 form 3– ions, when they form ions as opposed to sharing electrons
- Group 6 form 2– ions, when they form ions as opposed to sharing electrons
- Group 7 form 1– ions, when they form ions as opposed to sharing electrons
- Group 0 never form ions in compounds.

Figure 1 shows how the electrons are transferred between a magnesium atom and an oxygen atom during ionic bonding.

Another example of an ionic compound is calcium chloride. Each calcium atom (2,8,8,2) needs to lose two electrons but each chlorine atom (2,8,7) needs to gain only one electron. This means that two chlorine atoms react with every one calcium atom to form calcium chloride. So the formula of calcium chloride is $CaCl_2$.

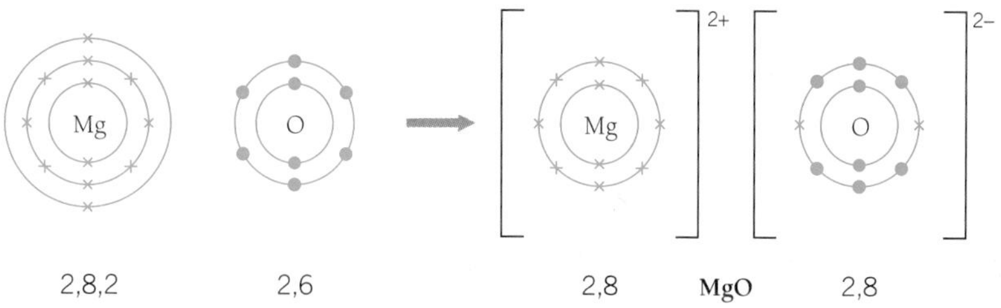

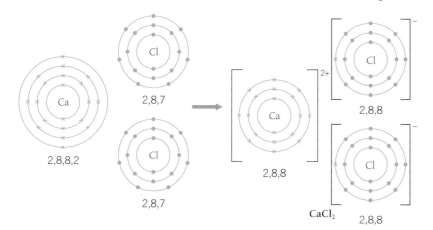

Figure 1 *When magnesium oxide, MgO, is formed, the reacting magnesium atoms lose two electrons and the oxygen atoms gain two electrons*

Figure 2 *The formation of calcium chloride, $CaCl_2$*

Ionic changes

Notice that the charges in the formula of an ionic compound cancel each other out, as the overall charge on the compound is zero. So aluminium oxide (made up of aluminium ions, Al^{3+}, and oxide ions, O^{2-}) has the formula Al_2O_3. The total charge of 6+ on the two aluminium ions is cancelled out by the total charge of 6– on the three oxide ions.

1 a Copy and complete the table:

Atomic number	Atom	Electronic structure of atom	Ion	Electronic structure of ion
9	F			2,8
3		2,1	Li^+	
16	S		S^{2-}	
20		2,8,8,2		

[4 marks]

2 Write down the general rules used to remember the charge on any ions formed by elements in:
   a Group 1, Group 2, and Group 3 [1 mark]
   b Group 5, Group 6, and Group 7. [1 mark]

3 Draw diagrams to show how you would expect the following elements to form ions together:
   a potassium and oxygen [3 marks]
   b aluminium and fluorine. [3 marks]

4 a Explain why potassium bromide is KBr but potassium oxide is K_2O. [3 marks]
   b Explain why magnesium oxide is MgO but magnesium chloride is $MgCl_2$. [3 marks]

5 Explain why metal atoms form positively charged ions whereas non-metal atoms form negatively charged ions. [6 marks]

Key points

- Ionic compounds are held together by strong forces of attraction between their oppositely charged ions. This is called ionic bonding.
- Besides the elements in Group 1 and Group 7, other elements that can form ionic compounds include those from Group 2 (forming 2+ ions) and Group 6 (forming 2– ions).

C3.4 Giant ionic structures

Learning objectives

After this topic, you should know:

- why ionic compounds have high melting points
- why ionic compounds conduct electricity when molten or dissolved in water.

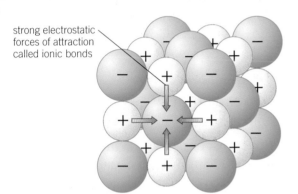

strong electrostatic forces of attraction called ionic bonds

3D model of a giant ionic lattice

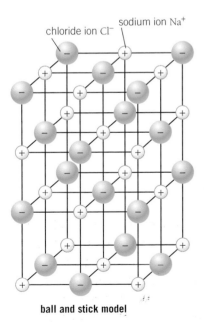

chloride ion Cl⁻
sodium ion Na⁺

ball and stick model

Figure 1 *The electrostatic forces of attraction between the oppositely charged ions in an ionic compound are very strong, operating in all directions. The regular arrangement of ions in the giant lattice results in ionic compounds forming crystals. The ball and stick model shows the 1 : 1 ratio of Na⁺ and Cl⁻ ions in NaCl*

You have already seen that an ionic compound consists of a giant structure of ions arranged in a lattice. The attractive electrostatic forces between the oppositely charged ions act in all directions and are very strong. This holds the ions in the lattice together very tightly. Look at the two models showing small parts of a giant ionic lattice in Figure 1. You will look at the limitations of using models like these in Topic C3.6.

It takes a lot of energy to break up a giant ionic lattice. There are lots of strong ionic bonds to break. To separate the ions you have to overcome all those electrostatic forces of attraction acting in all directions. This means that ionic compounds have high melting points and boiling points.

Once you have supplied enough energy to separate the ions from the lattice, they become mobile so can start to move around. This is when the ionic solid melts and becomes a liquid. The ions are free to move anywhere in this liquid. They are attracted to oppositely charged electrodes held in the molten compound. Therefore, they can carry their electrical charge through the liquid (Figure 2). A solid ionic compound cannot conduct electricity because its ions are held in fixed positions in the lattice. The ions in a solid ionic compound cannot move around. They can only vibrate 'on the spot'.

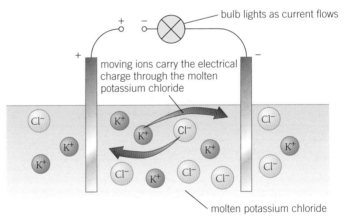

bulb lights as current flows

moving ions carry the electrical charge through the molten potassium chloride

molten potassium chloride

Figure 2 *Because the ions are mobile, a molten ionic compound can conduct electricity*

Many, but not all, ionic compounds will dissolve in water. When an ionic compound is dissolved in water, the lattice is split up by the water molecules. Then the ions are free to move around within the solution formed. They can carry their charge to oppositely charged electrodes in the solution. Just as molten ionic compounds will conduct electricity, solutions of ionic compounds will also conduct electricity. The ions are able to move to an oppositely charged electrode dipped in the solution.

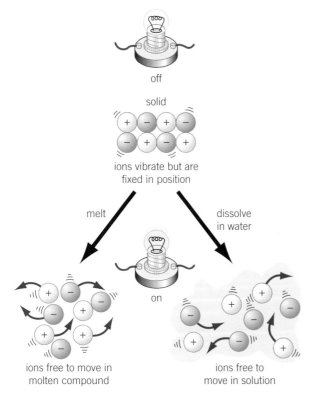

solid

ions vibrate but are
fixed in position

melt

dissolve
in water

on

ions free to move in
molten compound

ions free to
move in solution

Figure 3 *Ionic compounds do not conduct electricity in the solid
state but do when molten or when dissolved in water*

Ionic solid	Molten ionic compound	Ionic compound in solution
Ions are fixed in position in a giant lattice. They vibrate but cannot move around.

It does *not* conduct electricity. | High temperature provides enough energy to overcome the many strong attractive forces between ions. Ions are free to move around within the molten compound.

It *does* conduct electricity. | Water molecules separate ions from the lattice. Ions are free to move around within the solution.

It *does* conduct electricity. |

1 Why is seawater a better conductor of electricity than water from a
freshwater lake?   [1 mark]
2 Which of these ions will move to the positive electrode and which
will move to the negative electrode in a circuit like the one shown
in Figure 2?

lithium ions   chloride ions   bromide ions
calcium ions   sodium ions   zinc ions
oxide ions   iodide ions   barium ions   [4 marks]
3 Explain why ionic compounds have high melting points.   [4 marks]
4 Explain why ionic compounds conduct electricity only when
they are molten or dissolved in water.   [2 marks]
5 Predict which of these two ionic compounds has the higher melting
point – sodium oxide or aluminium oxide. Explain your answer.
                                                             [5 marks]

C3.5 Covalent bonding

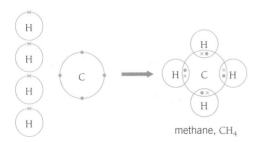

Reactions between metals and non-metals usually result in compounds with ionic bonding. However, many more compounds, including the vast majority of those making up your body, are formed in a very different way. When non-metals react together, their atoms share pairs of electrons to form molecules. This is called covalent bonding.

Simple molecules

The atoms of non-metals generally tend to gain electrons to achieve stable electron structures. When they react together, neither atom can give away electrons. So they get the electronic structure of a noble gas by sharing electrons. The atoms in the molecules are then held together by the shared pairs of electrons. These strong bonds between the atoms are called **covalent bonds**.

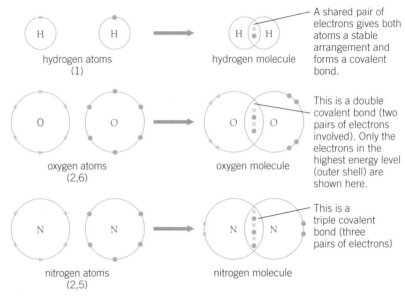

Figure 1 Most of the molecules in substances that make up living things are held together by covalent bonds between non-metal atoms

Figure 2 Atoms of hydrogen, oxygen, and nitrogen join together to form stable molecules. The atoms in H₂, O₂, and N₂ molecules are held together by strong covalent bonds

Sometimes in covalent bonding each atom has the same number of electrons to share (Figure 2) but this is not always the case. Sometimes the atoms of one element will need several electrons, whilst those of the other element only need one more electron for each atom to get a stable electronic structure. In this case, more atoms become involved in forming the molecule, such as in water, H_2O, and methane, CH_4 (Figure 3).

You can represent the covalent bonds in substances such as water, ammonia, and methane in a number of ways. Each way represents the same thing. The method chosen depends on what you want to show.

hydrogen chloride, HCl

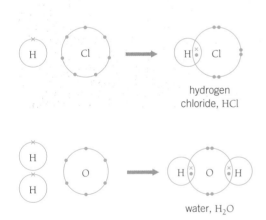

water, H_2O

methane, CH_4

Figure 3 The principles of covalent bonding remain the same however many atoms are involved – the atoms share one or more pairs of electrons to attain the electron structure of a noble gas, e.g., He (2), Ne (2,8), Ar (2,8,8)

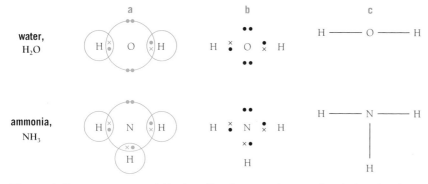

Figure 4 *You can represent the bonding in a covalent molecule by showing:*
a *the highest energy levels (or outer shells)*
b *the outer electrons in a dot and cross diagram*
c *the number of covalent bonds.*
Double bonds, as in the O_2 molecule in Figure 2, can be shown by two lines,
i.e., $O{=}O$ in an oxygen molecule

Giant covalent structures

Many substances containing covalent bonds consist of small molecules, for example, H_2O. However, some covalently bonded substances are very different. They have giant structures where huge numbers of atoms are held together by a network of covalent bonds. These giant covalent structures are sometimes referred to as macromolecules.

Diamond has a giant covalent structure. In diamond, each carbon atom forms four covalent bonds with its neighbours. This results in a rigid giant covalent lattice. You will look at the structure of diamond in more detail in Topic C3.7.

1 Which of these compounds will contain covalent bonds? Give the reason for your answer.
   hydrogen iodide   iron(II) chloride   lithium oxide
   sulfur dioxide   nitrogen(III) chloride   magnesium nitride
                                                         [2 marks]

2 Draw diagrams, showing all the electrons, to represent the covalent bonding between the following atoms:
   a two hydrogen atoms   [2 marks]
   b two chlorine atoms   [2 marks]
   c a hydrogen atom and a fluorine atom.   [2 marks]

3 Draw dot and cross diagrams to show the covalent bonds when:
   a a phosphorus atom bonds with three hydrogen atoms   [3 marks]
   b a carbon atom bonds with two oxygen atoms.   [3 marks]

4 Which noble gas electron structures do the atoms in a molecule of hydrogen chloride attain?   [2 marks]

5 A covalent bond consists of a pair of electrons, which occupy a space mainly between the nuclei of the two atoms bonded together. Suggest how this pair of electrons bonds the atoms to each other. (Hint: think of the electrostatic forces of attraction between oppositely charged particles.)   [3 marks]

Figure 5 *Diamonds owe their hardness to the way the carbon atoms are arranged in a giant covalent structure*

C3.6 Structure of simple molecules

Learning objectives

After this topic, you should know:

- the limitations of using models such as dot and cross, ball and stick, 2D diagrams, and 3D diagrams to represent molecules or giant structures
- why substances made of simple molecules have low melting points and boiling points
- why these substances do not conduct electricity.

Figure 2 *A molecular model of a methane molecule using a molecular model kit. Molecular model kits use plastic rods to join spheres together to construct molecules. These models have rigid bonds and solid atoms but actual bonds can vibrate and molecules can bend and twist*

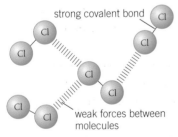

Figure 3 *Covalent bonds and the weak forces between molecules in chlorine gas. It is the weak intermolecular forces that are overcome when substances made of simple molecules melt or boil. The covalent bonds are not broken*

Using models

Models are used in everyday life to help us understand things. For example, the map of the London Underground is a simplified model of the maze of tunnels carrying trains beneath the streets. The diagrams of the covalent bonding shown in Figure 4 in Topic C3.5 are models used by chemists. The 'ball and stick' model can be added to these (Figure 1 in Topic C3.4). Figure 1 summarises some of the models used to represent methane, CH_4.

Like all models, each one is useful but has some limitations. In Figure 1, the 2D ball and stick model shows which atoms are bonded to each other but does not show the true shape of the molecule. It shows the H—C—H bond angles as 90°, whereas they are in fact 109.5°. The 3D formula attempts to show this on paper but a molecular model kit allows you to appreciate its tetrahedral shape more easily (Figure 2).

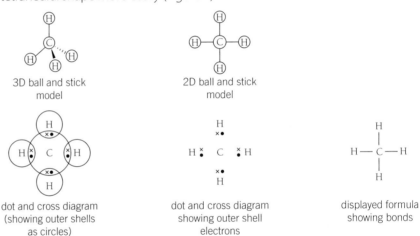

Figure 1 *The models used to help us understand the covalent bonding in a methane molecule*

Dot and cross diagrams show which atom the electrons in the bonds came from originally, but in reality all electrons are identical. Like all models drawn on paper, the electrons are in fixed positions between two atoms. However, scientists believe that the electrons in covalent bonds are constantly moving, but on average are found most of the time between the two nuclei of the atoms they are bonding together.

In giant structures, the models can never accurately reflect the many millions of atoms (or ions) bonded together in the giant lattices. However, they can represent a tiny fraction of a structure. They can indicate the chemical formula of a compound by the *simplest ratio* of the atoms or ions in models of their giant structures. For example, in a ball and stick model of sodium chloride (Figure 1 in Topic C3.4), there will be an equal number of Na^+ and Cl^- ions (ratio 1 : 1), so its formula is NaCl.

Intermolecular forces

Covalent bonds are very strong. So the atoms within each molecule are held very tightly together. However, each molecule tends to be quite separate from its neighbouring molecules. The force of attraction between the individual molecules in a covalent substance is relatively small – there are weak **intermolecular forces** between molecules (Figure 3). Overcoming these forces does not take much energy.

Intermolecular forces *increase* with the *size* of the molecules, so larger molecules have higher melting points and boiling points. For example, polymers are made from very long chain molecules, so the intermolecular forces are relatively high compared with smaller molecules. These stronger intermolecular forces make polymers solids at room temperature.

Polymers are made up of many small reactive molecules that bond to each other to form long chains. The simplest example is poly(ethene), made up from thousands of small ethene molecules, C_2H_4, reacting together.

We can represent the long polymer chains in poly(ethene) like this:

H H H H H H
| | | | | |
— C — C — C — C — C — C — etc.
| | | | | |
H H H H H H
poly(ethene)

Instead of showing all the covalent bonds, this can be abbreviated using the repeating unit of the polymer chain:

$\left(\begin{array}{cc} H & H \\ | & | \\ C - C \\ | & | \\ H & H \end{array}\right)_n$ where *n* is a large number

long chain of poly(ethene)

You have seen that ionic compounds will conduct electricity when they are liquids. Although a substance that is made up of simple molecules may be a liquid at room temperature, it will not conduct electricity (see Practical box).

Compounds made of simple molecules do not conduct electricity, even when they are molten or dissolved in water, unless they react with the water to form aqueous ions, as acid molecules do. This is because there is *no overall charge* on the simple molecules in a compound like sucrose. So their neutral molecules cannot carry electrical charge, unlike the ions in molten or aqueous ionic compounds.

Conductivity of a simple molecular compound

Sucrose is a sugar used to sweeten food and drinks. It is a compound made of individual molecules.

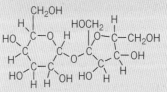

Figure 4 *A molecule of sucrose.*

Watch your teacher test the conductivity of solid sucrose, molten sucrose, and sucrose dissolved in water.

● What happens?
● Compare these results with what would happen in a similar experiment using sodium chloride instead of sucrose.

1 a Describe what is meant by intermolecular forces. [1 mark]
  b Look at Figure 4. What is the chemical formula of sucrose? [1 mark]

2 a Diamond has a very high melting point. Explain why. [2 marks]
  b Nitrogen gas has a very strong triple covalent bond holding the nitrogen atoms together in diatomic molecules. Explain why nitrogen has a boiling point of −196 °C. [2 marks]

3 A compound called sulfur hexafluoride, SF_6, is used to stop sparks forming inside electrical switches designed to control large currents. State why the properties of this compound make it particularly useful in electrical switches. [2 marks]

4 Explain why the melting point of hydrogen chloride is −115 °C, whereas sodium chloride's melting point is 801 °C. [4 marks]

Key points

● Substances made up of simple molecules have low melting points and boiling points.
● The forces between simple molecules are weak. These weak intermolecular forces explain why substances made of simple molecules have low melting points and boiling points.
● Simple molecules have no overall charge, so they cannot carry electrical charge. Therefore, substances made of simple molecules do not conduct electricity.
● Models are used to help understand bonding but each model has its limitations in representing reality.

C3.7 Giant covalent structures

Learning objectives

After this topic, you should know:

- the general properties of substances with giant covalent structures
- why diamond is hard and graphite is slippery
- why graphite can conduct electricity and thermal energy.

Figure 1 *Hard, shiny and transparent – diamonds make beautiful jewellery*

Study tip

Giant covalent structures are held together by covalent bonds throughout the lattice.

Did you know that diamond is a form of the element carbon, the same element contained in your pencil leads? Diamond is the hardest known natural substance. Artificial diamonds can be made by heating pure carbon to very high temperatures under enormous pressures. 'Industrial diamonds' made like this are embedded in the drills used by oil companies. They have to drill through layers of rock to get to the crude oil deep underground.

Many covalently bonded substances are made up of individual molecules. However, some substances, such as diamond, form very different structures. These do not have relatively small numbers of atoms arranged in simple molecules. They form huge networks of atoms held together by strong covalent bonds in **giant covalent structures**.

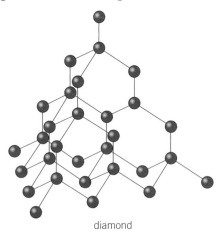

diamond

Figure 2 *A very small part of the structure of diamond. The giant covalent structure continues on and on in all directions*

As well as diamond, graphite and silicon dioxide (silica) also have giant covalent structures.

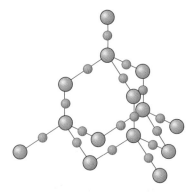

silicon dioxide, SiO_2

Figure 3 *Silicon dioxide has a giant covalent structure similar to that of diamond*

Having a giant covalent structure gives substances some very special properties:

- they have very high melting points and boiling points
- they are insoluble in water
- apart from graphite, they are hard and do not conduct electricity.

For example, as mentioned earlier, diamond is exceptionally hard and it has a boiling point of 4827 °C. Each carbon atom forms four strong covalent bonds, arranged in a perfectly symmetrical giant lattice.

Bonding in graphite

Carbon is not always found as diamonds. Another form is graphite, the form of carbon used in pencil lead. In graphite, carbon atoms are only bonded to three other carbon atoms. They form hexagons, which are arranged in giant layers. There are no covalent bonds between the layers, only weak intermolecular forces, so the layers can slide over each other quite easily. It is a bit like the effect of cards sliding off a pack of playing cards. This makes graphite a soft material that feels slippery to the touch.

As the carbon atoms in graphite's layers are arranged in hexagons, each carbon atom forms three strong covalent bonds (Figure 4). Carbon atoms have four electrons in their outer shell available for bonding. This leaves one spare outer electron on each carbon atom in graphite.

These mobile electrons can move freely along the layers of carbon atoms. The mobile electrons found in graphite are called **delocalised electrons**. They no longer belong to any one particular carbon atom. They behave rather like the electrons in a metallic structure (which you will look at in detail in Topic C3.9).

These delocalised electrons allow graphite to conduct electricity. The electrons will drift away from the negative terminal of a battery and towards its positive terminal when put into an electrical circuit. Diamond – and most other covalently-bonded substances – cannot conduct electricity. This is because their atoms have no free electrons, as all their outer shell electrons are involved in covalent bonding.

Graphite is also an excellent conductor of thermal energy. As more energy is transferred to the delocalised electrons, they move around faster and rapidly transfer the energy along the layers in the graphite.

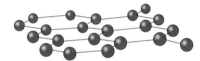

Figure 4 *The giant structure of graphite. When you write with a pencil, some layers of carbon atoms slide off the 'lead' and are left on the paper*

1   a Name two forms of the element carbon. [1 mark]
    b Find out what chemists call different forms of the same element in the same state. [1 mark]
2   List the general properties of a substance with a typical giant covalent structure. [2 marks]
3   Draw a rough sketch of the structure of graphite. Insert a plus and a minus sign to represent the terminals of a battery attached to the ends of the graphite.
    On your sketch use arrows to indicate the movement of the electrons when an electric current flows through the graphite. [2 marks]
4   Graphite is sometimes used to reduce the friction between two surfaces that are rubbing together.
    Explain how it does this. [2 marks]
5   Explain why graphite can conduct electricity but diamond cannot. 🖊 [6 marks]

Key points

- Some covalently-bonded substances have giant structures. These substances have very high melting points and boiling points.
- Graphite contains giant layers of covalently bonded carbon atoms. However, there are no covalent bonds between the layers. This means they can slide over each other, making graphite soft and slippery. The carbon atoms in diamond have a rigid giant covalent structure, making it a very hard substance.
- Graphite can conduct electricity and thermal energy because of the delocalised electrons that can move along its layers.

C3.8 Fullerenes and graphene

Learning objectives

After this topic, you should know:

- about the structure of the fullerenes and graphene
- recognise fullerenes and graphene from diagrams
- some uses of fullerenes.

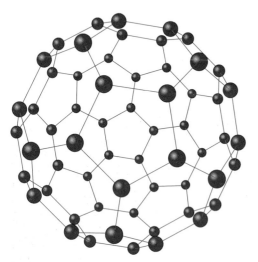

Figure 1 *The first fullerene to be discovered contained 60 carbon atoms but chemists can now make giant fullerenes that contain many thousands of carbon atoms. The name buckminsterfullerene was chosen for the C_{60} molecule after the Canadian architect Buckminster Fuller. He designed a similar shaped building in Montreal in 1967. The name is often abbreviated to bucky-ball*

Fullerenes

Apart from diamond and graphite, there are other structures that carbon atoms can form. In these structures the carbon atoms join together to make large hollow cages, which can have all sorts of shapes.

The ability of carbon to behave like this was not discovered until 1985. Radio astronomers had revealed that long chains of carbon atoms existed in outer space. When scientists tried experiments in the lab to recreate the conditions that might account for these carbon chains, they created a new molecule by chance. The molecule was made of 60 carbon atoms but how were the atoms arranged within each molecule? Analysis showed that all the carbon atoms in the new molecule were equivalent. There were no carbon atoms stuck at the ends of the molecule.

Professor Sir Harry Kroto, of Sussex University, solved the problem by suggesting a structure of hexagons and pentagons arranged in a sphere – just like the panels stitched together to make a football (Figure 1).

Since then, scientists have made many other new molecules. They can be shaped like footballs (as in the spherical C_{60} molecules), rugby balls, doughnuts, onions (spheres within spheres), and cones or tubes (open or closed at the ends). The general name for all these hollow-shaped molecules of carbon are the **fullerenes**. The structure of fullerenes is based on hexagonal rings of carbon atoms, as in graphite (Topic C3.7). However, they may also have rings of five (pentagonal) or seven (heptagonal) carbon atoms.

Cylindrical fullerenes called carbon nanotubes can also be produced. These fullerenes form incredibly thin cylinders, whose length is much greater than their diameter. They have very useful properties, such as:

- high tensile strength (leading to their use in reinforcing composite materials, such as those used in making tennis rackets)
- high electrical conductivity and high thermal conductivity (because their bonding is like the bonding in graphite, giving them delocalised

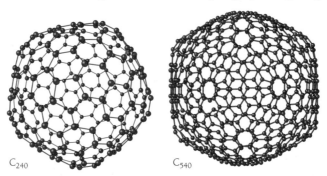

C_{240} C_{540}

Figure 2 *Other fullerenes. Some are bigger spheres than buckminsterfullerene*

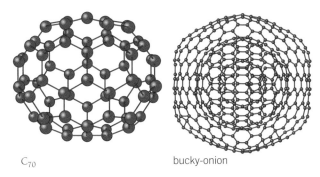

C₇₀   bucky-onion

Figure 3 *Some fullerenes are more elliptical in shape than C_{60}. The molecule with a ball inside a ball is nicknamed a 'bucky-onion'*

electrons, resulting in their use in the electronics industry).

Fullerenes could be used for drug delivery into the body. For example, the cage-like structures could be used as 'bucky-mules' to deliver drugs or radioactive atoms to treat cancer at very specific sites within the body. They can also be used as lubricants (as can graphite) and as catalysts because of the large surface area to volume ratio of their nanoparticles (Topic C3.11).

Graphene

If you could separate a single sheet of carbon atoms from graphite, you would get a layer of inter-locking hexagonal rings of carbon atoms (Figure 4 in Topic C3.7). It would be just one atom thick. Scientists at Manchester University managed to do this in 2004, basically by using a piece of sticky-tape. They stuck the tape across a piece of graphite, pulled it off, and looked at the tape under a powerful electron microscope. They had managed to isolate a 2D material – the thinnest ever made.

The new material is called graphene. It is an excellent conductor of thermal energy and electricity (even better than graphite), has a very low density, is the most reactive form of carbon, and pieces of it are incredibly strong for their mass. Graphene can be laid on a solid support and could one day be used to make quicker and more powerful computer chips. Its first large-scale application is likely to be in flexible electronic displays – just imagine watching a film on your coat sleeve.

1 a State a possible medical application of fullerene cages. [1 mark]
   b Why did chemists call a molecule of the new form of carbon discovered in 1985 a 'bucky-ball', and why did related molecules become known as fullerenes? [1 mark]
   c Write the chemical formula of the first fullerene discovered and describe the structure of its molecule. [1 mark]
2 Suggest which properties of graphene would make it useful in the manufacture of bullet-proof vests. [2 marks]
3 a Explain why graphene is such a good conductor of electricity. [5 marks]
   b Suggest why graphene could have many more applications in mobile electronics than graphite. [6 marks]

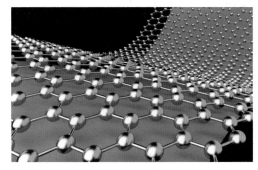

Figure 4 *A computer-generated image of the structure of graphene*

Key points

- As well as diamond and graphite, carbon also exists as fullerenes, which can form large cage-like structures and tubes, based on hexagonal rings of carbon atoms.
- The fullerenes are finding uses as a transport mechanism for drugs to specific sites in the body, as catalysts, and as reinforcement for composite materials.
- Graphene is a single layer of graphite and so is just one atom thick. Its properties, such as its excellent electrical conductivity, will help create new developments in the electronics industry in the future.

C3.9 Bonding in metals

Learning objectives

After this topic, you should know:

- how the atoms in metals are arranged
- how the atoms in metals are bonded to each other.

Figure 2 *Metal crystals, such as the zinc ones shown on this galvanised post give us evidence that metals are made up of atoms arranged in regular patterns*

Synoptic link

For more information on protecting iron from rusting, see Topic C15.1.

Metal crystals

The atoms in metals are built up layer upon layer in a regular pattern.

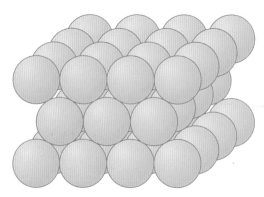

Figure 1 *The close-packed arrangement of copper atoms in copper metal*

This means that metals form crystals, although these are not always obvious to the naked eye. However, sometimes you can see them. You can see zinc crystals on the surface of some steel that has been dipped into molten zinc to prevent it from rusting. This is called galvanised steel. For example, look at the surface of galvanised lamp posts (Figure 2).

Growing silver crystals

You can grow crystals of silver metal by suspending a coiled length of copper wire in silver nitrate solution. Crystals of silver will appear on the wire quite quickly. However, for the best results they need to be left for several hours.

- Describe and explain your observations.

Safety: Wear eye protection.

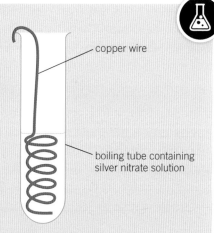

Figure 3 *Growing silver crystals.*

Survey of metallic crystals

Take a look around your school or college to see if you can find any galvanised steel. See if you can spot the metal crystals. You can also look for crystals on brass fittings that have been left outside and not polished.

Metallic bonding

Metals are another example of giant structures. You can think of a metal as a lattice of positively charged ions. The metal ions are arranged in regular layers, one on top of another.

The outer electrons from each metal atom can easily move throughout the giant structure. The outer electrons (in the highest occupied energy level) form a 'sea' of free-moving electrons surrounding the positively charged metal ions. Strong electrostatic attraction between the negatively charged electrons and the positively charged ions bond the metal ions to each other.

The electrons in the 'sea' of free-moving electrons are called **delocalised electrons**. They are no longer linked with any particular metal ion in the giant metallic structure. These delocalised electrons help to explain the properties of metals (Topic C3.10).

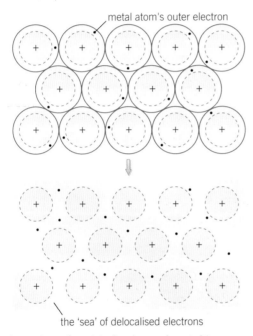

Figure 4 *A metal consists of positively charged metal ions surrounded by a 'sea' of delocalised electrons. This diagram shows a model of metallic bonding*

1 What can you deduce about the arrangement of the particles in a metal from the fact that metals form crystals? [1 mark]

2 a Why are the particles that make up a metal described as positively charged ions? [3 marks]
  b What are delocalised electrons? [2 marks]

3 Use the theory of metallic bonding to explain the bonding in magnesium metal. Make sure you mention delocalised electrons. (The atomic number of magnesium is 12.) ✍ [4 marks]

4 Using a model to explain metallic bonding, delocalised electrons could be thought of as a glue.
  Explain why thinking of delocalised electrons in a metal as a glue is a useful model but one which has a major drawback. [3 marks]

Key points

- The atoms in metals are closely packed together and arranged in regular layers.
- You can think of metallic bonding as positively charged metal ions, which are held together by electrons from the outermost shell of each metal atom. These delocalised electrons are free to move throughout the giant metallic lattice.

C3.10 Giant metallic structures

Learning objectives

After this topic, you should know:

- why metals can be bent and shaped without breaking
- why alloys are harder than pure metals
- why metals conduct electricity and thermal energy.

pure iron

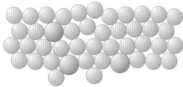

iron alloy

Figure 1 *The atoms in pure iron are arranged in layers, which can easily slide over each other. In alloys, the layers cannot slide so easily, because atoms of other elements distort the layers*

Metals can be hammered and bent into different shapes, and drawn out into wires. This is because the layers of atoms in a pure metal are able to slide easily over each other.

The atoms in a pure metal, such as iron, are held together in a giant metallic structure. The atoms are arranged in closely packed layers. This regular arrangement allows the atoms to slide over one another quite easily. This is why pure iron is relatively soft and easily bent and shaped.

Alloys are usually mixtures of metals. However, most steels contain iron with controlled amounts of carbon, a non-metal, mixed into its structure. The carbon atoms are a different size to the iron atoms. This makes it more difficult for the layers in the metal's giant structure to slide over each other. So, alloys are harder than the pure metals used to make them (Figure 1).

An alloy is a mixture of two or more elements, at least one of which is a metal.

Making models of metals

We can make a model of the structure of a metal by blowing small bubbles onto the surface of a soap solution to represent atoms.

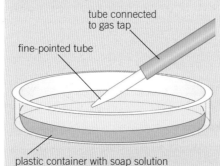

Figure 2 *Making a bubble raft to model the structure of a metal*

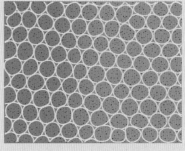

A regular arrangement of bubble 'atoms'

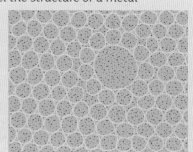

A larger bubble 'atom' disrupts the regular arrangement around it

- Why are models useful in science?

Metal cooking utensils are used all over the world because metals are good conductors of thermal energy and most have high melting points. Pans are usually made of steel (alloys of iron) but aluminium or copper are also used.

Wherever electricity is generated, it passes through metal wires (usually made of copper) to get to where it is needed because metals are also good conductors of electricity.

Explaining the properties of metals

The positive ions in a metal's giant structure are bonded to each other by a sea of delocalised electrons (Topic C3.9). These electrons are a bit like 'glue'. Their negative charge between the positively charged ions holds the metal ions in position by electrostatic forces of attraction. (Remember that opposite charges attract.)

However, unlike glue, the electrons are able to move throughout the whole giant lattice. Because they can move around and hold the metal ions together at the same time, the delocalised electrons enable the lattice to distort. When struck, the metal atoms can slip past one another without breaking up the metal's structure. The metals are malleable – they can be hammered into different shapes without cracking.

In Figure 3, the copper metal is being drawn into wires. As described above, the metallic bonding is maintained as layers of metal ions slide over each other. The metals are ductile – they can be drawn out into wires.

The high melting points of metals are explained by their giant structures. The electrostatic forces of attraction extend in all directions, as the electrons move freely between the positive metal ions in the giant lattices. It therefore takes a lot of energy to separate the metal ions from their fixed positions and break down the lattice, melting the metal.

Metals are also good conductors of thermal energy and electricity. This is because their delocalised electrons can readily flow through the giant metallic lattice. The electrical charge and thermal energy are transferred quickly through the metal by the free-moving delocalised electrons.

Figure 3 *Drawing copper out into wires depends on being able to make the layers of metal atoms slide easily over each other, without breaking the metal*

Figure 4 *Metals are essential in our lives – the delocalised electrons mean that they are good conductors of both thermal energy and electricity*

1 a Describe why metals can be bent, shaped, and pulled out into wires when forces are applied. [1 mark]
   b What word is used to describe a material, such as a metal, that can be:
      i hammered into shapes [1 mark]
      ii drawn out into wire? [1 mark]
2 Using your knowledge of metal structures, explain why alloying a metal can make the metal harder. [3 marks]
3 Explain why metals are good conductors of thermal energy and electricity, in terms of their structure and bonding. 🖉 [5 marks]
4 Explain why aluminium (from Group 3 in the periodic table) has a higher melting point than sodium (a Group 1 metal). 🖉 [6 marks]

Key points

- Metals can be bent and shaped because the layers of atoms (or positively charged ions) in a giant metallic structure can slide over each other.
- Alloys are harder than pure metals because the regular layers in a pure metal are distorted by atoms of different sizes in an alloy.
- Delocalised electrons in metals enable electricity and thermal energy to be transferred through a metal easily.

C3.11 Nanoparticles

Learning objective

After this topic, you should know:

- how to compare 'nano' dimensions to typical dimensions of atoms and molecules.

Nanoscience is a new and exciting area of science. 'Nano' is a prefix to a unit, like 'milli' or 'micro'. Whilst milli means one-thousandth and micro means one-millionth, 'nano' units are really small. Nano means one thousand-millionth or one billionth.

$$\textbf{1 nanometre (1 nm)} = 1 \times 10^{-9} \textbf{ metres}$$
$$(= \textbf{0.000 000 001 m} \text{ or } \textbf{a billionth of a metre})$$

So nanoscience is the science of really tiny things. It deals with structures that are just a few hundred atoms in size or even smaller, that measure between 1 and 100 nm. To give you an idea of these nanometre dimensions, a pin-head measures about 1 million nanometres and a human hair is about 80 000 nm wide.

Scientists need a scale and a common language to use when referring to tiny particles. As well as quoting the sizes of nanoparticles in nanometres (nm), for larger particles in the air, they also use micrometres (µm, where $1 \, \mu m = 1 \times 10^{-6} \, m$). The particles in air are often pollutants, pollen, or dust known as particulate matter (PM).

Scientists describe airborne particles with a diameter of 10 µm (i.e., 1×10^{-5} m) as PM_{10}. These are classified as coarse particles, and are often referred to as dust. Fine particles with a smaller diameter of between 0.1 µm (1×10^{-7} m or 100 nm) and 2.5 µm (i.e., 2.5×10^{-6} m or 2500 nm) are described as $PM_{0.1} - PM_{2.5}$. So nanoparticles can be 100 times smaller than even the finest dust. You can express this another way by saying that nanoparticles can be 2 orders of magnitude smaller than the finest dust particles (since $100 = 10^2$). This makes nanoparticles invisible in light. Look at Figure 2 below to see some scales comparing sizes of different types of particles.

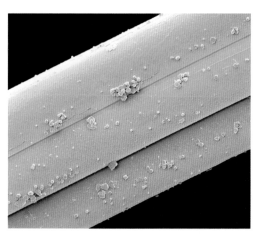

Figure 1 *Silver nanoparticles on fibres from an antibacterial wound dressing. The image was produced by a scanning electron microscope*

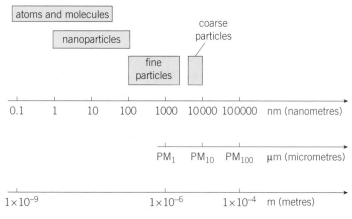

Figure 2 *The size of different types of particles. Nanoparticles are classified within the range of 1 to 100 nanometres in size. Getting larger, fine particles are sized between 100 nm and 2500 nm, and then coarse particles cover a range between 2500 nm and 10 000 nm.*

Materials may behave very differently on a very tiny scale, with sizes under 100 nm. When nanotechnologists arrange atoms and molecules to make particles on a nanoscale, their properties can be truly remarkable. This happens because nanoparticles have a huge proportion of their atoms or molecules at the surface of the particle compared with what you might normally consider a small particle, such as a grain of salt.

The surface area to volume ratio (SA : V) of particles gives a good indication of the proportion of particles at the surface. The higher the ratio, the greater the proportion of particles exposed at the surface. In fact, as the side of a cube decreases in size by a factor of 10, its surface area to volume ratio increases by 10.

Surface area to volume ratio

Look at the diagrams of the three cubes below and the calculations that follow:

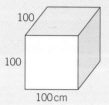

Synoptic link

To look at a surface area to volume calculation on a nanoscale, see Topic C8.2.

$SA = (100 \times 100) \times 6 \, cm^2$

$= 60\,000 \, cm^2$

$V = (100 \times 100 \times 100) \, cm^3$

$= 1\,000\,000 \, cm^3$

SA : V ratio

$= 60\,000 \, cm^2 : 1\,000\,000 \, cm^3$

$= 0.06 / cm$

$SA = (10 \times 10) \times 6 \, cm^2$

$= 600 \, cm^2$

$V = (10 \times 10 \times 10) \, cm^3$

$= 1000 \, cm^3$

SA : V ratio

$= 600 \, cm^2 : 1000 \, cm^3$

$= 0.6 / cm$

$SA = (1 \times 1) \times 6 \, cm^2$

$= 6 \, cm^2$

$V = (1 \times 1 \times 1) \, cm^3$

$= 1 \, cm^3$

SA : V ratio

$= 6 \, cm^2 : 1 \, cm^3$

$= 6 / cm$

This shows that as the side of a cube *decreases* in size by a factor of 10, its surface area to volume ratio *increases* by 10.

The exposure of such a large percentage of atoms at the surface of the nanoparticles makes them highly reactive compared to even powdered reactants.

The use of nanoparticles instead of traditional bulk materials should mean that smaller quantities are needed. Nanoparticles' high surface area to volume ratio makes them much more reactive than materials with normal particle sizes. This will result in a more sustainable approach in industry as less resources are used up, but there are some concerns about their environmental impact (Topic C3.12).

1 What is meant by nanoscience? [1 mark]

2 A nanoparticle has a diameter of 50 nm. Give this diameter in:
   a metres (m) [1 mark]
   b micrometres (μm). [1 mark]

3 Look at the Using maths box above. Show that the pattern in the surface area to volume ratio continues for a cube of side:
   a 0.1 cm [2 marks]
   b 10 m. [2 marks]

4 Explain why the properties of nanoparticles of a material may differ from the properties of the bulk material. [5 marks]

Key points

- Nanoscience is the study of small particles that are between 1 and 100 nanometres in size.
- Nanoparticles may have properties different from those for the same materials in bulk. This arises because nanoparticles have a high surface area to volume ratio, with a high percentage of their atoms exposed at their surface.
- Nanoparticles may result in smaller quantities of materials, such as catalysts, being needed for industrial processes.

C3.12 Applications of nanoparticles

Learning objectives

After this topic, you should know:

- some uses and benefits of nanoparticles
- some possible risks associated with the uses of nanoparticulate materials.

Figure 1 *Nanoparticles in sunscreen lotions will save many people from damaged skin and cancers caused by too much UV light. Zinc oxide nanoparticles have been found to have superior UV blocking properties compared to its use in lotions as a powder*

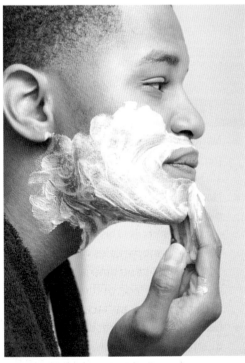

Figure 2 *Nanoparticles in cosmetic products can work deeper in the skin*

Nanoparticles at work

Here are some uses of nanoparticulate materials:

- Glass can be coated with titanium oxide nanoparticles. Sunshine triggers a chemical reaction that breaks down dirt that lands on the window. When it rains, the water spreads evenly over the surface of the glass, washing off the broken-down dirt.

- Titanium oxide and zinc oxide nanoparticles are also used in modern sunscreens. Scientists can coat nanoparticles of the metal oxide with a coating of silica. The thickness of the silica coating can be adjusted at an atomic level. These coated nanoparticles are more effective at blocking the Sun's rays than conventional UV absorbers.

- The cosmetics industry is one of the biggest users of this new technology. The nanoparticles in face creams are absorbed deeper into the skin. They are also used in sun creams and deodorants.

The delivery of active ingredients in cosmetics can also be applied to medicines. The latest techniques being developed use nanocages of gold to deliver drugs where they need to go in the body. Researchers have found that the tiny gold particles can be injected and absorbed by tumours. Tumours have thin, leaky blood vessels with holes large enough for the gold nanoparticles to pass into. However, they cannot get into healthy blood vessels. When a laser is directed at the tumour, energy is transferred to the gold nanoparticles and they warm up quickly. The temperature of the tumour increases enough to change the properties of its proteins but barely warms the surrounding tissue. This destroys the tumour cells without damaging healthy cells.

There is potential to use the gold nanocages to carry cancer-fighting drugs to the tumour at the same time. The carbon nanocages you met in Topic C3.8 can also be used to deliver drugs in the body. Incredibly strong, yet light, nanotubes are already being used to reinforce materials (Figure 3). The new materials are finding uses in sport, such as making very strong but light tennis racquets.

Silver nanoparticles are the most commonly engineered particles used by nanotechnologists. These are mainly used to inhibit the growth of microorganisms. They are used in fridges and are also put in the sprays used to clean operating theatres in hospitals. An increasingly common application is the use of silver nanoparticles for antimicrobial coatings. Many textiles, keyboards, wound dressings, and biomedical devices now contain coatings of silver nanoparticles. These continuously release a low level of silver ions to protect against bacteria (Figure 1 in Topic C3.11).

Future developments?

Nanotubes are now being developed that can be used as nanowires. This will make it possible to construct incredibly small electronic circuits. Nanotubes can be used to make highly sensitive sensors. For example, nanotube sensors have been made that can detect tiny traces of a gas present in the breath of asthmatics before an attack. This will let patients monitor and treat their own condition without having to visit hospitals to use expensive machines.

Nanowires would also help to make computers with vastly improved memory capacities and speeds.

Scientists in the US Army are developing nanotech suits – thin uniforms which are flexible and tough enough to withstand bullets and blasts. The uniforms would receive aerial views of the battlefield from satellites, transmitted directly to the soldier's brain. There would also be a built-in air-conditioning system to keep the body temperature normal. Inside the suit there would be a full range of nano-biosensors that could send medical data back to a medical team.

Possible risks

The large surface area of nanoparticles makes them very effective as catalysts and scientists are developing these for use in fuel cells. However, their large surface area also makes them dangerous. If a spark is made by accident near a large quantity of the catalyst, there could be a violent explosion.

If nanoparticles are used more and more, there is also going to be more risk of them finding their way into the atmosphere. In general, scientists believe that the health hazard increases as the diameter of the particles decreases. Breathing in tiny particles could damage the lungs. Nanoparticles could enter the bloodstream this way, or from their use in cosmetics, with unpredictable effects on our cells.

Although many scientists think there are few health risks, there are some worries about nanoparticles entering the environment, for example, after washing clothing impregnated with silver nanoparticles. This could affect aquatic life by accumulating in organisms over time.

More research, including long-term studies, needs to be done to find out about the effects of nanoparticles on health and the environment.

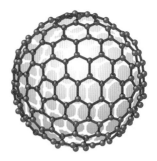
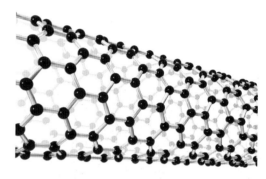

Figure 3 *Nanocages can carry drugs inside them and nanotubes can reinforce materials*

> ## Synoptic link
>
> To find out more about hydrogen fuel cells and their potential use in the future, see Topic C7.6.

1 Give two uses of silver nanoparticles. [2 marks]

2 a Give an advantage of using nanoparticles as catalysts. [1 mark]
  b Why are some people concerned about the use of nanoparticles as catalysts? [3 marks]

3 a Explain two uses of nanoparticles in cosmetic products for the skin. [4 marks]
  b How can nanoparticles possibly help to fight cancers? [2 marks]

4 In his book *Engines of Creation*, K. Eric Drexler speculates that one day we may invent a nanomachine that can reproduce itself. Then the world could be overrun by so-called 'grey goo'. Some people are so worried they have called for a halt in nanoscience research. Discuss this opinion. 🖊 [5 marks]

> ## Key points
>
> - New developments in nanoparticulate materials are very exciting and could improve many aspects of modern life.
> - The increased use of nanoparticles needs more research into possible issues that might arise in terms of health and the environment.

C3 Structure and bonding

Summary questions

1 Name the following changes:
   a liquid → solid [1 mark]
   b gas → liquid [1 mark]
   c solid → liquid [1 mark]
   d liquid → gas [1 mark]
   e solid → gas (in a single step). [1 mark]

2 Write a number to fill in the blanks in **a** to **e**:
   a The elements in Group …. in the periodic table all form ions with a charge of 1+. [1 mark]
   b The elements in Group …. in the periodic table all form ions with a charge of 2+. [1 mark]
   c The elements in Group …. in the periodic table all form ions with a charge of 1−. [1 mark]
   d The elements in Group …. in the periodic table all form ions with a charge of 2−. [1 mark]
   e The elements in Group …. in the periodic table have atoms which can form four covalent bonds. [1 mark]

3 This table contains data about some different substances:

Substance	Melting point in °C	Boiling point in °C	Electrical conductor
ammonia	−78	−33	solid – poor liquid – poor
magnesium oxide	2852	3600	solid – poor liquid – good
lithium chloride	605	1340	solid – poor liquid – good
silicon dioxide	1610	2230	solid – poor liquid – poor
hydrogen bromide	−88	−67	solid – poor liquid – poor
graphite	3652	4827	solid – good liquid – good

   a Make a table with the following headings:
   Giant covalent, Giant ionic, Simple molecules
   Now write the name of each substance above in the correct column. [3 marks]
   b Which substances are gases at 25 °C? [2 marks]
   c One of these substances behaves in a surprising way. Which one and why? [1 mark]
   d Draw a diagram to show the electron structures of a lithium atom and a chlorine atom and their ions in lithium chloride. (The atomic number of Li = 3, Cl = 17.) [4 marks]

4 a Which of the following substances will have covalent bonding?

   hydrogen iodide   **chlorine(VII) oxide**
   silver chloride   **phosphorus(V) fluoride**
   calcium bromide   **silver nitrate** [3 marks]

   b Explain how you decided on your answers in part **a**. [1 mark]
   c What type of bonding will the remaining substances in the list have? [1 mark]
   d What is the chemical formula of:
   i hydrogen iodide [1 mark]
   ii calcium bromide? [1 mark]

5 Copy and complete the following table with the formula of each ionic compound formed. (The first one is done for you.)

	chloride, Cl^-	oxide, O^{2-}	sulfate, SO_4^{2-}	phosphate (V), PO_4^{3-}
potassium, K^+	KCl			
magnesium, Mg^{2+}				
iron(III), Fe^{3+}				

[4 marks]

6 Draw a diagram which shows the bonding in:
   a hydrogen, H_2 [3 marks]
   b carbon dioxide, CO_2. [3 marks]

7 Silver nanoparticles can be incorporated into the fibres used to make hiking socks.
   a Why do manufacturers of the socks use silver nanoparticles? [1 mark]
   b Why might some people be worried about wearing the socks? [2 marks]
   c How might the silver nanoparticles find their way into the environment? Describe two possible mechanisms. [2 marks]
   d The silver nanoparticles have a diameter of 3.5×10^{-8} m. Give their diameter in nanometres. [1 mark]

8 Look at the ball and stick model of sodium chloride in Topic C3.4 and explain how you can use it to work out its chemical formula. [2 marks]

Practice questions

01.1 Carbon nanoparticles are used in very small amounts in deodorants.

Tick (✓) one other use of nanoparticles.

extraction of aluminium	
sun cream	
test for oxygen	

[1 mark]

01.2 Why are only very small amounts of carbon nanoparticles needed? Tick (✓) the correct box

Nanoparticles are highly flammable	
Nanoparticles are very reactive	
Nanoparticles have a very high surface area to volume ratio	

[1 mark]

01.3 Suggest why some people may be concerned about the use of nanoparticles in deodorants. [1 mark]

01.4 **Table 1** shows the size of a nanoparticle of carbon and the size of a coarse particle of carbon.

Table 1

Particle	Size of particle in metres
nanoparticle	1×10^{-5}
coarse particle	1×10^{-9}

How many times bigger is a coarse particle of carbon than a nanoparticle of carbon?

Tick (✓) the correct answer

10 times bigger	
1000 times bigger	
10 000 times bigger	
10 000 000 times bigger	

[1 mark]

02 Properties of five different substances **A, B, C, D,** and **E** are shown in **Table 2**.

Table 2

	Melting point in °C	Boiling point in °C	Conduction of electricity when solid	Conduction of electricity when liquid
Substance **A**	0	100	does not conduct	does not conduct
Substance **B**	1538	2862	conducts	conducts
Substance **C**	801	1413	does not conduct	conducts
Substance **D**	1610	2230	does not conduct	does not conduct
Substance **E**	−7	59	does not conduct	does not conduct

Which substance **A, B, C, D,** or **E**:

02.1 is a liquid at 20 °C? [1 mark]

02.2 has a giant covalent structure? [1 mark]

02.3 has a giant ionic structure? [1 mark]

02.4 The properties of another three substances **F, G,** and **H** are shown in **Table 3**.

Table 3

	Melting point in °C	Boiling point in °C	Conduction of electricity when solid	Conduction of electricity when liquid
Substance **F**	1064	1947	conducts	conducts
Substance **G**	−220	−188	does not conduct	does not conduct
Substance **H**	302	337	does not conduct	does not conduct

Substance **F** is a metal. Describe the structure of a metal and explain why a metal can conduct electricity when it is solid. [3 marks]

02.5 Substance **G** and substance **H** are elements in Group 7 of the periodic table. They are both simple molecules with the formula **G**₂ and **H**₂. Explain why substance **H** has a higher boiling point than substance **G**. [2 marks]

03 The electronic structures of an atom of magnesium and an atom of fluorine are shown in **Figure 1**.

Figure 1

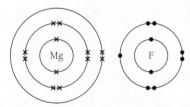

03.1 Describe in terms of electrons what happens when magnesium reacts with fluorine to form the ionic compound magnesium fluoride MgF_2. [3 marks]

03.2 The structure of magnesium fluoride can be represented by the ball and stick model shown in **Figure 2**.

Figure 2

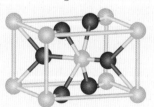

Explain why an ionic substance such as magnesium fluoride will conduct electricity when it is molten. [2 marks]

03.3 The ball and stick model is **not** a true representation of the structure of an ionic compound.

Give **one** reason why. [1 mark]

Learning objectives

After this topic, you should know:

- what is meant by the relative atomic mass of an element
- how to calculate the relative atomic mass of an element and the relative formula mass of a compound
- **H** how to calculate the number of moles, given the mass (or the mass, given the number of moles) of substance.

Calculating relative atomic mass

You can calculate the relative atomic mass A_r of an element given the percentage abundance of its isotopes, for example, copper has two isotopes, ^{63}Cu (abundance = 69%) and ^{65}Cu (31%).

To work out the relative atomic mass of copper from this data, imagine you had 100 copper atoms. 69 copper atoms would have a relative mass of 63, and the other 31 copper atoms would have a mass of 65. Then calculate the mean relative mass of these 100 atoms:

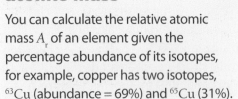

$$A_r \text{ of Cu} = \frac{(69 \times 63) + (31 \times 65)}{100} = 63.5$$

Relative formula mass

You also need to know how to work out the relative formula mass of more complex ionic compounds such as aluminium sulfate, $Al_2(SO_4)_3$.

Aluminium has an A_r of 27, the A_r of sulfur is 32, and the A_r of oxygen is 16. In this case, you must multiply any atoms *within* the brackets by the subscript number *after* the brackets. This means that the M_r of aluminium sulfate is:

$(27 \times 2) + (32 \times 3) + (16 \times 12)$
$= 54 + 96 + 192 = 342$

Relative atomic masses

The mass of a single atom is so tiny that it would not be practical to use it in experiments or calculations. So instead of working with the actual masses of atoms, the *relative* masses of atoms of different elements are used. These are called **relative atomic masses** A_r.

On any relative scale you need a standard reference point to compare against. In quoting relative atomic masses, the atom of carbon-12, $^{12}_6C$, is used as a standard atom. Carbon-12 is given a 'mass' of exactly 12 units because it has six protons and six neutrons. You can then compare the masses of atoms of all the other elements with this standard carbon atom. For example, hydrogen has a relative atomic mass of 1, as most of its atoms have a mass that is $\frac{1}{12}$ of the mass of a $^{12}_6C$ atom.

The A_r takes into account the proportions of any isotopes of the element found naturally. So it is the mean (average) relative mass of the isotopes of an element compared with the standard carbon atom. That is why chlorine has a relative atomic mass of 35.5, although you could never have half a proton or neutron in an atom.

Relative formula masses

You can use the A_r of the various elements to work out the **relative formula mass** M_r of compounds. This is true whether the compounds are made up of molecules or ions.

A simple example is a substance such as sulfuric acid, H_2SO_4. Hydrogen has an A_r of 1, the A_r of sulfur is 32, and the A_r of oxygen is 16. This means that the M_r of sulfuric acid is:

$(1 \times 2) + 32 + (16 \times 4) = 2 + 32 + 64 = 98$

In the case of molecular substances, such as H_2SO_4, the relative formula mass can also be referred to as the relative molecular mass.

The mole

Saying or writing 'relative atomic mass in grams' or 'relative formula mass in grams' is rather clumsy. So chemists have a shorthand word to describe this amount of substance – a **mole**. The abbreviation used for mole is mol. The relative atomic mass in grams of carbon (i.e., 12 g of carbon) is a mole of carbon atoms. One mole is simply the relative atomic mass or relative formula mass of any substance expressed in grams.

A mole of any substance always contains the same number of atoms, molecules, or ions. This is a huge number and its value is called the **Avogadro constant**. In standard form, it is written as 6.02×10^{23} per mole. In fact, if you had as many soft drink cans as there are atoms, molecules, or ions in a mole, they would cover the surface of the Earth to a depth of 200 miles.

Higher

Higher

Moles from masses

Chemists prefer to use the mole when describing relative numbers of particles (atoms, molecules, or ions) in a certain mass of substance. They use the equation:

$$\text{number of moles} = \frac{\text{mass (g)}}{A_r} \quad \text{or} \quad \frac{\text{mass (g)}}{M_r}$$

Masses from moles

Sometimes you will have to work out the mass of a substance from a given number of moles.

By re-arranging:

$$\text{number of moles} = \frac{\text{mass (g)}}{A_r} \quad \text{or} \quad \frac{\text{mass (g)}}{M_r}$$

you can calculate the mass of a certain number of moles of substance using the equation:

mass (g) = number of moles $\times A_r$ **or number of moles** $\times M_r$

Synoptic links

For more information on standard form, see Topic 1.7 and Maths skills M1b. To revise significant figures, see Maths skills M2a.

Synoptic links

To revise isotopes, look back to Topic C1.7.

Worked example: Calculating moles

How many moles of sulfuric acid molecules are there in 4.9 g of sulfuric acid?

Solution

number of moles of H_2SO_4 molecules $= \frac{4.9}{98} = 0.20 \, \text{mol}$

The answer is given as 0.20 mol, as opposed to 0.2 mol. This is because the data in the question was provided to 2 significant figures, so the answer should also be given to 2 significant figures.

Worked example: Calculating mass

What is the mass of 7.5×10^{-3} moles of aluminium sulfate?

Solution

mass of $Al_2(SO_4)_3 = (7.5 \times 10^{-3}) \times 342 = 2.6 \, \text{g}$

The number that appears on your calculator is 2.565. This is quoted to 4 significant figures. In the question, 7.5×10^{-3} moles, is only given to 2 significant figures, so the answer should reflect this. Hence 2.565 is rounded up to 2.6.

1   What is the relative atomic mass of an element? [2 marks]

2   What is the relative formula mass of:
   a MgF_2 (A_r values: Mg = 24, F = 19) [1 mark]
   b $C_6H_{12}O_6$ (A_r values: C = 12, H = 1, O = 16)? [1 mark]

3   **H**   a   How many moles of helium atoms are there in 0.02 g of helium? [1 mark]
      b   How many moles of sulfur atoms are there in:
         i   9.6 g of sulfur [1 mark]
         ii   16 tonnes of sulfur (where 1 tonne = 1000 kg)? [1 mark]

4   **H**   What is the mass of:
   a   50 moles of calcium carbonate, $CaCO_3$ [1 mark]
   b   0.05 moles of hydrogen, H_2 [1 mark]
   c   0.6 moles of phosphorus, P_4? [1 mark]

5   Why can you have relative atomic masses, which are not whole numbers, e.g., the A_r of chlorine, Cl, is 35.5. [1 mark]

Key points

- The masses of atoms are compared by measuring them relative to atoms of carbon-12.
- You can work out the relative formula mass of a compound by adding up the relative atomic masses of the elements in it, in the ratio shown by its formula.
- **H** One mole of any substance is its relative formula mass, in grams.
- **H**  $\text{number of moles} = \frac{\text{mass (g)}}{A_r} \text{ or } \frac{\text{mass (g)}}{M_r}$
- **H** Avogadro constant is 6.02×10^{23} per mole.

C4.2 Equations and calculations

Learning objectives

After this topic, you should know:

- what balanced symbol equations tell you about chemical reactions
- how to use balanced symbol equations to calculate masses of reactants and products.

Fractions in equations

Sometimes you will see fractions written in balanced equations. Even though you cannot have half an atom, ion, or molecule, you can read these equations as half a mole of the substance. These fractions are usually used in combustion reactions, such as:

$$C_2H_6 + 3\frac{1}{2}O_2 \rightarrow 2CO_2 + 3H_2O$$

Here you have two moles of carbon atoms, six moles of hydrogen atoms, and seven moles of oxygen atoms on both sides of the balanced equation.

By multiplying all the large numbers (multipliers) in the equation by 2 you will get all whole numbers and maintain a balanced equation:

$$2C_2H_6 + 7O_2 \rightarrow 4CO_2 + 6H_2O$$

Chemical equations can be very useful. When you want to know how much of each substance is involved in a chemical reaction, you can use the balanced symbol equation.

Think about what happens when hydrogen molecules, H_2, react with chlorine molecules, Cl_2. The reaction makes hydrogen chloride molecules, HCl:

$$H_2 + Cl_2 \rightarrow HCl \qquad \text{(not balanced)}$$

This equation shows the reactants and the product – but it is not balanced.

Here is the balanced equation:

$$H_2 + Cl_2 \rightarrow 2HCl$$

This balanced equation tells you that 'one hydrogen molecule reacts with one chlorine molecule to make two hydrogen chloride molecules. The balanced equation also tells you the number of moles of each substance involved. It tells you that 1 mole of hydrogen molecules reacts with 1 mole of chlorine molecules to make 2 moles of hydrogen chloride molecules.

one hydrogen molecule	one chlorine molecule	two hydrogen chloride molecules
H_2 +	Cl_2 →	$2HCl$
1 mole of hydrogen molecules	1 mole of chlorine molecules	2 moles of hydrogen chloride molecules

Using balanced equations to work out reacting masses

The balanced equation above is really useful because you can use it to work out what mass of hydrogen and chlorine react together. You can also calculate how much hydrogen chloride is made (Worked example 1).

Worked example 1

What masses of reactants and products are involved in the balanced symbol equation:

$$H_2 + Cl_2 \rightarrow 2HCl$$

Solution

To do this, you need to know that the A_r for hydrogen is 1 and the A_r for chlorine is 35.5.

A_r of hydrogen = 1	mass of 1 mole of $H_2 = 2 \times 1 = 2\,g$
A_r of chlorine = 35.5	mass of 1 mole of $Cl_2 = 2 \times 35.5 = 71\,g$
M_r of HCl = (1 + 35.5)	mass of 1 mole of HCl = 36.5 g

The balanced equation tells you that one mole of hydrogen reacts with one mole of chlorine to give two moles of hydrogen chloride molecules. So turning this into masses you get:

1 mole of hydrogen molecules, H_2	$= 1 \times 2$	$= 2\,g$
1 mole of chlorine molecules, Cl_2	$= 1 \times 71$	$= 71\,g$
2 moles of hydrogen chloride molecules, $2HCl$	$= 2 \times 36.5$	$= 73\,g$

More reacting mass calculations

These calculations are important when you want to know the mass of chemicals that react together.

Worked example 2

Sodium hydroxide reacts with chlorine gas to make bleach. This reaction happens when chlorine gas is bubbled through a solution of sodium hydroxide. The balanced symbol equation for the reaction is:

$$2NaOH \quad + \quad Cl_2 \quad \rightarrow NaOCl + NaCl + H_2O$$

sodium hydroxide   chlorine   bleach   salt   water

If you have a solution containing 100.0 g of sodium hydroxide, what mass of chlorine gas do you need to convert it to bleach?

Solution

A_r values: hydrogen = 1, oxygen = 16, sodium = 23, chlorine = 35.5

Mass of 1 mole of	
NaOH	**Cl$_2$**
$= 23 + 16 + 1 = 40\,g$	$= 35.5 \times 2 = 71\,g$

The table shows that 1 mole of sodium hydroxide has a mass of 40 g.

So 100.0 g of sodium hydroxide is $\frac{100}{40} = 2.5$ moles.

The balanced symbol equation tells you that for every 2 moles of sodium hydroxide you need 1 mole of chlorine to react with it.

So you need $\frac{2.5}{2} = 1.25$ moles of chlorine.

The table shows that 1 mole of chlorine has a mass of 71 g.

So you will need $1.25 \times 71 = 88.75\,g$ of chlorine to react with 100.0 g of sodium hydroxide.

The answer 88.75 g is given to 4 significant figures. This is to be consistent with the data supplied in the question, as you started with 100.0 g of sodium hydroxide.

The number of significant figures to which the relative atomic masses are quoted does not need to be taken into account in chemical calculations.

1 $2HCl$ can have two meanings. What are they? [1 mark]

2 Magnesium burns in oxygen with a bright white flame:
$2Mg(s) + O_2(g) \rightarrow 2MgO(s)$
What mass of oxygen will react exactly with 6.0 g of magnesium?
(A_r values: O = 16, Mg = 24) [2 marks]

3 a An aqueous solution of hydrogen peroxide, H_2O_2, decomposes to form water and oxygen gas. Write a balanced symbol equation, including state symbols, for this reaction. [3 marks]

b When hydrogen peroxide decomposes, what mass of hydrogen peroxide is needed in solution to produce 1.6 g of oxygen gas? [2 marks]

4 When a small lump of calcium metal, Ca, is added to water, it reacts giving off hydrogen gas. A solution of calcium hydroxide, $Ca(OH)_2$, is also formed in the reaction.

a Write a balanced symbol equation, including state symbols, for the reaction. [3 marks]

b Calculate how much calcium metal must be added to an excess of water to produce 3.7 g of calcium hydroxide. [2 marks]

C4.3 From masses to balanced equations

Learning objectives

After this topic, you should know:

- how to balance an equation, given the masses of reactants and products
- why a limiting quantity of a reactant affects the amount of product it is possible to obtain (in terms of amounts in moles or masses in grams).

In Topic C4.2 you saw how to use a balanced chemical equation to calculate the mass of reactants and products in a reaction. Alternatively, if you have the masses of the substances involved in a reaction, you can work out the ratio of the number of moles of each reactant and product (called the stoichiometry of the reaction). The simplest whole-number ratio gives you the balanced equation.

Worked example 1

Sodium nitrate, $NaNO_3$, decomposes on heating to give sodium nitrite, $NaNO_2$, and oxygen gas, O_2.

When 8.5 g of sodium nitrate is heated in a test tube until its mass is constant, 6.9 g of sodium nitrite is produced.

a What mass of oxygen must have been given off in the reaction?

b Find the ratio of reactants and products involved in the reaction, and show how these can be used to produce the balanced symbol equation for the decomposition of sodium nitrate:

(A_r values: Na = 23, N = 14, O = 16)

Solution

a You know that the total mass of reactants = total mass of products (from the Law of conservation of mass). So if the mass of oxygen is x g:

sodium nitrate $\rightarrow$ sodium nitrite + oxygen

$\qquad 8.5\,g \qquad = \qquad 6.9\,g \qquad + \quad x\,g$

$(8.5 - 6.9)\,g = x\,g$

$1.6\,g$ = mass of oxygen

b From the masses given in the question and our answer to part **a**, you can work out the numbers of moles of each reactant and product:

First of all, you will need to calculate the relative formula masses M_r of the reactants and products using the A_r values provided:

M_r of $NaNO_3 = [23 + 14 + (16 \times 3)] = 85$

M_r of $NaNO_2 = 69$

M_r of $O_2 = 32$

Then use the equation from Topic C4.1 to convert masses to moles:

$$\text{number of moles} = \frac{\text{mass}}{M_r}$$

$$\frac{\text{moles of}}{NaNO_3} = \frac{8.5}{85} \qquad \frac{\text{moles of}}{NaNO_2} = \frac{6.9}{69} \qquad \frac{\text{moles}}{\text{of } O_2} = \frac{1.6}{32}$$

$$= 0.1\,mol \qquad\qquad = 0.1\,mol \qquad\qquad = 0.05\,mol$$

Synoptic link

You first met the Law of conservation of mass in Topic C1.2.

Then find the simplest whole-number ratio of the numbers of moles of $NaNO_3 : NaNO_2 : O_2$

moles of $NaNO_3 : NaNO_2 : O_2$

$$0.1 : 0.1 : 0.05$$

Dividing the ratio by the smallest number gives:

$$2 : 2 : 1$$

So the balanced equation is:

$$2NaNO_3 \rightarrow 2NaNO_2 + O_2$$

Limiting reactants

More often than not, when you carry out a reaction in experiments, you do not use the exact amounts of reactants as predicted in the balanced equation. One of the reactants will be in excess. For example, if you add dilute hydrochloric acid to magnesium ribbon, hydrogen gas is given off and you see bubbles rising from the magnesium. If the reaction stops (no more bubbles of gas appear) but there is still magnesium ribbon – the magnesium is in excess. The reason the reaction stops is that all the acid has been used up. In this case the hydrochloric acid is called the **limiting reactant**. What would you expect to see at the end of the reaction if the magnesium was the limiting reactant?

The reactant that gets used up first in a reaction is called the limiting reactant (or limiting reagent).

You can work out which is the limiting reactant from the balanced equation, if you know the number of moles of reactants you start with.

Worked example 2

If you have 4.8 g of magnesium ribbon reacting in a solution of dilute hydrochloric acid containing 7.3 g of HCl, which reactant is the limiting reactant?

(A_r values: Mg = 24, H = 1, Cl = 35.5)

Solution

The balanced equation for the reaction is:

$$Mg(s) + 2HCl(aq) \rightarrow MgCl_2(aq) + H_2(g)$$

You are only interested in the reactants in this question.

number of moles $= \dfrac{mass}{A_r}$ or $\dfrac{mass}{M_r}$

You start with 4.8 g of Mg, which is $\dfrac{4.8}{24}$ moles = 0.2 mol

and 7.3 g of HCl, which is $\dfrac{7.3}{(1 + 35.5)}$ moles $= \dfrac{7.3}{36.5} = 0.2$ mol

From the balanced equation, you see that 1 mole of Mg will react with 2 moles of HCl.

Therefore 0.2 mol of Mg will need 0.4 mol of HCl to react completely. In this case, we have not got 0.4 mol of HCl – we only have 0.2 mol – so the dilute hydrochloric acid is the limiting reactant (and the magnesium is in excess).

Key points

- You can deduce balanced symbol equations from the masses (and hence the ratio of the numbers of moles) of substances involved in a chemical reaction.
- The reactant that gets used up first in a reaction is called the limiting reactant. This is the reactant that is NOT in excess.
- Therefore, the amounts of product formed in a chemical reaction are determined by the limiting reactant.

1 State what we mean by a limiting reactant in a chemical reaction.

[1 mark]

2 When copper metal reacts with oxygen gas, black copper oxide, CuO, is formed. In an experiment it was found that when copper reacted completely with oxygen, 6.35 g of copper reacted with 1.60 g of oxygen gas, O_2, to form 7.95 g of copper oxide.

  a Calculate the number of moles of each reactant and product.

[3 marks]

  b Show how this relates to the balanced symbol equation for the reaction. [2 marks]

3 Aluminium reacts with iron(III) oxide, Fe_2O_3, to give iron metal and aluminium oxide, Al_2O_3.

  a Write a balanced symbol equation for this reaction. [3 marks]

  b In an experiment, 32.0 g of iron(III) oxide was reacted with 16.2 g of aluminium.
     Which of the two reactants is the limiting reactant? Show your working. [2 marks]

  c Calculate the maximum mass of iron that could be collected at the end of this experiment. [2 marks]

C4.4 The yield of a chemical reaction

Learning objectives

After this topic, you should know:

- what is meant by the yield of a chemical reaction
- what factors can affect the yield
- how to calculate the theoretical yield of a reaction
- how to calculate the percentage yield of a chemical reaction.

A   +   2B   ⟶   C
(reactants)   (product)

Figure 1 *The reaction A + 2B → C (reactants to product)*

Many of the substances that you use every day have to be made from other chemicals. This may involve using complex chemical reactions. Examples include the plastics and composites used in your phones and computers, the ink in your pen or printer, and the artificial fibres in your clothes. All of these are made using chemical reactions.

Imagine a reaction: **A** + 2**B** → **C**

If you need 1000 kg of **C**, you can work out how much **A** and **B** you need. All you need to know is the relative formula masses of **A**, **B**, and **C** and the balanced symbol equation.

If you carry out the reaction, it is unlikely that you will get as much of **C** as you worked out. This is because our calculations assumed that all of the reactants **A** and **B** would be turned into product **C**. The mass of product that a chemical reaction produces is called its **yield**.

It is useful to think about reactions in terms of their **percentage yield**. This compares the mass of product that the reaction *actually* produces with the maximum theoretical mass that it could *possibly* produce, as predicted from the balanced symbol equation. This maximum mass possible is known as the theoretical yield of a reaction. Here is the formula you use to calculate percentage yield:

$$\text{percentage yield} = \frac{\text{actual mass of product produced}}{\text{maximum theoretical mass of product possible}} \times 100\%$$

So, if the maximum mass of product predicted for a reaction is 1000 kg, but only 800 kg is collected, the percentage yield is (800 kg/1000 kg) × 100% = 80%.

Calculating percentage yield

Higher

Worked example

Limestone is made mainly of calcium carbonate. In the production of calcium oxide, crushed lumps of limestone are heated in a rotating lime kiln. The calcium carbonate decomposes to make calcium oxide, and carbon dioxide gas is given off. A company processes 200 tonnes of limestone a day. It collects 98.0 tonnes of calcium oxide, the useful product. What is the percentage yield of the reaction in the kiln, assuming limestone contains only calcium carbonate? (A_r values: Ca = 40, C = 12, O = 16)

Solution

calcium carbonate → calcium oxide + carbon dioxide

$CaCO_3(s)$ → $CaO(s)$ + $CO_2(g)$

Work out the relative formula masses of $CaCO_3$ and CaO:

M_r of $CaCO_3$ = 40 + 12 + (16 × 3) = 100

M_r of CaO = 40 + 16 = 56

So the balanced symbol equation tells you that 100 tonnes of $CaCO_3$ could make 56 tonnes of CaO, assuming a 100% yield.

Therefore 200 tonnes of $CaCO_3$ could make a maximum of (56×2) tonnes of $CaO = 112$ tonnes (the theoretical yield in this case).

$$\text{So percentage yield} = \frac{\text{mass of product produced}}{\text{maximum mass of product possible}} \times 100\%$$

$$= \frac{98.0}{112} \times 100\% = 87.5\%$$

Factors affecting percentage yield

You can explain why the percentage yield of calcium oxide from the decomposition of calcium carbonate is less than 100%, as some of the limestone is lost as dust in the crushing process and in the rotating kiln. There will also be some other mineral compounds in the limestone. It is not 100% calcium carbonate, as you assumed in our calculation. Very few chemical reactions have a yield of 100% because:

- The reaction may be reversible – as products form they react to re-form the reactants again. You show reversible reactions using this symbol $\rightleftharpoons$ instead of the normal arrow between reactants and products. Chemists can manipulate reversible reactions by the conditions they choose in the reaction vessels in chemical plants.

- Some reactants may react to give unexpected or unwanted products in alternative reactions.

- Some of the product may be lost in handling or left in the apparatus.

- The reactants may not be pure (as in the case of the lime kiln).

- Some of the desired product may be lost during its separation from the reaction mixture.

1 State why it is good for the environment if industry finds ways to make products using high yield reactions and processes that waste as little energy as possible. [2 marks]

2 List the factors that can affect the percentage yield of a reaction. [5 marks]

3 Ammonia gas, NH_3, is made by heating the gases nitrogen and hydrogen under pressure in the presence of an iron catalyst:
$$N_2(g) + 3H_2(g) \rightleftharpoons 2NH_3(g)$$
If 7.0 g of nitrogen are reacted with excess hydrogen and 1.8 g of ammonia is collected, what is the percentage yield? [3 marks]

4 Sodium hydrogencarbonate, $NaHCO_3$, can be converted into sodium carbonate, Na_2CO_3, by heating. This is a thermal decomposition reaction in which water vapour and carbon dioxide are also products of the reaction. A student started with 16.8 g of sodium hydrogencarbonate and collected 9.20 g of sodium carbonate.
   a Write a balanced symbol equation for the thermal decomposition. [1 mark]
   b Calculate the percentage yield the student obtained. [3 marks]

Synoptic links

For more information about reversible reactions, see Topics C8.6–8.9.

Also see an example of how chemists influence reversible reaction in industry in Topic C15.5.

Key points

- The yield of a chemical reaction describes how much product is made.
- The percentage yield of a chemical reaction tells you how much product is made compared with the maximum amount that could be made (100%).
- Factors affecting the yield of a chemical reaction include product being left behind in the apparatus, reversible reactions not going to completion, some reactants may produce unexpected reactions, and losses in separating the products from the reaction mixture.

C4.5 Atom economy

Learning objectives

After this topic, you should know:

- how to calculate the atom economy of a reaction to form a desired product from the balanced equation
- why atom economy is important in industrial processes.

Synoptic links

To find out more about the Earth's natural resources and the impact of industrial processes on the environment, see Topic C14.1 and Topic C14.5.

Sustainable production

Chemical companies use reactions to make the products they sell. Ideally, they want to use reactions with high percentage yields (that also happen at a reasonable rate). Making a product more efficiently means making less waste. As much product as possible should be made from the reactants.

Chemical factories (or chemical plants) are designed by chemical engineers. They design a plant to work as safely and economically as possible. It should waste as little energy and raw materials as possible. This helps the company to make money. It is better for the environment too, as it conserves limited resources. It also reduces the pollution produced whenever fossil fuels are used as sources of energy.

As part of this move to a 'greener' chemical industry, atom economy (also known as atom utilisation) is now considered an important issue. The atom economy of a reaction is a measure of the extent to which the atoms in the starting materials (the reactants) end up in the desired product. Any remaining atoms end up in other products. Ideally the chemical company can use these other products for some useful purpose. If not, there will be extra expense to get rid of the waste products. If the waste products are in any way hazardous, they will have to be treated to make them harmless before being released into the air, waterways, or landfill sites.

Calculating atom economy

The equation used to work out the atom economy of a particular reaction is:

$$\text{percentage atom economy} = \frac{\text{relative formula mass of the desired product from equation}}{\text{sum of the relative formula masses of the reactants from equation}} \times 100\%$$

In some reactions, all the atoms in the reactants appear in the products, and the ideal '100% atom economy' is achieved. Examples include:

$$N_2(g) + 3H_2(g) \rightleftharpoons 2NH_3(g)$$

and

$$C_2H_4(g) + H_2O(g) \rightleftharpoons C_2H_5OH(g)$$

Both these reactions are reversible, but to counter this any reactants emerging from the reaction vessel (unreacted or re-formed from the product) are continuously recycled back into the vessel.

Any reactions that form other products as well as the desired product will fall short of 100% atom economy.

Worked example

When lead is extracted from its ore, galena, the first stage is the roasting of lead sulfide, PbS, to convert it to lead oxide, PbO. The lead oxide is then reduced to form lead metal. The balanced symbol equation for the first stage, with state symbols is:

$$2PbS(s) + 3O_2(g) \rightarrow 2PbO(s) + 2SO_2(g)$$

Calculate the percentage atom economy of this first stage in the process of extracting lead.

(A_r values: Pb = 207, S = 32, O = 16)

Solution

$$\% \text{ atom economy} = \frac{\text{relative formula mass of the desired product from equation}}{\text{sum of the relative formula masses of the reactants from equation}} \times 100\%$$

$$= \frac{M_r (2PbO)}{[M_r (2PbS) + M_r (3O_2)]} \times 100\%$$

$$= \frac{2 \times (207 + 16)}{[2 \times (207 + 32)] + [3 \times (16 \times 2)]} \times 100\%$$

$$= \frac{446}{(478 + 96)} \times 100\%$$

$$= 77.7\%$$

In the Worked example above, the other product of the reaction is sulfur dioxide gas. This gas is a serious pollution threat. It causes acid rain, so it has to be removed from waste gases by treating with a slurry of a base, such as calcium oxide or calcium carbonate. The product formed in this neutralisation reaction is calcium sulfate, which can be used in the construction industry but it is sometimes cheaper to dump it in landfill sites.

Synoptic links

For more information on neutralisation reactions using insoluble bases, see Topic C5.5.

1 Write down the formula that chemists use to calculate the percentage atom economy of a reaction. [1 mark]

2 Calculate the atom economy of the thermal decomposition of calcium carbonate. [2 marks]

3 A chemical company is setting up a plant to manufacture the compound called chloroethane, C_2H_5Cl. Their chemists can make it in two ways, either from ethene (Reaction 1) or ethanol (Reaction 2):
   Reaction 1: $C_2H_4 + HCl \rightarrow C_2H_5Cl$
   Reaction 2: $C_2H_5OH + HCl \rightarrow C_2H_5Cl + H_2O$
   a Calculate the percentage atom economy of Reaction 2. [2 marks]
   b Many factors have to be considered before deciding which route to take to make chloroethane. However, on the basis of atom economy, explain which reaction is preferable. [2 marks]
   c Suggest two factors that the chemical company should consider before making their decision, and give a reason for each. [4 marks]

Key points

- It is important to maximise atom economy in industrial processes to conserve the Earth's resources and minimise pollution.
- The atom economy of a reaction uses its balanced equation to compare the relative formula mass of the desired product with the sum of the relative formula masses of the reactants. It is usually expressed as a percentage.

C4.6 Expressing concentrations

Learning objectives

After this topic, you should know:

- the concentration of solutions can be expressed in grams per dm³ (g/dm³)
- how the mass of a solute and the volume of a solution is related to the concentration of the solution.

Figure 1 *The orange squash is getting less concentrated going left to right (the darker colour indicates more squash is in the same volume of its solution)*

Figure 2 *Volumetric flasks are used to make up solutions. They have a graduation mark around their narrow necks. Water is added to the solute until the bottom of its meniscus (the curve at the surface of the solution when viewed from the side) is level with the mark*

What is the concentration of a solution?

When you make a drink of orange squash, sometimes you put too much squash in the glass – sometimes you add too much water. Then you can add more water or add more squash until the colour looks right for you. A chemist would say that you are adjusting the **concentration** of the solution.

Chemists often carry out their reactions in solution. The solvent is usually water but can be other liquids, such as ethanol.

To record, interpret, and communicate their results, they need to express the concentration of the solutions they use. Other chemists should be able to repeat published experiments to verify data. So chemists quote the amount of substance (solute) dissolved in a certain volume of the solution. The units they use to express the concentration of a solution can be grams per decimetre cubed (g/dm³). A decimetre cubed (1 dm³) is equal to 1000 cm³.

Calculating concentrations

If you know the mass of solute dissolved in a certain volume of solution, you can work out its concentration.

As an equation:

$$\text{concentration (g/dm}^3) = \frac{\text{amount of solute (g)}}{\text{volume of solution (dm}^3)}$$

If you are working in centimetres cubed (cm³), convert the volume to dm³ by dividing it by 1000, and use the equation above. Alternatively, substitute your data in cm³ into the following equation:

$$\text{concentration (g/dm}^3) = \frac{\text{amount of solute (g)}}{\text{volume of solution (cm}^3)} \times 1000$$

As an example, imagine that you make a solution of sodium hydroxide in water. You dissolve exactly 40.0 g of sodium hydroxide in enough water to make exactly 3.00 dm³ of solution. You can calculate the concentration of the solution in g/dm³:

$$\frac{40.0\ g}{3.0\ dm^3} = 13.3\ g/dm^3.$$

Worked example: Concentration of a solution

50 g of sodium hydroxide is dissolved in water to make up 200 cm³ of solution. What is its concentration, given in g/dm³? (Remember that 1 dm³ = 1000 cm³.)

Solution

To find the concentration of the solution, you should use the equation:

$$\text{concentration g/dm}^3 = \frac{\text{amount of solute (g)}}{\text{volume of solution (cm}^3)} \times 1000$$

So, $\frac{50\ g}{200\ cm^3} = 0.25$

$0.25\ g/cm^3 \times 1000 = 250\ g/dm^3$

If you know the concentration of a given volume of solution you can calculate the amount of solute in the solution.

Worked example: Calculating mass of solute

A solution of sodium chloride has a concentration of 200 g/dm³. What is the mass of sodium chloride in 700 cm³ of the solution?

Solution

First, you need to convert 700 cm³ into dm³.

$$\frac{700}{1000} = 0.7\,\text{dm}^3$$

Then rearrange the equation to make amount of solute (g) the subject.

amount of solute (g) = concentration (g/dm³) × volume of solution (dm³)

So, 200 g/dm³ × 0.7 = 140 g

You can increase the concentration of an aqueous solution by:

- adding more solute and dissolving it in the same volume of its solution
- evaporating off some of the water from the solution so you have the same mass of solute in a smaller volume of solution.

1 Calculate the concentration in g/dm³ for:
  a 50 g of sodium chloride in 2.5 dm³ of water. [1 mark]
  b 1.8 g of sodium carbonate in 862 cm³ of water. [1 mark]
2 A technician made up a solution of potassium hydroxide, KOH, by placing 7.00 g of solid potassium hydroxide into a volumetric flask and added water up to 100 cm³ mark. She then stoppered the flask and shook the solution until the potassium hydroxide had dissolved completely. What was the concentration of the solution in g/dm³? [1 mark]

3 **H** Explain how the mass of a solute and the volume of water affect the concentration of a solution. [2 marks]
4 A student had a solution of sodium chloride with a concentration of 93.6 g/dm³ sodium.
Calculate the mass of sodium chloride dissolved in 25.0 cm³ of the solution. [2 marks]

Key points

- concentration (g/dm³) =
  $$\frac{\text{amount of solute (g)}}{\text{volume of solution (dm}^3)}$$
- To calculate the mass of solute in a certain volume of solution of known concentration:
  1 Calculate the mass (in grams) of the solute there is in 1 dm³ (1000 cm³) of solution.
  2 Calculate the mass (in grams) of solute in 1 cm³ of solution.
  3 Calculate the mass (in grams) of solute there is in the given volume of the solution.
- **H** A more concentrated solution has more solute in the same volume of solution than a less concentrated solution.

C4.7 Titrations

Learning objectives

After this topic, you should know:

- how to accurately measure the amount of acid and alkali that react together completely
- how to determine when the reaction is complete.

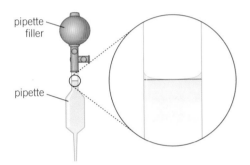

Figure 1 *A volumetric pipette and pipette filler. Fill the pipette until the bottom of the meniscus (curved surface of the solution) lines up with the mark. Allow the liquid to run out of the pipette and touch the tip on the side of the conical flask to drain out the solution. It is normal for a tiny amount of solution to remain in the pipette*

An acid and an alkali (a soluble base) react together and neutralise each other. They form a salt and water in the process.

Suppose you mix a strong acid and a strong alkali. The solution made will be neutral only if you add exactly the right quantities of acid and alkali.

If you start off with more acid than alkali, then the alkali will be neutralised. However, the solution left after the reaction will be acidic, not neutral. That is because some acid will be left over – the acid is *in excess*. If you have more alkali than acid to begin with, then all the acid will be neutralised and the solution left will be alkaline.

You can measure the exact volumes of acid and alkali needed to react with each other using a technique called **titration**. The point at which the acid and alkali have reacted completely is called the **end point** of the reaction. You judge when the end point is reached by using an acid/base indicator.

Carrying out a titration

In this experiment you can carry out a titration. You will find out how much acid is needed to completely react with an alkali.

1. Measure a known volume of alkali into a conical flask using a volumetric **pipette** (Figure 1). Before doing this, you should first wash the pipette with distilled water, and then with some of the alkali.

2. Now add a few drops of acid/base indicator to the solution in the conical flask and swirl.

3. Rinse a **burette** with distilled water, and then with some of the acid, allowing some acid to pass through the tap. Then pour the acid you are going to use into the burette. The burette has markings on it to enable you to measure volumes accurately. Burettes are graduated every 0.1 cm^3, so with care you can measure volumes of solution to the nearest 0.05 cm^3.

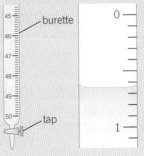

Figure 2 *A burette – use the bottom of the meniscus to read the scale. The reading here is 0.65 cm^3*

4 Record the reading on the burette. Then open the tap to release a small amount of acid into the flask. Swirl the flask to make sure that the two solutions are mixed.

5 Keep on repeating step 4 until the indicator in the flask changes colour. This shows when the alkali in the flask has completely reacted with the acid added from the burette. Record the reading on the burette and work out the volume of acid that has run into the flask. This volume is known as a titre. On your first go at doing this you will probably run too much acid into the flask, so treat this titre as a rough estimate of how much acid is needed.

6 Repeat the whole process at least three times. Discard any anomalous results (usually the first rough titre). Alternatively, keep repeating the titration until you get two results within 0.1 cm³ of each other. These precise results are called **concordant**. Then calculate a mean value to give the most accurate results possible.

Higher

7 Now you can use your results to calculate the concentration of the alkali in mol/dm³ (see the Worked example Concentration from titrations in Topic C4.9).

Safety: Wear eye protection. Chemicals in this practical may be harmful or irritant.

Go further

In A Level Chemistry you will have lots of opportunity to hone your titration skills. As well as acid/base titrations, you will also investigate the stoichiometry of redox reactions using this technique.

Key points

- Ⓗ Titration is used to measure accurately what volumes of acid and alkali react together completely.
- The point at which a reaction between an acid and an alkali is complete is called the end point of the reaction.
- You use an acid/base indicator to show the end point of the reaction between an acid and an alkali.
- To calculate the concentration of a solution in mol/dm³, given the mass of solute in a certain volume:
  - Calculate the mass (in grams) of solute in 1 cm³ of solution.
  - Calculate the mass (in grams) of solute in 1000 cm³ of solution.
  - Convert the mass (in grams) to moles.

1 a What word is used for the curved surface of a liquid in a measuring cylinder? [1 mark]

b i Name two measuring instruments used to measure volumes of liquid in a titration. [2 marks]

ii What is the correct way to read the volume from the level of liquid in one of these measuring instruments? [2 marks]

2 a Describe how to carry out a titration between dilute nitric acid of known concentration and sodium hydroxide solution of unknown concentration. ✏ [6 marks]

b Write a balanced symbol equation, including state symbols, for the reaction in part **a**. [1 mark]

3 a Name two acid/base indicators that can be used in the titration between dilute hydrochloric acid and sodium hydroxide solution. [2 marks]

b What makes a good indicator for a titration between an acid and an alkali? [1 mark]

c Describe how you obtain concordant results. [2 marks]

C4.8 Titration calculations

Learning objectives

After this topic, you should know:

- how to calculate the number of moles or the mass of solute in a given volume of solution of known concentration.
- how to calculate the amount of acid or alkali needed in a neutralisation reaction
- how to calculate an unknown concentration from reacting volumes of two solutions, where one concentration is known.

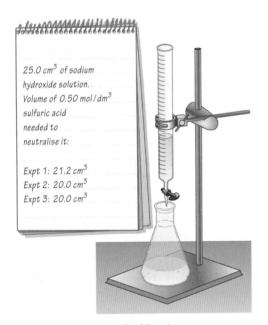

25.0 cm³ of sodium hydroxide solution.
Volume of 0.50 mol/dm³ sulfuric acid needed to neutralise it:

Expt 1: 21.2 cm³
Expt 2: 20.0 cm³
Expt 3: 20.0 cm³

Figure 1 *From results like these, you can calculate the unknown concentration of a solution – in this case the sodium hydroxide solution. In titrations you usually put the alkaline solution in the conical flask and the dilute acid in the burette*

Although chemists now have many sophisticated measuring instruments, some analytical chemists still perform titrations in their daily routine to find the concentrations of solutions.

Concentration in mol/dm³

You can also work out the concentration of a solution in mol/dm³ by using the equations on the previous page. Imagine that you make a solution of sodium hydroxide in water. You dissolve exactly 40 g of sodium hydroxide in water to make exactly 1 dm³ of solution. You know how to work out the mass of 1 mole of sodium hydroxide, NaOH, that is, its relative formula mass in grams. You add up the relative atomic masses of sodium ($A_r = 23$), oxygen ($A_r = 16$), and hydrogen ($A_r = 1$):

23 + 16 + 1 = 40 g = mass of 1 mole of NaOH.

Therefore you know that the solution contains 1 mole of sodium hydroxide in 1 dm³ of solution. So the concentration of sodium hydroxide in the solution is 1 mol/dm³.

When tackling calculations of concentrations in mol/dm³ you can either:

- substitute data provided into the concentration equations
- work through the concentration calculation in logical steps, after first converting the mass given to moles.

> **Worked example: Mass of a solute in solution**
>
> What mass of potassium sulfate, K_2SO_4, is there in 25 cm³ of a 2.0 mol/dm³ solution? (A_r values: K = 39, S = 32, O = 16)
>
> **Solution**
>
> In 1 dm³ of solution there would be 2 moles of K_2SO_4.
>
> The mass of 1 mole of K_2SO_4 is $(2 \times 39) + 32 + (4 \times 16)$ g = 174 g
>
> So in 1000 cm³ of solution there would be:
>
> $(174 \times 2) = 348$ g of K_2SO_4.
>
> Therefore in 1 cm³ of solution there are $\frac{348}{1000}$ g of K_2SO_4.
>
> So in 25 cm³ of solution there are $\frac{348}{1000} \times 25$ g = 8.7 g of K_2SO_4
>
> There are **8.7 g** of K_2SO_4 in 25 cm³ of 2.0 mol/dm³ potassium sulfate solution.

Titrations and concentration

In a titration, you always have one solution with a concentration that you know accurately. This goes in the burette. Then you can place the other solution, with an unknown concentration, in a conical flask. This is done using a volumetric pipette. This ensures you know the volume of this solution accurately. The result from the titration is used to calculate the number of moles of the substance dissolved in the solution in the conical flask.

Worked example: Concentration from titrations

In a titration experiment, 20.0 cm³ of the potassium hydroxide solution was placed in a conical flask. A few drops of phenolphthalein were added to indicate the end point of the reaction. It was titrated against dilute hydrochloric acid with a concentration of 1.00 mol/dm³. The titration was repeated until two concordant results (within 0.1 cm³ of each other) were obtained. In the experiment it was found that the potassium hydroxide solution reacted completely with exactly 12.5 cm³ of the dilute hydrochloric acid added from a burette.

What was the concentration of the potassium hydroxide solution in mol/dm³?

Solution

The balanced symbol equation for this reaction is:

$$KOH(aq) + HCl(aq) \rightarrow KCl(aq) + H_2O(l)$$

This equation tells you that 1 mole of KOH reacts with 1 mole of HCl. The concentration of the HCl is 1.00 mol/dm³, so:

- 1.00 mole of HCl is dissolved in 1000 cm³ of the dilute acid
- $\dfrac{1.00}{1000}$ moles of HCl are dissolved in 1 cm³ of acid

Therefore $\left(\dfrac{1.00}{1000}\right) \times 12.5$ moles of HCl are dissolved in 12.5 cm³ of acid.

So there are 0.0125 moles of HCl dissolved in 12.5 cm³ of the dilute acid. The balanced equation tells you that the KOH and the HCl react together in the ratio 1 : 1. So in this titration 0.0125 moles of HCl will react with exactly 0.0125 moles of KOH.

So there must have been 0.0125 moles of KOH in the 20.0 cm³ of solution in the conical flask originally.

Now you can calculate the concentration of KOH in the solution of unknown concentration in the flask.

You need to calculate the number of moles of KOH in 1 dm³ (1000 cm³) of solution.

0.0125 moles of KOH are dissolved in 20.0 cm³ of solution, so:

- $\dfrac{0.0125}{20}$ moles of KOH are dissolved in 1 cm³ of solution

Therefore, there will be $\left(\dfrac{0.0125}{20}\right) \times 1000$ moles of KOH dissolved in 1000 cm³ of the solution.

The concentration of the potassium hydroxide solution is 0.625 mol/dm³.

You can convert a concentration from mol/dm³ to g/dm³.

Worked example: Concentration in g/dm³

Convert the concentration 0.625 mol/dm³ into g/dm³.

Solution

In Topic C4.1 you saw that mass = moles × relative formula mass in grams. The relative formula mass of KOH is 39 + 16 + 1 = 56.

So the mass of 0.625 moles of KOH is (0.625 × 56) g = 35.0 g

The concentration of the potassium hydroxide solution is 35.0 g/dm³. (The answer is given to 3 significant figures.)

Key points

- You can use titration to find the unknown concentration of a solution.
- You need to know the accurate concentration of one solution, then once the end point is established, the balanced equation gives you the number of moles in a certain volume of solution.
- This value is multiplied up to give the concentration in moles per decimetre cubed (which can be converted to grams per decimetre cubed if necessary).

1 A solution of potassium nitrate has a concentration of 0.2 mol/dm³. Calculate how many moles of potassium nitrate are dissolved in 10 cm³ of the solution. [1 mark]

In a titration, a 12.5 cm³ sample of nitric acid, HNO₃, reacted exactly with 10.0 cm³ of 0.40 mol/dm³ potassium hydroxide solution. Answer questions 2 to 6 below:

2 Write down a balanced symbol equation, including state symbols, for this reaction. [1 mark]

3 Calculate the number of moles of potassium hydroxide used. [1 mark]

4 How many moles of nitric acid react? [1 mark]

5 Calculate the concentration of the nitric acid in mol/dm³. [1 mark]

6 Calculate the concentration of the nitric acid in g/dm³. [1 mark]

C4.9 Volumes of gases

Learning objectives

After this topic, you should know:

- how to calculate the volume of a gas at room temperature and pressure from its mass and relative formula mass
- how to calculate volumes of gaseous reactants and products from a balanced equation and a given volume of a gaseous reactant or product.

Figure 1 *Crash-test dummies are used to gauge the best volume of gas to cushion the effects of a collision. A sensor detects the sudden deceleration and completes an electrical circuit to ignite the chemicals in the air-bag*

Go further

In A Level Chemistry, you will use the Ideal Gas equation to calculate volumes, pressures, and temperatures of gases under conditions that differ from room temperate and pressure. You will also find out why real gases do not behave like ideal gases.

Having air-bags in cars has saved many lives. When a crash takes place, the chemicals in the bags react together to rapidly give off nitrogen gas, N_2. The reacting chemicals have to release just the right volume of gas to make the air-bag act like a cushion, reducing the damage caused to the occupants of the car.

The designers of air-bags can calculate the volume of gas produced from the balanced equation for the reaction. This tells them how many moles of gas are made, as it has been found that equal numbers of moles of any gas occupy the same volume.

The volume of 1 mole of any gas is 24 dm³ (24 000 cm³) at room temperature and pressure (r.t.p), that is, at 20 °C and 1 atmosphere.

24 dm³ per mole is known as the molar gas volume.

Using the molar gas volume you can say that:

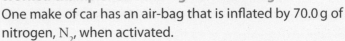
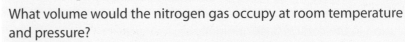

$$\text{number of moles of gas} = \frac{\text{volume of gas (dm}^3)}{24\,\text{dm}^3} = \frac{\text{volume of gas (cm}^3)}{24\,000\,\text{cm}^3}$$

Worked example: Calculating volume of a gas

One make of car has an air-bag that is inflated by 70.0 g of nitrogen, N_2, when activated.

What volume would the nitrogen gas occupy at room temperature and pressure?

(A_r of N = 14)

Solution

First of all you have to find out how many moles of nitrogen gas are in 70.0 g of N_2.

You have seen in Topic C4.1 that:

$$\text{number of moles} = \frac{\text{mass}}{\text{relative formula mass (in g)}}$$

The relative formula mass of $N_2 = (14 \times 2) = 28$.

So the number of moles of N_2 gas $= \dfrac{70.0}{28} = 2.5$ mol.

To find the volume that 2.5 mol of N_2 gas will occupy, you need to rearrange the equation:

$$\text{number of moles of gas} = \frac{\text{volume of gas (dm}^3)}{24\,\text{dm}^3}$$

To get:

volume of gas (dm³) = no. of moles × 24 dm³

So the volume of nitrogen gas = 2.5 × 24 dm³ = 60.0 dm³.

Calculating volumes of gaseous reactants or products

You can use a balanced symbol equation to find the numbers of moles of reactants and products in reactions involving gases.

This is straightforward if the reaction involves more than one gas and you are given the volume of one of the gases. The ratio of the numbers of moles in the balanced equation gives you the ratio of the volume of gases involved. This is because the same number of moles of any gas occupies the same volume. For example hydrogen and chlorine react to make hydrogen chloride:

$$H_2(g) + Cl_2(g) \rightarrow 2HCl(g)$$

1 mole of hydrogen gas reacts with 1 mole of chlorine gas to give 2 moles of hydrogen chloride gas. So the ratio of moles of $H_2(g) : Cl_2(g) : HCl(g)$ is $1 : 1 : 2$. So if you start with 50 cm³ of $H_2(g)$, it will react with 50 cm³ of $Cl_2(g)$ to give 100 cm³ of $HCl(g)$.

The problems are more complex if you have reactions of solids and/or solutions that produce a gas. However, working logically from the numbers of moles in the balanced equation will allow you to solve them.

Worked example: Calculating mass

The nitrogen gas produced in an air-bag is formed in two reactions. In the first reaction, a solid called sodium azide, NaN_3, is ignited and decomposes, producing the majority of the gas to fill the air-bag:

$$2NaN_3(s) \rightarrow 2Na(s) + 3N_2(g)$$

What mass of sodium azide would be needed to produce 48.0 dm³ of nitrogen gas at room temperature and pressure?

(A_r of N = 14, Na = 23)

Solution

You want to make 48.0 dm³ of nitrogen gas, so converting that to moles using:

$$\text{number of moles of gas} = \frac{\text{volume of gas (dm}^3)}{24\,\text{dm}^3} = \frac{48.0\,\text{dm}^3}{24\,\text{dm}^3}$$
$$= 2\text{ moles of }N_2(g)$$

The balanced equation for the reaction shown above tells you that:

2 moles of $NaN_3(s) \rightarrow$ 2 moles of $Na(s)$ + 3 moles of $N_2(g)$

So the ratio we need from the equation is:

2 mol $NaN_3(s)$: 3 mol $N_2(g)$

Dividing this ratio by 2 means that $\frac{2}{3}$ mol $NaN_3(s)$ would give 1 mol $N_2(g)$, and then multiplying by 2 (so we get two moles of N_2 gas) means that $\frac{4}{3}$ mol $NaN_3(s)$ would give 2 mol $N_2(g)$.

Using mass = number of moles × relative formula mass, we get the mass of $\frac{4}{3}$ moles of $NaN_3(s) = \frac{4}{3} \times [23 + (14 \times 3)]\,g = 86.7\,g$

Key points

- A certain volume of gas always contains the same number of gas molecules under the same conditions.
- The volume of 1 mole of any gas at room temperature and pressure is 24 dm³ (24 000 cm³).
- You can use the molar gas volume and balanced symbol equations to calculate volumes of gaseous reactants or products.

1 What is meant by the molar gas volume? [1 mark]

2 a How many moles of gas are present in:
   i 36 dm³ of carbon dioxide, $CO_2(g)$ [1 mark]
   ii 10 000 dm³ of hydrogen, $H_2(g)$? [1 mark]
  b What volume of gas would be occupied by:
   i 36.0 g of helium, $He(g)$ [2 marks]
   ii 13.8 g of nitrogen dioxide, $NO_2(g)$? [2 marks]
  c What mass of gas is present in 48 cm³ of oxygen, O_2? (Take care with the units!) [2 marks]

3 When methane gas burns completely in air, it forms carbon dioxide and water:
   $CH_4(g) + 2O_2(g) \rightarrow CO_2(g) + 2H_2O(g)$
   What volume of oxygen gas is needed to burn 150 dm³ of methane? [1 mark]

4 Calcium reacts with dilute hydrochloric acid vigorously, giving off hydrogen gas:
   $Ca(s) + 2HCl(aq) \rightarrow CaCl_2(aq) + H_2(g)$
   What volume of hydrogen would be produced when 0.80 g of calcium is added to excess dilute acid? [3 marks]

C4 Chemical calculations

Summary questions

1 Calculate the relative formula mass M_r of each of the following compounds:

   a H_2S [1 mark]   **e** Na_2CO_3 [1 mark]
   b SO_2 [1 mark]   **f** $Al_2(SO_4)_3$ [1 mark]
   c C_2H_4 [1 mark]   **g** $NaAl(OH)_4$ [1 mark]
   d $NaOH$ [1 mark]

H 2 How many moles of:

   a Ag atoms are there in 27 g of silver   [1 mark]
   b Fe atoms are there in 0.056 g of iron [1 mark]
   c P_4 molecules are there in 6.2 g of phosphorus?   [1 mark]

3 In a lime kiln, calcium carbonate is decomposed to calcium oxide:

$$CaCO_3 \rightarrow CaO + CO_2$$

1500 tonnes of calcium carbonate gave 804 tonnes of calcium oxide. Calculate the percentage yield for the process.   [2 marks]

4 **a** Ethene gas, C_2H_4, reacting with steam, H_2O, to form ethanol gas, C_2H_5OH, is a reversible reaction. Write the balanced symbol equation for this reaction, including state symbols.   [1 mark]

   b If 14.00 g of ethene reacts with excess steam to produce 17.25 g of ethanol, what is the percentage yield of the reaction?   [2 marks]

H 5 Sulfur is mined in Poland and is brought to Britain in ships. The sulfur is used to make sulfuric acid. Sulfur is burnt in air to produce sulfur dioxide. Sulfur dioxide and air are passed over a heated catalyst to produce sulfur trioxide. Water can be added to sulfur trioxide to produce sulfuric acid. The reactions are:

$$S + O_2 \rightarrow SO_2$$
$$2SO_2 + O_2 \rightleftharpoons 2SO_3$$
$$SO_3 + H_2O \rightarrow H_2SO_4$$

(A_r values: H = 1, O = 16, S = 32)

   a How many moles of sulfuric acid could be produced from one mole of sulfur?   [1 mark]

   b Calculate the maximum mass of sulfuric acid that can be produced from 64 kg of sulfur.   [2 marks]

   c In an industrial process, the mass of sulfuric acid that was produced from 64.00 kg of sulfur was 188.16 kg. Use your answer to part **b** to calculate the percentage yield of this process.   [1 mark]

   d Suggest two reasons why the yield of the industrial process was less than the maximum yield.   [2 marks]

6 A student placed 12.5 cm³ of potassium hydroxide solution of an unknown concentration into a conical flask using a volumetric pipette. The potassium hydroxide reacted with exactly 20.0 cm³ of 0.50 mol/dm³ hydrochloric acid, which was added from a burette.

   a Write the balanced symbol equation, including state symbols, for this reaction.   [3 marks]

H   **b** **i** How many moles of hydrochloric acid are in 20.0 cm³ of 0.50 mol/dm³ hydrochloric acid?   [1 mark]

       ii How many moles of potassium hydroxide will this react with?   [1 mark]

H   **c** What is the concentration of the potassium hydroxide solution:

       i in moles per dm³   [1 mark]
       ii in grams per dm³?   [2 marks]

7 Magnesium carbonate, $MgCO_3$, decomposes when heated in a similar reaction to calcium carbonate.

   a Write a balanced symbol equation for the decomposition of magnesium carbonate.   [1 mark]

H   **b** When a sample of magnesium carbonate was decomposed, 1.10 g of carbon dioxide gas was given off. What volume of carbon dioxide gas was given off (as measured at room temperature and pressure)?   [2 marks]

H   **c** What mass of magnesium carbonate was decomposed in the reaction?   [1 mark]

H   **d** Magnesium carbonate also reacts with dilute hydrochloric acid:

$$MgCO_3(s) + 2HCl(aq) \rightarrow$$
$$MgCl_2(aq) + H_2O(l) + CO_2(g)$$

       i What is the easiest way to tell when this reaction has finished?   [1 mark]

       ii 8.4 g of magnesium carbonate was added to 25 cm³ dilute hydrochloric acid with a concentration of 0.20 mol/dm³. Show which reactant is the limiting reactant.   [4 marks]

       iii What volume of carbon dioxide gas was given off in the reaction described in part **ii**, at r.t.p.?   [1 mark]

8 Calculate the relative atomic mass of:

   a bromine (made up of 50% ^{79}Br and 50% ^{81}Br)   [1 mark]

   b iron (with 5.8% ^{54}Fe, 91.8% ^{56}Fe, 2.1% ^{57}Fe and 0.3% ^{58}Fe).   [2 marks]

Practice questions

01 This question is about the change in mass when chemical reactions take place in a crucible.
A crucible has a loose lid so that gases can get in or get out. A diagram of a crucible is shown in **Figure 1**.

Figure 1

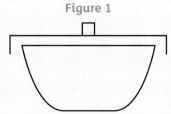

01.1 A student heated a piece of magnesium in a crucible. The magnesium reacts as shown in the equation. Balance the equation.
……. $Mg(s) + O_2(g) \rightarrow$ ……. $MgO(s)$ [2 marks]

01.2 The student recorded the masses shown in **Table 1**.

Table 1

	Mass in g
Mass of crucible at the start of the reaction	0.24
Mass of crucible at the end of the reaction	0.40

Explain why the mass increased. [2 marks]

01.3 The student heated the crucible again at the end of the reaction. What could the student do to make sure the reaction was complete? [2 marks]

01.4 Another student heated lithium carbonate in a crucible. The lithium carbonate reacts as shown in the equation.
$Li_2CO_3(s) \rightarrow Li_2O(s) + CO_2(g)$
Use the equation to predict whether the mass would increase or decrease. Explain your answer.
[3 marks]

H **02** This question is about the combustion of hydrocarbons.

02.1 0.010 moles of hydrocarbon **Z** are burnt completely in an excess of oxygen. The equation for the reaction is below.
$C_xH_y + O_2(g) \rightarrow$ …….$CO_2(g) +$ …….H_2O
1.76 g of carbon dioxide and 0.90 g of water are produced.
Use this information to work out the balancing numbers for CO_2 and H_2O.
Relative atomic masses: C = 12; H = 1; O = 16
[4 marks]

02.2 Use your answer from **02.1** to work out the identity of hydrocarbon **Z**. Tick (✓) the correct answer

C_2H_4	
C_3H_8	
C_4H_{10}	
C_8H_{18}	

[1 mark]

02.3 In a different experiment, a student dissolved 0.22 g of carbon dioxide in 400 cm³ of pure water. Calculate the concentration in mol/dm³ of the solution produced. [3 marks]

H **03** This question is about the manufacture of ammonia
Ammonia is made from nitrogen and hydrogen as shown in the equation
$N_2(g) + 3H_2(g) \rightleftharpoons 2NH_3(g)$
Two molecules of ammonia are shown in **Figure 2**

Figure 2

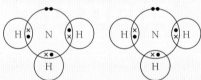

Ammonia is a gas at room temperature because it has a low boiling point.

03.1 Explain why ammonia has a low boiling point. You should refer to **Figure 2** in your answer. [2 marks]

03.2 84 tonnes of nitrogen were mixed with 30 tonnes of hydrogen.
Relative atomic masses: N = 14; H = 1.
1 tonne = 1 000 000 g.
Calculate the number of moles of nitrogen and the number of moles of hydrogen, and show that nitrogen is the limiting reactant. [3 marks]

03.3 Calculate the maximum mass of ammonia that can be produced from 84 tonnes of nitrogen.
[3 marks]

03.4 Only 18 tonnes of ammonia were produced. Use your answer from **03.3** to calculate the percentage yield.
Give your answer to one decimal place. [3 marks]

03.5 The percentage atom economy for this reaction is 100%.
Explain why. You do not need to do a calculation.
[3 marks]

2 Chemical reactions and energy changes

In the early 19th century, people began experimenting with chemical reactions in a systematic way, organising their results logically. Gradually they began to predict exactly what new substances would be formed and used this knowledge to develop a wide range of different materials and processes. They could extract important resources from the Earth, for example, by using electricity to decompose ionic substances. This is how reactive metals such as aluminium and sodium were discovered.

Energy changes are also an important part of chemical reactions. Transfers of energy take place due to the breaking and formation of bonds. The heating or cooling effects of reactions are used in a range of everyday applications.

Key questions

- How can we extract metals from their ores?
- How can we make and prepare pure, dry samples of salts?
- How can we decompose ionic compounds to get useful products?
- Why do chemical reactions always involve transfers of energy?

Making connections

- Reaction profile diagrams will be used in **C8 Rates and equilibrium** to explain the effect of catalysts on the rate of a chemical reaction.
- The calculations of energy changes using bond energy values relies on you drawing the 2D structures of the molecules involved in the reaction, which was covered in **C3 Structure and bonding**.
- Displacement reactions and the use of electrolysis will be applied in **C14 The Earth's resources**.

I already know...

I will learn...

how to define acids and alkalis in terms of neutralisation reactions.

how to represent neutralisation using an ionic equation.

how to use the pH scale for measuring acidity and alkalinity.

how to calculate the concentration of hydrogen ions in a solution given its whole number pH value.

about displacement reactions and the reactions of acids with metals to produce a salt plus hydrogen.

to interpret displacement reactions and the reaction between an acid and a metal in terms of reduction and oxidation.

the reactions of acids with alkalis to produce a salt plus water.

to calculate the concentration of an unknown acid or alkali from experimental results.

combustion and rusting are examples of oxidation reactions.

to identify and describe oxidation and reduction reactions in terms of electron transfer.

that chemical reactions are exothermic and endothermic.

to use bond energy values to calculate the approximate energy change accompanying a reaction.

Required Practicals

Practical		Topic
1	Making a copper salt	C5.5
	Making a salt from a metal carbonate	C5.6
3	Investigating the electrolysis of a solution	C6.4
4	Investigating temperature changes	C7.1

Learning objectives

After this topic, you should know:

- how some common metals react with water and with dilute acid
- how to deduce an order of reactivity of metals based on experimental results
- explain reduction and oxidation in terms of loss or gain of oxygen
- predict reactions of unfamiliar metals given information about their relative reactivity.

Figure 1 *Metals are important materials in transportation*

Metals are important in all your lives. For example, in transport metals are used to make bicycles, cars, ships, trains, and aeroplanes. In this chapter you will look at the chemistry involved in getting some of these metals from their raw materials, **ores**. Ores are rocks from which it is economical to extract the metals that they contain. With new techniques, metals can now be extracted from rock that was once thought of as waste.

Most metals in ores are chemically bonded to other elements in compounds. Many of these metals have been **oxidised** (i.e., have oxygen added) by oxygen in the air to form their oxides. For example:

$$\text{iron} + \text{oxygen} \rightarrow \text{iron(III) oxide}$$

$$4Fe(s) + 3O_2(g) \rightarrow 2Fe_2O_3(s)$$

So to extract the metals from their oxides, the metal oxide must be **reduced** (i.e., have oxygen removed). To understand how this is done, you will need to know about the **reactivity series** of metals.

The reactivity series is a list of metals in order of their reactivity, with the most reactive metals at the top and the least reactive ones at the bottom.

Metals plus water

You can start putting the metals in order of reactivity by looking at their reactions with water. Most metals do not react vigorously with water. Metals such as copper, which does not react at all with water, can be used to make water pipes. However, there is a great range in reactivity between different metals. For example, in Topic C2.3 you have seen how the alkali metals in Group 1 react with water. They react vigorously, giving off hydrogen gas and leaving alkaline solutions.

The reactivity of the metals increases going down Group 1, so of the metals you have observed the order of reactivity is:

1 potassium (most reactive)

2 sodium

3 lithium.

Magnesium lies somewhere between lithium and copper in the reactivity series. If magnesium is left in a beaker of water, it takes several days to collect enough gas to test with a lighted spill. The resulting 'pop' shows that the gas is hydrogen.

Metals plus dilute acid

You have now seen how a range of metals react with water, and that you can use your observations of these reactions to place the metals into an order of reactivity. However, where the reactions are very slow, the task of ordering the metals is difficult. With these metals you can look at their reactions with dilute acid to arrive at an order of reactivity.

Table 1 summarises the reactions of some important metals with water and dilute acid.

Table 1 *The reactivity series. Note: aluminium is protected by a layer of aluminium oxide, so will not undergo the reactions below unless the oxide layer is removed. That is why this fairly reactive metal can be used outside, for example, in sliding patio doors, without corroding.*

Order of reactivity	Reaction with water	Reaction with dilute acid
potassium	fizz, giving off hydrogen gas, leaving an alkaline solution of metal hydroxide	explode
sodium		
lithium		
calcium	very slow reaction	fizz, giving off hydrogen gas and forming a salt
magnesium		
aluminium		
zinc		
iron		
tin	slight reaction with steam	react slowly with warm acid
lead		
copper	no reaction, even with steam	no reaction
silver		
gold		

1 Write the word equation and balanced symbol equation for:
  a lithium reacting with water [3 marks]
  b zinc reacting with dilute hydrochloric acid. [3 marks]

2 A student added a piece of magnesium ribbon to dilute sulfuric acid.
  a List three ways she could tell that a chemical reaction was taking place. [2 marks]
  b Write down the general equation that describes the reaction between a metal and an acid. [1 mark]
  c Write the word equation and balanced symbol equation, including state symbols, for the reaction between magnesium and dilute sulfuric acid. [3 marks]

3 Explain the following facts in terms of chemical reactivity.
  a Gold, silver, and platinum are used to make jewellery. [2 marks]
  b Potassium, lithium, and sodium are stored in jars of oil. [2 marks]
  c Food cans are plated with tin, but not with zinc. [2 marks]

4 Explain why aluminium can be used outdoors, for example, for window frames, even though it is quite high in the reactivity series. [2 marks]

Ⓗ 5 Using the balanced equation for the oxidation of iron shown on the previous page, calculate the mass of iron(III) oxide formed when 2.80 g of iron is completely oxidised. [3 marks]

Synoptic link

For more about the reaction of metals with dilute acid, see Topic C5.4.

Metals and acid

You are given coarse-grained filings of the metals copper, zinc, iron, and magnesium to put into an order of reactivity according to their reactions with dilute hydrochloric acid.

Plan a test to put the metals in order of reactivity based on your observations. Your plan should include the quantities of reactants you intend to use. (You will have access to a balance and measuring cylinders.)

Let your teacher check your plan before you start your tests.

● Give a brief outline of your method, including how you will make it as fair a test as possible. Identify any hazards.

● Record your results in a suitable table.

● Put the metals in order of reactivity according to your observations.

● Evaluate your investigation.

Safety: Wear eye protection.

Key points

● The metals can be placed in order of reactivity by their reactions with water and dilute acid.
● Hydrogen gas is given off if metals react with water or dilute acids. The gas 'pops' with a lighted spill.

C5.2 Displacement reactions

Learning objectives

After this topic, you should know:

- the position of carbon and hydrogen in the reactivity series and how to predict displacement reactions
- how the reactivity of metals is related to the tendency of the metal to form its positive ion
- **H** how to write ionic equations for displacement reactions
- **H** how to identify in a given reaction, symbol equation, or half equation which species are oxidised and which are reduced, in terms of electron transfer.

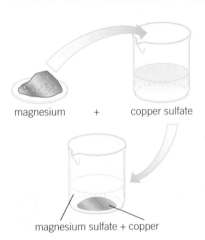

magnesium   +   copper sulfate

magnesium sulfate + copper

Figure 1 *Magnesium displaces copper from copper(II) sulfate solution*

Displacing a metal from solution

Set up the test tube as shown:

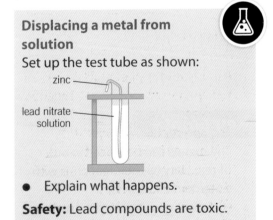

zinc

lead nitrate solution

- Explain what happens.

Safety: Lead compounds are toxic. Wear eye protection.

You have now seen how to use the reactions of metals with water and dilute acid to get an order of reactivity. You can also judge reactivity by putting the metals 'into competition' with each other. One metal starts off as atoms of the element and the other metal as positive ions in a solution of one of its salts. For example, you might have magnesium metal, Mg(s), and copper(II) ions, $Cu^{2+}(aq)$, in a solution of copper(II) sulfate.

A more reactive metal will displace a less reactive metal from an aqueous solution of one of its salts.

In this case, magnesium is more reactive than copper. Therefore the copper ions will be displaced from solution to form copper metal, Cu(s). In this reaction, the magnesium metal forms aqueous magnesium ions, $Mg^{2+}(aq)$, and dissolves into the solution. This is a **displacement reaction**.

magnesium + copper(II) sulfate → magnesium sulfate + copper

The balanced symbol equation, including state symbols, is:

$$Mg(s) + CuSO_4(aq) \rightarrow MgSO_4(aq) + Cu(s)$$

An **ionic equation** shows only the atoms and ions that change in a reaction. In this reaction the sulfate ions, $SO_4^{2-}(aq)$, remain the same, so do not appear in the ionic equation. So the correct ionic equation is:

$$Mg(s) + Cu^{2+}(aq) \rightarrow Mg^{2+}(aq) + Cu(s)$$

Higher

This shows that magnesium atoms have a greater tendency to form positive ions than copper atoms.

Zinc is more reactive than lead – it is higher up the reactivity series. Therefore, zinc displaces lead from its solution:

zinc   +   lead nitrate   →   zinc nitrate   +   lead
$$Zn(s) + Pb(NO_3)_2(aq) \rightarrow Zn(NO_3)_2(aq) + Pb(s)$$

You would see the lead metal forming as crystals on the zinc.

Hydrogen and carbon in the reactivity series

You can include the non-metals hydrogen and carbon in the reactivity series using displacement reactions. You can think of the metal plus acid reactions as displacement of hydrogen ions, $H^+(aq)$, from solution. Copper cannot displace the hydrogen from an acid, whereas lead can. So hydrogen is positioned between copper and lead. Carbon can be used in the extraction of metals from their oxides (another type of displacement reaction). However, it can only do this for metals below aluminium in the reactivity series. It does not displace aluminium from aluminium oxide, but can displace zinc from zinc oxide. So carbon is placed between aluminium and zinc in the series (Figure 2 in Topic C5.3).

Oxidation and reduction

In Topic C5.1 you saw oxidation was defined as the chemical addition of oxygen, and reduction as the removal of oxygen. A wider definition involves the transfer of electrons rather than oxygen atoms.

Oxidation is the loss of electrons.   Reduction is the gain of electrons.

You can apply these definitions to displacement reactions in solution. Take, as an example, the displacement of copper(II) ions by iron. This reaction is used in industry to extract copper metal from copper sulfate solution. The iron added to the copper sulfate solution is cheap scrap iron. The **ionic equation** for the reaction is:

$$Fe(s) + Cu^{2+}(aq) \rightarrow Fe^{2+}(aq) + Cu(s)$$

You can use **half-equations** to show what happens to each reactant:

$$Fe(s) \rightarrow Fe^{2+}(aq) + 2e^-$$

The iron atoms lose two electrons to form iron(II) ions. This is oxidation (the loss of electrons). The iron atoms have been oxidised.

The two electrons from iron are gained by the copper(II) ions as they form copper atoms:

$$Cu^{2+}(aq) + 2e^- \rightarrow Cu(s)$$

This is reduction (the gain of electrons). The copper(II) ions have been reduced.

This is why these displacement reactions are also known as *redox* reactions (*reduction-oxidation*).

1 Which of the following pairs of metals and solutions will result in a reaction? If a reaction is predicted, write a word equation and a balanced symbol equation.
   a iron + zinc sulfate   [1 mark]
   b zinc + copper sulfate   [2 marks]
   c magnesium + iron(II) chloride   [2 marks]

2 Explain why carbon can reduce zinc oxide, but magnesium oxide cannot.   [2 marks]

3 Hydrogen gas is used in the reduction of tungsten oxide, WO_3, to extract tungsten metal.
   a What can you deduce about tungsten metal?   [1 mark]
   b Write a balanced symbol equation for this reaction.   [1 mark]
 H 4 a Write the ionic equation, including state symbols, for the reaction between zinc and iron(II) sulfate.   [3 marks]
      b Explain, in terms of the transfer of electrons which species is oxidised and which is reduced in the reaction.   [4 marks]
      c 3.25 g of zinc powder is added to $50\,cm^3$ of a $0.50\,mol/dm^3$ iron(II) sulfate solution. Which is the limiting reactant, zinc or iron(II) sulfate, and which is present in excess? Show how you arrived at your answer.   [3 marks]

Predicting reactions

You will be provided with small samples of magnesium, copper, zinc, and iron, plus their sulfate solutions. Draw a table to record possible combinations of metal + metal sulfate solution.

Predict which metals and solutions will react (enter a tick in the table), and which will not (enter a cross). Then try out the reactions on a spotting tile.

Safety: Wear eye protection. Some solutions are harmful.

Study tip

You can use the phrase **OILRIG** to remember the definition of oxidation and reduction reactions:

Oxidation **I**s **L**oss of electrons

Reduction **I**s **G**ain of electrons

Go further

You can use the redox reactions of metals in cells or batteries to power your mobile devices, as you will discover if you study A Level Chemistry.

Synoptic link

For more about an industrial application of the displacement of copper ions from solution by iron, see Topic C14.4.

Key points

- A more reactive metal will displace a less reactive metal from its aqueous solution.
- The non-metals hydrogen and carbon can be given positions in the reactivity series on the basis of displacement reactions.
- **H** Oxidation is the loss of electrons.
- **H** Reduction is the gain of electrons.

C5.3 Extracting metals

Figure 1 *The Angel of the North stands 20 metres tall. It is made of steel, which contains a small amount of copper*

Metals have been important to people for thousands of years. You can follow the course of history by the materials people used. After the Stone Age came the Bronze Age (copper/tin) and then on to the Iron Age.

Where do metals come from?

Metals are found in the Earth's crust. Most metals are combined chemically with other chemical elements, often with oxygen or sulfur. This means that the metal must be chemically separated from its compounds before it can be used.

When there is enough of a metal or metal compound in a rock to make it worth extracting the metal, the rock is called a **metal ore**. Ores are mined from the ground. Some need to be concentrated before the metal is extracted and purified. For example, copper ores are ground up into a powder. Then they are mixed with water and a chemical that makes the copper compound repel water. Air is then bubbled through the mixture and the copper compound floats on top as a froth. The rocky bits sink and the concentrated copper compound is scraped off the top. It is then ready to have its copper extracted.

Whether it is worth extracting a particular metal depends on:

- how easy it is to extract it from its ore
- how much metal the ore contains
- the changing demands for a particular metal.

These factors can change over time. For example, a new, cheaper method might be discovered for extracting a metal. You might also discover a new way to extract a metal efficiently from rock which contains only small amounts of a metal ore. An ore that was once thought of as 'low grade' could then become an economic source of a metal.

A few metals, such as gold and silver, are so unreactive that they are found in the Earth as the metals (elements) themselves. They exist in their native state.

Sometimes a nugget of gold is so large it can simply be picked up. At other times, tiny flakes have to be physically separated from sand and rocks by panning.

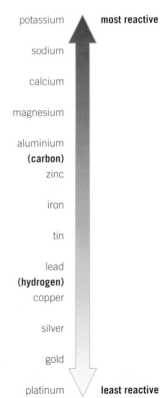

Figure 2 *This reactivity series shows the position of the non-metals carbon and hydrogen in the series*

Reduction of oxide by carbon

The way that a metal is extracted depends on its place in the reactivity series. The reactivity series lists the metals in order of their reactivity (Figure 2, which includes the non-metals carbon and hydrogen). As you have seen in Topic C5.2, a more reactive metal will displace a less reactive metal from its compounds. Carbon (a non-metal) will also displace less reactive metals from their oxides. Carbon is used to extract some metals from their ores in industry.

You can find many metals, such as copper, lead, iron, and zinc, combined with oxygen as metal oxides. Because carbon is more reactive than each of these metals, carbon is used to extract the metals from their oxides.

You must heat the metal oxide with carbon. The carbon removes the oxygen from the metal oxide to form carbon dioxide. The metal is also formed, as the element:

$$\text{metal oxide} + \text{carbon} \rightarrow \text{metal} + \text{carbon dioxide}$$

For example:

$$\text{lead oxide} + \text{carbon} \xrightarrow{\text{heat}} \text{lead} + \text{carbon dioxide}$$
$$2PbO(s) + C(s) \rightarrow 2Pb(l) + CO_2(g)$$

The removal of oxygen from a compound is called chemical reduction.

Reduction of oxide by hydrogen

Another metal extracted by reduction of its oxide is tungsten, W. The non-metal used as the reducing agent is hydrogen, not carbon, even though carbon would be cheaper. This is because the carbon forms a compound, tungsten carbide, with the metal formed by reduction. The tungsten obtained from reduction of its oxide by hydrogen is very pure:

$$\text{tungsten oxide} + \text{hydrogen} \xrightarrow{\text{heat}} \text{tungsten} + \text{water (as steam)}$$
$$WO_3(s) + 3H_2(g) \rightarrow W(s) + 3H_2O(g)$$

The metals that are more reactive than carbon are not extracted from their ores by reduction with carbon. Instead, they are extracted by **electrolysis** of the molten metal compound.

1 Define the term metal ore. [2 marks]

2 Why is gold found as the metal rather than combined with other elements in compounds? [1 mark]

3 Platinum is found in its native state. What does this tell you about its reactivity? Give a use of platinum that depends on this chemical property. [2 marks]

4 Zinc oxide, ZnO, can be reduced to zinc by heating it in a furnace with carbon. Carbon monoxide, CO, gas is given off in the reaction. The zinc formed is molten at the temperature in the furnace.
   a Write a word equation for the reduction of zinc oxide by carbon in the furnace, labelling what is reduced and what is oxidised. [1 mark]
   b Write a balanced symbol equation, including state symbols, for the reaction in part a. [3 marks]

Reduction by carbon

Heat some copper oxide with carbon powder in a test tube, gently at first, then more strongly.

Empty the contents into an evaporating dish.

You can repeat the experiment with lead oxide and carbon if you have a fume cupboard to work in.

● Explain your observations. Include a word equation and a balanced symbol equation with state symbols.

Safety: Wear eye protection. Wash hands after experiments. Copper oxide is harmful and lead oxide is toxic.

Synoptic link

To find out more about the use of electrolysis to extract reactive metals, see Topic C6.3.

Key points

● A metal ore contains enough of the metal to make it economic to extract the metal. Ores are mined and might need to be concentrated before the metal is extracted and purified.

● Gold and some other unreactive metals can be found in their native state.

● The reactivity series helps you decide the best way to extract a metal from its ore. The oxides of metals below carbon in the series can be reduced by carbon to give the metal element.

● Metals more reactive than carbon *cannot* be extracted from their ores using carbon. They are extracted by electrolysis of the molten metal compound (see Chapter C6).

C5.4 Salts from metals

Learning objectives

After this topic, you should know:

- the reactions of magnesium, zinc, and iron with hydrochloric and sulfuric acids, and how to collect the salts formed
- **(H)** why these reactions are classed as redox reactions
- **(H)** how to identify which species are oxidised and which are reduced in given chemical equations, in terms of electron transfer.

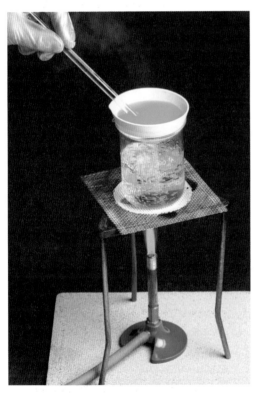

Figure 1 *Heating by using a water bath is a gentler way to evaporate water from the salt solution than heating the solution directly. Steam from a beaker of water heated below the dish heats the dish and its contents. The slower the water evaporates from the solution, the larger the crystals of the salt collected at the end of the experiment*

Acids, metals, and salts

You used the reactions of metals with dilute acid as one of the reactions to deduce the reactivity series in Topic C5.1. Reactions between metals and acids can only occur when the metal is more reactive than the hydrogen in the acid. For example, iron reacts with dilute acids but silver does not. All acids contain hydrogen, which is released as hydrogen ions when the acid is dissolved in water (Topic C5.7).

Whenever a reaction does take place between a metal and an acid, a **salt** is formed. A salt is the general name for *a compound formed when the hydrogen in an acid is wholly, or partially, replaced by metal (or ammonium) ions.*

So one way you can make salts is by reacting acids directly with metals that are more reactive than hydrogen.

metal	+	acid	$\rightarrow$	a salt	+	hydrogen
iron	+	hydrochloric acid	$\rightarrow$	iron(II) chloride	+	hydrogen
$Fe(s)$	+	$2HCl(aq)$	$\rightarrow$	$FeCl_2(aq)$	+	$H_2(g)$

If the metal is very reactive, the reaction with acid is too violent to be carried out safely. So alkali metals are never added to acid.

Pure, dry crystals of a salt (iron(II) chloride in the equation above) can be obtained from the solution. Some of the water is evaporated from the solution by heating it until the point of crystallisation is reached. At this point, the solution is saturated and crystals will appear at the edge of a salt solution being heated in an evaporating dish. This point can also be tested by dipping a glass rod into the hot salt solution, removing it, and seeing if crystals form in the solution left on the rod as it cools down.

In order to prepare the best samples of salt crystals, the salt solution should then be left at room temperature for the remaining water to evaporate slowly. You can remove any small amounts of solution left on the crystals by dabbing on filter papers, then leaving to dry.

The salt that you make depends on the metal you use, as well as on the acid. So magnesium metal will always make salts containing magnesium ions, Mg^{2+}. Zinc metal will always make zinc salts, containing Zn^{2+} ions.

The acid used provides the negative ions present in all salts:

- the salts formed when you react a metal with hydrochloric acid, HCl, are always *chlorides* (containing Cl^- ions)
- sulfuric acid, H_2SO_4, makes *sulfates* (containing SO_4^{2-} ions)
- nitric acid, HNO_3, always makes *nitrates* (containing NO_3^- ions).

Explaining the reaction between a metal and an acid

Higher

In the reaction between magnesium, Mg, and dilute sulfuric acid, H_2SO_4, hydrogen ions will be displaced from solution by magnesium. This happens because magnesium is more reactive than hydrogen. Magnesium has a stronger tendency to form positive ions than hydrogen has, so the following reaction takes place:

$$Mg(s) + H_2SO_4(aq) \rightarrow MgSO_4(aq) + H_2(g)$$

You can summarise this reaction as an ionic equation (see Topic C5.2):

$$Mg(s) + 2H^+(aq) \rightarrow Mg^{2+}(aq) + H_2(g)$$

The sulfate ions in the solution, $SO_4^{2-}(aq)$, do not change in the reaction, so are not included in the ionic equation. They are called spectator ions.

Then you can look more closely at the ionic equation by dividing it into two half equations (see Topic C5.2). You can see what happens to the magnesium atoms when they change into positive magnesium ions:

$$Mg(s) \rightarrow Mg^{2+}(aq) + 2e^-$$

A magnesium atom loses its two electrons from its outer shell. It gives these electrons to two hydrogen ions from the acidic solution, $2H^+(aq)$, forming two H atoms. These bond to each other (sharing a pair of electrons in a covalent bond) to make a molecule of hydrogen gas, H_2:

$$2H^+(aq) + 2e^- \rightarrow H_2(g)$$

In Topic C5.2, you found that oxidation is loss of electrons and that reduction is gain of electrons (remember OILRIG). So you can conclude that electrons have been transferred from magnesium atoms to hydrogen ions in the reaction:

● The magnesium atoms have lost electrons, so magnesium atoms have been *oxidised* in the reaction.

● The hydrogen ions have gained electrons, so hydrogen ions have been *reduced* in the reaction.

The reaction of a metal with an acid is always a redox reaction, because the metal atoms always donate electrons to the hydrogen ions, displacing hydrogen as a gas and leaving the metal ions in the solution.

1 Write the general equation for the reaction 'acid + metal'. [1 mark]

2 a Why can't copper sulfate be prepared by adding copper metal to dilute sulfuric acid? [1 mark]

 b Why is potassium chloride never prepared by reacting potassium metal and dilute hydrochloric acid together? [1 mark]

3 Write a balanced symbol equation, including state symbols, for:

 a iron + sulfuric acid [3 marks]

 b zinc + hydrochloric acid [3 marks]

Ⓗ 4 Using the reaction of zinc with dilute hydrochloric acid:

 a Write an ionic equation for the reaction, with state symbols. [2 marks]

 b From your answer to part a, construct two half equations to show the electron transfers taking place. [2 marks]

 c Explain why this is a redox reaction. [5 marks]

Planning to make a salt

Given zinc powder and $10\,cm^3$ of dilute hydrochloric acid or dilute sulfuric acid, plan an experiment to make and collect pure, dry crystals of the salt.

● Write a word equation (and balanced symbol equation, with state symbols, if possible) for the reaction chosen.

● Think how you can make sure that all the dilute acid has reacted. How will you tell when the reaction is finished?

● Then write a step-by-step method that another student could follow to collect a sample of the salt crystals from adding zinc to dilute acid.

Safety: Do not try out your plan before your teacher has checked it.

Key points

● A salt is a compound formed when the hydrogen in an acid is wholly or partially replaced by metal or ammonium ions.

● Salts can be made by reacting a suitable metal with an acid. The metal must be above hydrogen in the reactivity series, but not dangerously reactive.

● The reaction between a metal and an acid produces hydrogen gas as well as a salt. A sample of the salt made can then be crystallised out of solution by evaporating off the water.

● Ⓗ The reaction between a metal and an acid is an example of a redox reaction. The metal atoms lose electrons and are oxidised, and hydrogen ions from the acid gain electrons and are reduced.

C5.5 Salts from insoluble bases

Learning objectives

After this topic, you should know:

- the reaction between an acid and a base
- how to prepare pure, dry crystals of the salts formed in neutralisation reactions between acids and insoluble bases
- how to predict products from given reactants
- how to use the formulae of common ions to deduce the formulae of salts.

Acids and bases are an important part of your understanding of chemistry. They play an important part inside you and all other living things. You have probably neutralised acids with bases before in your science lessons. In fact, you have probably neutralised acids in your mouth today when you brushed your teeth with toothpaste

When you neutralised acid previously in scientific experiments you will have used an alkali, such as sodium hydroxide solution, $NaOH(aq)$. Alkalis are part of a larger class of compounds called bases. Bases are compounds that can neutralise acids, and alkalis are those bases that are soluble in water, such as the hydroxides of Group 1 metals. Many bases are metal oxides, such as sodium oxide or copper oxide.

When you react an acid with a base, a salt and water are formed.

The general equation which describes this **neutralisation** reaction is:

$$\text{acid} \; + \; \text{base} \; \longrightarrow \; \text{a salt} \; + \; \text{water}$$

The oxide of a transition metal, such as iron(III) oxide, is an example of a base you can use to make a salt in this way:

hydrochloric acid	+	solid iron(III) oxide	$\longrightarrow$	iron(III) chloride solution	+	water
$6HCl(aq)$	+	$Fe_2O_3(s)$	$\longrightarrow$	$2FeCl_3(aq)$	+	$3H_2O(l)$

Formulae of salts

Salts are made up of positive metal ions (or ammonium ions, NH_4^+) and a negative ion from an acid. The positive ions can come from a metal, a base, or a carbonate. A carbonate reacts with an acid to form a salt, water, and carbon dioxide gas.

Like all ionic compounds, salts have no overall charge, as the sum of the charges on their ions equals zero. So once you know the charges on the ions that make up a salt, you can work out its formula.

The charges on common positive ions	The charges on common negative ions
ions of Group 1 metals = +1 (e.g., Li^+, Na^+, K^+)	ions of Group 7 non-metals = −1 (e.g., F^-, Cl^-, Br^-, I^-)
ions of Group 2 metals = +2 (e.g., Mg^{2+}, Ca^{2+})	nitrate ions = −1, NO_3^-
aluminium ion = +3, Al^{3+}	sulfate ions = −2, SO_4^{2-}
ammonium ion = +1, NH_4^+	
transition metals = variable (size of positive charge given by the roman numeral in the name, e.g., copper(II) ion, Cu^{2+}, or iron(III) ion, Fe^{3+})	

Making a copper salt

You can make copper sulfate crystals from copper(II) oxide (an insoluble base) and sulfuric acid. The equation for the reaction is:

sulfuric acid + copper(II) oxide → copper(II) sulfate + water

$$H_2SO_4(aq) + CuO(s) \rightarrow CuSO_4(aq) + H_2O(l)$$

1

2

warm gently

Add insoluble copper oxide to sulfuric acid and stir. Warm gently on a tripod and gauze (do not boil).

The solution turns blue as the reaction occurs, showing that copper sulfate is being formed. Excess black copper oxide can be seen.

3

4

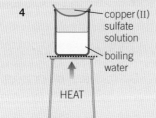

copper(II) sulfate solution

boiling water

HEAT

When the reaction is complete, filter the solution to remove excess copper oxide.

You can evaporate the water so that crystals of copper sulfate start to form. Stop heating when you see the first crystals appear at the edge of the solution. Then leave for the rest of the water to evaporate off slowly. This will give you larger crystals. Any small excess of solution on the crystals can be removed by dabbing between filter papers (do not touch the solution), then leaving to dry.

● What does the copper sulfate look like? Draw a diagram if necessary.

Safety: Wear eye protection. Chemicals in this practical are harmful. Make sure you only warm the acid gently – do not boil it!

1 Write the general word equation for an acid–base reaction. [1 mark]

2 Write the word equation and balanced symbol equation for the reaction between zinc oxide and dilute hydrochloric acid. [5 marks]

3 What is the formula of each of the following salts?
  a sodium bromide [1 mark] **b** magnesium fluoride [1 mark]
  c potassium nitrate [1 mark] **d** aluminium sulfate [1 mark]

4 Describe in detail how you could prepare a sample of copper sulfate crystals from its solution. [4 marks]

5 **a** Write a balanced symbol equation, including state symbols, for the reaction of lithium oxide (in excess) and dilute sulfuric acid. [2 marks]

  b If 15 cm³ of 2.0 mol/dm³ sulfuric acid was used, calculate the maximum mass of the lithium salt that could be formed (ignoring any water of crystallisation in the salt). [3 marks]

Key points

● When an acid reacts with a base, a neutralisation reaction occurs.
● The reaction between an acid and a base produces a salt and water.
● The sum of the charges on the ions in a salt add up to zero. This enables you to work out the formula of salts, knowing the charges on the ions present.
● A pure, dry sample of the salt made in an acid–base reaction can be crystallised out of solution by evaporating off most of the water, and drying with filter papers if necessary.

C5.6 Making more salts

Learning objectives

After this topic, you should know:

- the reactions of acids with alkalis
- the reactions of acids with carbonates
- how to make pure, dry samples of a named soluble salt from information provided.

Figure 1 *Ammonium nitrate, NH_4NO_3, made from ammonia and nitric acid is used as a fertiliser*

There are two other important reactions you can use to make salts:

- reacting solutions of an acid and an alkali together
- reacting an acid with a carbonate (usually added as the solid).

Acid + alkali

When an acid reacts with an alkali, a neutralisation reaction takes place.

Hydrochloric acid reacting with sodium hydroxide solution is an example:

acid	**+**	**alkali**	$\rightarrow$	**a salt**	**+**	**water**
$HCl(aq)$	**+**	$NaOH(aq)$	$\rightarrow$	$NaCl(aq)$	+	$H_2O(l)$
hydrochloric acid	+	sodium hydroxide solution	$\rightarrow$	sodium chloride	+	water

You can think about neutralisation in terms of $H^+(aq)$ ions from the acid reacting with $OH^-(aq)$ ions from the alkali. The ions react to form water molecules. You can show this in an ionic equation. Remember that the ionic equation for a reaction just shows the ions, atoms, and molecules that change when the new products are formed:

$$H^+(aq) + OH^-(aq) \rightarrow H_2O(l)$$

You can make ammonium salts, as well as metal salts, by reacting an acid with an alkali. Ammonia reacts with water to form a weakly alkaline solution:

$$NH_3(aq) + H_2O(l) \rightleftharpoons NH_4^+(aq) + OH^-(aq)$$

Ammonia solution reacts with an acid (for example, dilute nitric acid):

acid	+	ammonia solution	$\rightarrow$	an ammonium salt	+	water
$HNO_3(aq)$	+	$NH_4^+(aq) + OH^-(aq)$	$\rightarrow$	$NH_4NO_3(aq)$	+	$H_2O(l)$
nitric acid	+	ammonia solution	$\rightarrow$	ammonium nitrate	+	water

When you react an acid with an alkali, you need to be able to tell when the acid and alkali have completely reacted. It is not obvious by just observing the reaction. There is no gas given off during the reaction (that would stop when the acid has been neutralised). Also, there is no excess insoluble base visible in the reaction mixture when excess has been added. So you need to use an acid/base indicator to help decide when the reaction is complete. You saw how to do this by titration in Topic C4.7.

To collect a pure, dry sample of crystals of the salt you would:

- carry out the titration with the indicator added to see how much acid reacts completely with the alkali
- run that volume of acid into the solution of alkali again, but this time without the indicator
- then crystallise and dry the crystals of salt from the reaction mixture, as described previously in Topics C5.4 and C5.5.

Synoptic link

For more about making fertilisers, see Topic C15.7.

Synoptic link

You first looked at acid/base indicators and titrations in Topic C4.7.

Acids + carbonates

Buildings and statues made of limestone suffer badly from damage by acid rain. You might have noticed statues where the fine features have been lost. Limestone, which is quarried from the ground, is mostly calcium carbonate. This reacts with acid, giving off carbon dioxide gas in the reaction.

Metal carbonates react with acids to give a salt, water, and carbon dioxide. The general equation is:

acid + a carbonate → a salt + water + carbon dioxide

For calcium carbonate, the reaction with hydrochloric acid is:

calcium carbonate + hydrochloric acid → calcium chloride + water + carbon dioxide

The balanced symbol equation, including state symbols, is:

$$CaCO_3(s) + 2HCl(aq) \rightarrow CaCl_2(aq) + H_2O(l) + CO_2(g)$$

Making a salt from a metal carbonate

Metal carbonates are generally insoluble in water (except the carbonates of the Group 1 metals, such as sodium carbonate).

- Think of two ways in which you could decide when an acid has been completely neutralised by an insoluble carbonate.

- You will be given a choice of magnesium carbonate or copper carbonate, plus dilute hydrochloric acid or sulfuric acid. Choose which combination you will react together to make a salt.

- Name the salt, write an equation, and describe a method for preparing a pure, dry sample of crystals of your salt.

Safety: Do not start any practical work before your teacher has checked your planned method.

Figure 2 _Powdered limestone is used to raise the pH of acidic soils or lakes affected by acid rain, making use of the reaction between calcium carbonate and acid_

1 a State the general equation for the neutralisation reaction between an acid and an alkali. [1 mark]
  b Write an ionic equation to show what happens to the $H^+(aq)$ and $OH^-(aq)$ ions in this neutralisation reaction. [1 mark]
  c Write down the general equation for the reaction between an acid and a carbonate. [1 mark]

2 a Write a detailed method to show how to make lithium chloride (a soluble salt) from an acid and an alkali – look back to Topic C4.7 for help, if necessary.  [6 marks]
  b Write a balanced symbol equation, including state symbols, for the reaction in part **a**. [3 marks]

3 a Barium is a Group 2 metal. Give the balanced symbol equation with state symbols for the reaction between barium carbonate and dilute nitric acid. [3 marks]
  H b Suggest an ionic equation, including state symbols, that can summarise the reaction of any insoluble carbonate with dilute acid. [3 marks]

Key points

- An indicator is needed when a soluble salt is prepared by reacting an alkali with an acid.
- The titration can be repeated without the indicator to make a salt, then a pure, dry sample of its crystals prepared.
- A carbonate reacts with an acid to produce a salt, water, and carbon dioxide gas.

C5.7 Neutralisation and the pH scale

Learning objectives

After this topic, you should know:

- why solutions are acidic or alkaline
- how to use universal indicator or a wide-range indicator to measure the approximate pH of a solution
- how to use the pH scale to identify acidic or alkaline solutions
- how to investigate pH changes when a strong acid neutralises a strong alkali.

Figure 1 *Acids and bases are all around you, in many of the things used in everyday life – and in your bodies too*

Figure 2 *pH probes or sensors can measure pH values accurately*

When you dissolve a substance in water, you make an aqueous solution, as happens with the washing powder in a washing machine. Whether the solution formed is acidic, alkaline, or neutral depends on which substance you have dissolved. A solution of washing powder is slightly alkaline.

- Soluble hydroxides are called **alkalis**. Their solutions are alkaline. An example is sodium hydroxide solution.

- **Bases**, which include alkalis, are substances that can neutralise acids. Metal oxides and metal hydroxides are bases. Examples include iron oxide and copper hydroxide, which are both insoluble in water.

- **Acids** include citric acid, sulfuric acid, and ethanoic acid. All acids taste very sour, although many acids are far too dangerous to put in your mouth. Ethanoic acid (in vinegar) and citric acid (in citrus fruit and fizzy drinks) are acids that are weak enough to be edible.

- Pure water is **neutral**; it is neither acidic nor alkaline.

One acid that you use in science labs is hydrochloric acid. This is formed when the gas hydrogen chloride, HCl, dissolves in water:

$$HCl(g) \xrightarrow{\text{water}} H^+(aq) + Cl^-(aq)$$

All acids release $H^+(aq)$ ions into solution when added to water. It is these excess $H^+(aq)$ ions that make a solution acidic.

Because alkalis are bases that dissolve in water and form solutions, they are the bases often used in experiments. Sodium hydroxide solution is often found in school labs. You get sodium hydroxide solution when you dissolve solid sodium hydroxide in water:

$$NaOH(s) \xrightarrow{\text{water}} Na^+(aq) + OH^-(aq)$$

All alkalis form aqueous hydroxide ions, $OH^-(aq)$, when added to water. It is these excess aqueous hydroxide ions, $OH^-(aq)$, that make a solution alkaline.

Measuring acidity or alkalinity

Indicators are substances which change colour when you add them to acids and alkalis. Litmus is a well-known indicator (red in acid and blue in alkali), but there are many more.

You use the **pH scale** to show how acidic or alkaline a solution is. The scale runs from 0 (most acidic) to 14 (most alkaline). You can use universal indicator to find the pH of a solution. It is a special indicator made from a number of dyes. It turns a range of colours as the pH changes. Anything in the middle of the pH scale (pH 7) is neutral – neither acidic nor alkaline (Figure 3).

Alternatively, you can use a pH meter, which has a glass probe attached to dip in the solution being tested. Some electronic pH sensors give a digital display of the pH directly, and others can be attached to data-loggers and computers to monitor and record pH changes over time.

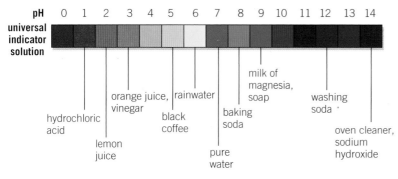

Figure 3 *The pH scale tells you how acidic or alkaline a solution is*

Greater than/less than

You can use the mathematical symbols > (read as 'is greater than') and < ('is less than') when interpreting pH values. You can say:

pH < 7 indicates an acidic solution, i.e., pH values less than 7 are acidic.

pH > 7 indicates an alkaline solution, i.e., pH values greater than 7 are alkaline.

Obtaining a pH curve

Collect 20 cm³ of sodium hydroxide solution in a small beaker. Measure and record its pH using a pH sensor.

Then add dilute hydrochloric acid to a burette.

● Predict the change of pH you will get in the experiment as you add the hydrochloric acid to the sodium hydroxide solution. You can sketch a predicted line on a graph of pH value against volume of dilute HCl added:

Now add the dilute hydrochloric acid from the burette, 1 cm³ at a time, to the acid. Stir after each addition of acid and take the pH of the solution in the beaker.

Record your data in a table and show it on a graph or use the computer to display the data collected.

● Evaluate your prediction.

Safety: wear eye protection.

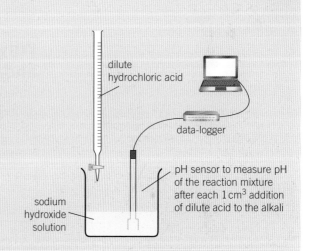

1 a What distinguishes alkalis from other bases? [1 mark]
  b What do all alkalis have in common? [1 mark]
  c Using an equation, including state symbols, show the change that happens when potassium hydroxide, KOH, dissolves in water. [2 marks]
2 a What ions do all acids produce in aqueous solution? [1 mark]
  b Using an equation, including state symbols, show the change that happens when hydrobromic acid, HBr, dissolves in water. [2 marks]
3 How could you use universal indicator paper as a way of distinguishing between distilled water, sodium hydroxide solution, and ethanoic acid solution? [3 marks]
4 Describe the way the pH changes when a strong acid is added slowly to a strong alkali. [3 marks]
5 Compare the advantages and disadvantages of the use of universal indicator paper or a pH sensor and data-logger to find the pH of a solution. [3 marks]

Key points

● Acids are substances that produce H⁺(aq) ions when you add them to water.
● Bases are substances that will neutralise acids.
● An alkali is a soluble hydroxide. Alkalis produce OH⁻(aq) ions when you add them to water.
● You can use the pH scale to show how acidic or alkaline a solution is.
● Solutions with pH values less than 7 are acidic, pH values more than 7 are alkaline, and a pH value of 7 indicates a neutral solution.

C5.8 Strong and weak acids

Learning objectives

After this topic, you should know:

- how to use and explain the terms dilute and concentrated, and weak and strong in relation to acids
- how the concentration of hydrogen ions in a solution affects the numerical value of pH (whole number values of pH only).

Why are some acids called dilute solutions, and others concentrated solutions?

Some acids are safer to use than others. The acids you have used in experiments in this chapter are all strong acids, made safer to use by using dilute solutions of them. For example, a solution of $6.0\,mol/dm^3$ hydrochloric acid is relatively concentrated and would have to be labelled as corrosive. However, when that solution is diluted by adding water to give a solution of $0.10\,mol/dm^3$ hydrochloric acid, it only requires labelling as irritant. Dilute the strong acid enough and it will eventually be harmless.

So a concentrated solution of a solute contains a greater amount of that solute in the same volume of solution than a more dilute solution.

Why are some acids called strong acids, and others weak acids?

Other acids, such as the weak acid citric acid, are not harmful when in dilute or even concentrated solutions.

Examples of strong acids	Examples of weak acids
hydrochloric acid	ethanoic acid (found in vinegar)
nitric acid	citric acid (found in citrus fruits)
sulfuric acid	carbonic acid (found in rainwater, fizzy drinks)

Carbon dioxide gas, CO_2, is given off more slowly when a metal carbonate reacts with a carboxylic acid (ethanoic acid), as compared with hydrochloric acid of the same concentration. Carboxylic acids are called **weak acids**, as opposed to **strong acids** such as hydrochloric acid.

The pH of a $0.1\,mol/dm^3$ solution of hydrochloric acid, a strong acid, is 1.0 (page 95). Yet a $0.1\,mol/dm^3$ solution of ethanoic acid (a weak acid) has a pH of only 2.9. The solution of ethanoic acid is not as acidic, even though the two solutions have the same concentration. So why is this?

Acids must dissolve in water before they show their acidic properties. That is because in water all acids ionise (split up). Their molecules split up to form $H^+(aq)$ ions and negative ions. It is the $H^+(aq)$ ions that all acidic solutions have in common. For example, in hydrochloric acid, the HCl molecules all ionise in water:

$$HCl(aq) \longrightarrow H^+(aq) + Cl^-(aq)$$

So strong acids ionise *completely* in solution. However, in weak acids, most of the molecules stay as they are; they do not release H^+ ions into the solution. Only a relatively small proportion of the acidic molecules will ionise in their solutions. The reaction is reversible, unlike the ionisation of a strong acid. So as the molecules of the weak acid split up to form H^+ ions and negative ions, the ions recombine to form the original molecules again. A position of **equilibrium** is reached in which both whole

Comparing ethanoic acid and hydrochloric acid

Ethanoic acid is a weak acid and hydrochloric acid is a strong acid.

Write down your observations in these two tests to compare solutions of ethanoic acid with hydrochloric acid, both with the same concentration.

a Take the pH of the solutions of both acids.

b Add a little sodium carbonate to solutions of both acids.

- Why did you use the same concentrations of each acid in the experiment?

Safety: Wear eye protection.

molecules (the majority) and their ions (the minority) are present. So, in ethanoic acid you get:

$$CH_3COOH(aq) \rightleftharpoons CH_3COO^-(aq) + H^+(aq)$$
ethanoic acid   ethanoate ions   hydrogen ions

Therefore, given two aqueous solutions of equal concentration, the strong acid will have a higher concentration of $H^+(aq)$ ions than the solution of the weak acid. So, a weak acid has a higher pH value (and hence reacts more slowly with a metal carbonate).

How are pH values related to the concentration of $H^+(aq)$ ions?

On the previous page you saw that the pH of a 0.10 mol/dm³ solution of dilute hydrochloric acid is 1.0. Remember that all acids contain $H^+(aq)$ ions. In this solution of hydrochloric acid, the concentration of hydrogen ions, $H^+(aq)$, is also 0.10 mol/dm³, because HCl(aq) ionises completely in solution.

If you make that solution 10 times more dilute by adding more water, you will find that the 0.010 mol/dm³ solution of HCl(aq) has a pH value of 2.0. So when the concentration of the $H^+(aq)$ is reduced by a factor of 10, the pH value goes up by one unit. Diluting by 10 again, to form a 0.0010 mol/dm³ solution, the HCl(aq) solution has a pH value of 3.0. Repeating the dilution yet again, to make 0.000 10 mol/dm³ solution, the HCl(aq) solution has a pH value of 4.0. Putting these data in a table helps you see the pattern:

Concentration of $H^+(aq)$ ions in mol/dm³	pH value
0.10	1.0
0.010	2.0
0.0010	3.0
0.00010	4.0

As the concentration of hydrogen ions, $H^+(aq)$, decreases by a factor of 10 (i.e., an order of magnitude), the pH value increases by one unit.

Put another way:

As the concentration of $H^+(aq)$ ions increases by a factor 10, the pH value decreases by one unit.

1 Equal amounts of sodium carbonate are added to beakers containing ethanoic acid and dilute nitric acid, of equal concentration. Describe the difference seen. [1 mark]

2 a What will be the pH of a 1.0×10^{-5} mol/dm³ solution of dilute hydrochloric acid? [1 mark]
   b A solution of sodium chloride is neutral. What will be the concentration of hydrogen ions in the solution? Give your answer in mol/dm³ as a decimal and in standard form. [2 marks]

3 Propanoic acid is a carboxylic acid. Explain in detail why propanoic acid, C_2H_5COOH, is described as a weak acid, whereas nitric acid, HNO_3, is a strong acid. 🖊 [4 marks]

4 Explain why it is possible to have a very dilute solution of a strong acid, with a higher pH value than a concentrated solution of a weak acid. 🖊 [4 marks]

Synoptic links ∞

For more information on the structure of the family of weak acids called carboxylic acids, see Topic C10.2.

For more information about reversible reactions at equilibrium, see Topic C8.8.

pH values

The pH value can be related to the concentration of $H^+(aq)$ ions expressed in standard form.

Using the values in the table below, with concentrations in standard form, you get:

Concentration of $H^+(aq)$ ions in mol/dm³	pH value
1.0×10^{-1}	1.0
1.0×10^{-2}	2.0
1.0×10^{-3}	3.0
1.0×10^{-4}	4.0

So, looking at the concentration of $H^+(aq)$ ions, if you take minus the value of the power to which 10 is raised, it gives you the pH value of the solution.

Go further

If you study A Level Chemistry, you will learn how to calculate the pH values of solutions of strong acids or alkalis, as well as the pH values of weak acids, knowing the concentration of their solutions.

Key points

- Aqueous solutions of weak acids, such as carboxylic acids, have a higher pH value than solutions of strong acids with the same concentration.
- As the pH decreases by one unit, the hydrogen ion concentration of the solution increases by a factor of 10 (i.e., one order of magnitude).

C5 Chemical changes

Summary questions

1 Imagine that a new metal, given the symbol X, has been discovered. It lies between calcium and magnesium in the reactivity series.

   a Describe the reaction of excess X with dilute sulfuric acid and give word and balanced symbol equations (metal X forms +2 ions). [5 marks]

   b Explain why you cannot be sure how X will react with cold water. [2 marks]

   c X is added to a solution of copper(II) nitrate.
      i Explain what you would expect to see happen, including a balanced symbol equation with state symbols. [3 marks]
      H ii Write an ionic equation, including state symbols, showing what happens in the change. [3 marks]
      H iii Explain which species is oxidised and which is reduced, using ionic half-equations. [4 marks]

   d Another new metal, Y, does not react with water but there is a slight reaction with warm dilute acid.
      i Where would you place metal Y in the reactivity series? [1 mark]
      ii Explain what you would expect to happen if metal Y was added to magnesium sulfate solution. [3 marks]

2 a Identify which of the pairs of substances will react.
      i carbon + copper(II) oxide [1 mark]
      ii iron + zinc nitrate [1 mark]
      iii iron + magnesium oxide [1 mark]
      iv magnesium + copper(II) sulfate [1 mark]

   b Write balanced symbol equations, including state symbols, for the pairs that will react. [6 marks]

3 The reaction between aluminium powder and iron(III) oxide is used in the rail industry.

   a i Write a word equation and a balanced symbol equation for the reaction that takes place. [3 marks]
      ii What do you call this type of reaction? [1 mark]

   b Compare the reaction described above with the reaction you would expect to see between powdered aluminium and copper(II) oxide. [1 mark]

   c Predict what would happen if you heat a mixture of aluminium oxide and iron. [1 mark]

   d Explain why the uses of aluminium metal are surprising, given its position in the reactivity series. [2 marks]

4 Lead is often found in the ore called galena, which contains lead sulfide, PbS. Before reduction with carbon, the lead sulfide must be converted into lead oxide, PbO, by roasting the ore. This reaction with oxygen from the air also produces sulfur dioxide gas.

   a Write a word equation for the roasting of lead sulfide. [1 mark]

   b Write a balanced symbol equation, including state symbols, for the reaction in part **a**. [3 marks]

   c Lead oxide is reduced by heating with carbon. Write a balanced symbol equation for this reaction, assuming CO_2 is produced. [3 marks]

5 Nickel(II) sulfate crystals can be made from an insoluble oxide base and sulfuric acid.

   a i Name the insoluble base that can be used to make nickel(II) sulfate. [1 mark]
      ii Write a balanced symbol equation, including state symbols, to show the reaction. [3 marks]
      iii What type of reaction is shown in part **ii**? [1 mark]

   b Describe how you could obtain crystals of nickel(II) sulfate from the reaction in part **a ii**. [6 marks]

6 Write balanced symbol equations, including state symbols, to describe the reactions below.

   a Lithium hydroxide solution (in excess) and dilute sulfuric acid. [3 marks]

   b Iron(III) oxide (an insoluble base) and dilute nitric acid. [3 marks]

   c Zinc metal and dilute hydrochloric acid. [3 marks]

H 7 You are given a $0.50 \, mol/dm^3$ solution of nitric acid, HNO_3 – a strong acid, and a $0.50 \, mol/dm^3$ solution of methanoic acid, HCOOH – a weak acid.

   a i What does 'a $0.50 \, mol/dm^3$ solution' mean? [1 mark]
      ii Calculate the concentration of each acid in g/dm^3. Give your answers to 3 significant figures. [2 marks]

   b i What can you predict about the relative pH values of the two solutions? [1 mark]
      ii Explain your prediction in part **i**. [3 marks]

   c Name two other weak acids not mentioned in this question. [2 marks]

   d i The pH of a $0.10 \, mol/dm^3$ solution of nitric acid is 1.0. Give this concentration of nitric acid in standard form. [2 marks]
      ii Calculate the pH of a $0.0010 \, mol/dm^3$ solution of nitric acid. [1 mark]

Practice questions

01 This question is about the reactivity of metals.
A student dropped five different metals **A**, **B**, **C**, **D** and **E** in water.
Four of the metals reacted to produce a metal hydroxide and hydrogen gas.
The reactions of the five metals are shown in **Figure 1**.

Figure 1

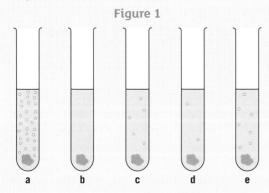

a   b   c   d   e

01.1 Use the information in **Figure 1** to put metals **A**, **B**, **C**, **D**, and **E** in order of reactivity. [2 marks]

01.2 Describe the test for hydrogen. [2 marks]

01.3 Give two variables that should be controlled. [2 marks]

01.4 **A** is a Group 2 metal.
Tick (✓) the formula for the hydroxide of **A**: [1 mark]

AOH		A_2OH	
$A(OH)_2$		$A(OH)_3$	

02 This question is about the preparation of copper sulfate crystals.
Copper metal does not react with dilute sulfuric acid. Copper sulfate can be prepared by reacting copper oxide, CuO, with dilute sulfuric acid, H_2SO_4.

02.1 Suggest why copper metal does not react with dilute sulfuric acid. [2 marks]
A student used this method to make crystals of copper sulfate.
Step 1: Place dilute sulfuric acid into a beaker. Heat for one minute.
Step 2: Add one spatula of copper oxide powder.
Step 3: Pour the copper sulfate solution into an evaporating basin and heat until all the water has evaporated.

02.2 Give a reason why the student heated the sulfuric acid in **Step 1**. [1 mark]

02.3 At the end of the method the student tested the solution with universal indicator paper. The solution was still acidic. Give a colour for universal indicator in acid. [1 mark]

02.4 Give an improvement to **Step 2** to ensure all the acid is used up. [2 marks]

02.5 How can **Step 3** be improved to produce large crystals of copper sulfate? [2 marks]

03 The colours of three metals and three metal sulfates are shown in **Table 1**.

Table 1

	Colour of metal	Formula of the sulfate	Colour of metal sulfate
iron, Fe	grey	$FeSO_4$	pale green solution
copper, Cu	brown	$CuSO_4$	blue solution
magnesium, Mg	silver	$MgSO_4$	colourless solution

Use your knowledge of the reactivity series and the information in **Table 1** to give the observations when the following are mixed together.

03.1 Iron nail with copper sulfate solution. [2 marks]

03.2 Magnesium ribbon with iron sulfate solution. [2 marks]

03.3 Explain why no observable change was seen when a piece of copper was dropped into magnesium sulfate solution. [2 marks]

Ⓗ 03.4 Write an ionic equation for the reaction between iron and copper sulfate solution in **03.1**. [2 marks]

03.5 Write a half equation to show the reduction of Fe^{2+} ions in **03.2**. Use the half equation to explain why Fe^{2+} ions are reduced. [2 marks]

04 Magnesium carbonate reacts with dilute nitric acid to form the soluble salt magnesium nitrate. The equation for the reaction is shown below.
$$MgCO_3(s) + 2HNO_3(aq) \rightarrow Mg(NO_3)_2(aq) + H_2O(l) + CO_2(g)$$

04.1 Plan a method to produce pure, dry crystals of magnesium nitrate. [6 marks]

Ⓗ 04.2 Magnesium also reacts with dilute ethanoic acid. Nitric acid is a strong acid and ethanoic acid is a weak acid.
What difference would you expect to see if magnesium carbonate was reacted with ethanoic acid of the same concentration as the nitric acid? [2 marks]

Ⓗ 04.3 The formula of ethanoic acid is CH_3COOH. Write an equation to show the ionisation of ethanoic acid. [2 marks]

Ⓗ 04.4 A solution of ethanoic acid has a pH of 4 and a solution of nitric acid has a pH of 1.
How many times greater is the concentration of H^+ ions in the nitric acid compared to the concentration of hydrogen ions in the ethanoic acid? [1 mark]

C 6 Electrolysis
6.1 Introduction to electrolysis

Learning objectives

After this topic, you should know:

- what happens in electrolysis
- the type of substances that can be electrolysed
- the products of electrolysis.

Electrolysis of zinc chloride

- When does the bulb light up?
- What is observed at each electrode?

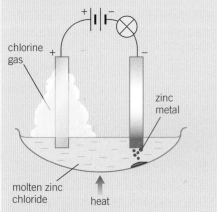

Figure 1 *Passing electricity through molten zinc chloride. Zinc metal forms on one electrode and pale green chlorine gas is given off at the other as the electrolyte (zinc chloride) is broken down by electrolysis.*

Safety: Warn asthmatics. Wear chemical splash-proof eye protection. Chlorine is toxic. Carry out in a fume cupboard. Zinc chloride is corrosive.

Synoptic links

Look back at Topic C3.3 and Topic C3.4 to remind yourself about ionic bonding and the properties of ionic compounds.

The word electrolysis means 'breaking down using electricity'. In electrolysis you use an electric current to break down an ionic compound. The compound that is broken down by electrolysis is called the **electrolyte**.

To set up an electrical circuit for electrolysis, you have two electrodes that dip into the electrolyte, with a gap between them. The electrodes are conducting rods. One of these is connected to the positive terminal of a power supply. This positive electrode is called the **anode**. The other electrode is connected to the negative terminal and is called the **cathode**.

The electrodes are often made of an unreactive (or **inert**) substance, such as graphite or sometimes expensive platinum metal. This is so the electrodes do not react with the electrolyte or the products made in electrolysis.

During electrolysis, positively charged ions move to the cathode (negative electrode). At the same time, the negative ions move to the anode (positive electrode), as opposite charges attract.

When the ions reach the electrodes, they lose their charge and become elements. At the electrodes, gases may be given off or metals deposited. This depends on the compound used and whether it is molten or dissolved in water (see Topic C6.2).

The first person to explain electrolysis was Michael Faraday. He worked on this and many other problems in science nearly 200 years ago.

Figure 1 shows how electricity breaks down zinc chloride into zinc and chlorine:

$$\text{zinc chloride} \rightarrow \text{zinc} + \text{chloride}$$
$$\text{ZnCl}_2(\text{l}) \rightarrow \text{Zn}(\text{s/l}) + \text{Cl}_2(\text{g})$$

Zinc chloride is an ionic compound. Ionic compounds do not conduct electricity when they are solid, as their ions are in fixed positions in their giant lattice. However, once an ionic compound is melted, the ions are free to move around within the hot liquid and carry their charge towards the electrodes.

The positive zinc ions, Zn^{2+}, move towards the cathode (negative electrode). At the same time, the negatively charged chloride ions, Cl^-, move towards the anode (positive electrode).

Notice the state symbols in the equation. They tell you that the zinc chloride is molten, so it is a *liquid* at the temperature in the evaporating dish, *solid* zinc coats the tip of one electrode (and melts if the temperature of the electrolyte reaches 420°C), and chlorine *gas* is given off at the other.

Electrolysis of solutions

Many ionic substances have very high melting points, so it takes a lot of energy to melt them and free the ions to move to electrodes in electrolysis. However, some ionic substances dissolve in water and when this happens, the ions also become free to move around.

When electrolysing ionic compounds in solution, and not as molten compounds, it is more difficult to predict what will be formed. This is because water also forms ions, so the products at each electrode are not always exactly what you expect (see Topic C6.2). In electrolysis only metals of very low reactivity, below hydrogen in the reactivity series, are deposited from their aqueous solutions.

When you electrolyse an aqueous solution of copper(II) bromide, copper ions, Cu^{2+}, move to the cathode (negative electrode). The bromide ions, Br^-, move to the anode (positive electrode). Copper(II) bromide is split into its elements at the electrodes (Figure 2):

$$copper(II)\ bromide\ \rightarrow\ copper\ +\ bromine$$
$$CuBr_2(aq)\ \rightarrow\ Cu(s)\ +\ Br_2(aq)$$

The state symbols in the equation tell you that the copper bromide is dissolved in water, the copper is formed as a solid, and the bromine formed dissolves in the water.

Covalent compounds cannot usually be electrolysed unless they react (or ionise) in water to form ions. For example, acids in water always contain $H^+(aq)$ ions plus negatively charged aqueous ions.

1 a Define electrolysis. [1 mark]
   b What do you call the substance broken down by electrolysis? [1 mark]
   c What type of bonding is present in compounds that can be electrolysed? [1 mark]
2 Predict the products formed at the cathode and anode when the following compounds are melted and then electrolysed:
   a zinc iodide [1 mark]
   b lithium bromide [1 mark]
   c iron(III) fluoride [1 mark]
   d sodium oxide [1 mark]
   e potassium chloride. [1 mark]
3 Write a balanced symbol equation, including state symbols, for the electrolytic decomposition of molten sodium chloride. [3 marks]
4 a Which of the following solutions would deposit a metal at the cathode during electrolysis?
      KBr(aq)   CaCl$_2$(aq)   FeBr$_2$(aq)
      CuCl$_2$(aq)   Na$_2$SO$_4$(aq)   AgNO$_3$(aq) [2 marks]
   b How did you decide on your answer to part a? [1 mark]
5 Solid ionic substances do not conduct electricity. Using words and diagrams, explain why they conduct electricity when molten or in aqueous solution, but not when solid. [3 marks]

Synoptic links

Remind yourself of the relative reactivity of metals and hydrogen in Topic C5.1 and Topic C5.2.

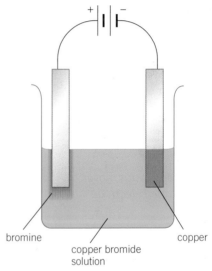

bromine

copper

copper bromide solution

Figure 2 *If you dissolve copper bromide in water, you can decompose it by electrolysis. Copper metal, Cu(s), is formed at the cathode (negative electrode). Brown bromine, Br$_2$(aq), appears in solution around the anode (positive electrode)*

Key points

- Electrolysis breaks down a substance using electricity.
- Ionic compounds can only be electrolysed when they are molten or dissolved in water. This is because their ions are then free to move and carry their charge to the electrodes.
- In electrolysis, positive ions move to the cathode (negative electrode), while negative ions move to the anode (positive electrode).

C6.2 Changes at the electrodes

Learning objectives

After this topic, you should know:

- what happens to the ions during electrolysis
- how water affects the products of electrolysis
- **H** how you can represent the reactions at each electrode using half equations.

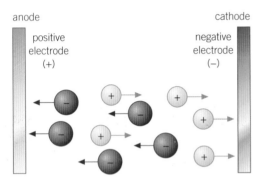

Figure 1 *An ion always moves towards the oppositely charged electrode*

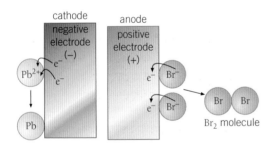

Figure 2 *Changes at the cathode and the anode in the electrolysis of molten lead bromide*

During electrolysis, mobile ions move towards the electrodes. The direction they move in depends on their charge. As you saw in Topic C6.1, positive ions move towards the cathode (negative electrode) and negative ions move towards the anode (positive electrode).

When ions reach an electrode, they either lose or gain electrons, depending on their charge (Figure 1).

Negatively charged ions *lose* electrons to become neutral atoms. Positively charged ions *gain* electrons to become neutral atoms.

The easiest way to think about this is to look at an example.

Think about the electrolysis of molten lead bromide, $PbBr_2$ (Figure 2). The lead ions, Pb^{2+}, move towards the negative electrode (cathode). When they get there, each ion gains two electrons from the power supply to become a neutral lead atom.

Gaining electrons is called *reduction*. The lead ions are *reduced*. Reduction is simply another way of saying gaining electrons.

The negatively charged bromide ions, Br^-, move towards the positive electrode (anode). Once there, each ion loses its one extra electron to become a neutral bromine atom. Two bromine atoms then form a covalent bond to make a bromine molecule, Br_2.

Losing electrons is called *oxidation*. The bromide ions are *oxidised*. Oxidation is another way of saying 'losing electrons'.

Half equations

You represent what is happening at each electrode using **half equations**.

At the cathode (negative electrode) you get reduction of a positive ion:

$$Pb^{2+} + 2e^- \rightarrow Pb$$

At the anode (positive electrode) you get oxidation of a negative ion:

$$2Br^- \rightarrow Br_2 + 2e^-$$

Sometimes half equations at the anode are written to show the electrons being removed from negative ions, like this:

$$2Br^- - 2e^- \rightarrow Br_2$$

You can write the half equation for negative ions either way. They both show the same oxidation of the negatively charged ions.

The effect of water

In aqueous solutions, electrolysis is more complex, because of the ions formed by water as it ionises:

$$H_2O(l) \rightleftharpoons H^+(aq) + OH^-(aq)$$

water   hydrogen ions   hydroxide ions

There is a rule for working out what will happen. Remember that if two elements can be produced at an electrode, the less reactive element will usually be formed. In aqueous solutions, there will usually be positively charged metal ions and $H^+(aq)$ ions (from water), which are attracted to the cathode (negative electrode).

Figure 3 shows what happens in the electrolysis of a solution of a potassium compound. Hydrogen is less reactive than potassium, so hydrogen is produced at the cathode rather than potassium metal.

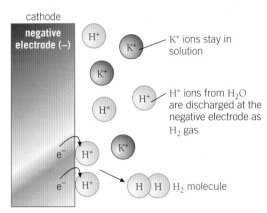

Figure 3 *Here is the cathode in the electrolysis of a solution of a potassium compound. Hydrogen is less reactive than potassium, so hydrogen gas is given off at the electrode*

Higher

At the cathode (–):

$$2H^+(aq) + 2e^- \rightarrow H_2(g)$$

So what happens at the anode in the electrolysis of aqueous solutions? Hydroxide ions, $OH^-(aq)$, from water are often discharged. When hydroxide ions are discharged, you see oxygen gas given off at the anode (positive electrode).

Higher

At the anode (+):

$$4OH^-(aq) \rightarrow 2H_2O(l) + O_2(g) + 4e^-$$

This happens unless the solution contains a reasonably high concentration of a halide (Group 7) ion, such as $Cl^-(aq)$. In this case, the halide ion is discharged and the halogen is formed:

Higher

At the anode (+):

$$2Cl^-(aq) \rightarrow Cl_2(g) + 2e^-$$

So the order of discharge at the anode (starting with the easiest) is:

halide ion > hydroxide > all other negatively charged ions

1 Predict what is formed at each electrode in the electrolysis of:
   a molten lithium oxide [1 mark]
   b copper chloride solution [1 mark]
   c sodium sulfate solution. [1 mark]

2 a i How do negatively charged ions become neutral atoms in electrolysis? [1 mark]
     ii Identify the process described in part **ai** [1 mark]
   b i How do positively charged ions become neutral atoms in electrolysis? [1 mark]
     ii Identify the process described in part **bi** [1 mark]

3 Ⓗ Copy and, where necessary, balance the following half equations:
   a $Cl^- \rightarrow Cl_2 + e^-$ [1 mark]
   b $Br^- \rightarrow Br_2 + e^-$ [1 mark]
   c $Mg^{2+} + e^- \rightarrow Mg$ [1 mark]
   d $Al^{3+} + e^- \rightarrow Al$ [1 mark]
   e $K^+ + e^- \rightarrow K$ [1 mark]
   f $H^+ + e^- \rightarrow H_2$ [1 mark]
   g $O^{2-} \rightarrow O_2 + e^-$ [1 mark]
   h $OH^- \rightarrow O_2 + H_2O + e^-$ [1 mark]

Synoptic link

For more information on reduction and oxidation in terms of electron transfer, look back to Topic C5.2.

Key points

- In electrolysis, the ions move towards the oppositely charged electrodes.
- At the negative electrode (cathode), positive ions gain electrons, so are reduced.
- At the positive electrode (anode), negative ions lose their extra electrons, so are oxidised.
- When electrolysis happens in aqueous solution, the less reactive element, either hydrogen or the metal, is usually produced at the cathode. At the anode, you get either:
  - oxygen gas given off, from discharged hydroxide ions produced from water, or
  - a halogen produced if the electrolyte is a solution of a halide.

C6.3 The extraction of aluminium

Learning objectives

After this topic, you should know:

- why some metals are extracted with carbon and others by electrolysis
- the process of extracting aluminium from its ore
- **H** the half equation at each electrode during the electrolysis of aluminium oxide.

Figure 1 *Aluminium alloys have a low density, but are very strong*

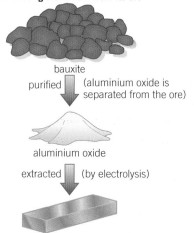

extracting aluminium from its ore

bauxite
purified ↓ (aluminium oxide is separated from the ore)

aluminium oxide
extracted ↓ (by electrolysis)

aluminium metal

Figure 2 *Extracting aluminium from its ore. This process requires a lot of energy. The purification of aluminium oxide from the ore makes aluminium hydroxide. This is separated from the impurities, but then must be heated to turn it back to pure aluminium oxide. Then even more energy is needed to melt and electrolyse the oxide*

You already know that aluminium is a very important metal. The uses of the metal or its alloys include:

- pans
- overhead power cables
- aeroplanes
- cooking foil
- drink cans
- window and patio door frames
- bicycle frames and car bodies.

Aluminium is quite a reactive metal (look back at the reactions of metals and the reactivity series in Topic C5.1 and Topic C5.3). It is less reactive than magnesium, but more reactive than zinc or iron. Carbon is not reactive enough to use in its extraction, as it cannot displace aluminium from its compounds, so you must use electrolysis. The compound electrolysed is aluminium oxide, Al_2O_3.

You get aluminium oxide from the ore called bauxite. The ore is mined by open cast mining, digging it directly from the surface. Bauxite contains mainly aluminium oxide. However, it is mixed with other rocky impurities, so the first step is to separate aluminium oxide from the ore. The impurities contain a lot of iron(III) oxide. This makes the waste solution from the separation process a rusty brown colour. The brown wastewater has to be stored in large lagoons.

Electrolysis of aluminium oxide

The electrolysis of aluminium oxide requires a lot of energy. Once purified, the aluminium oxide must be melted. This enables the ions to move to the electrodes.

Unfortunately, aluminium oxide has a very high melting point of 2050 °C. However, chemists have found a way of saving at least some energy. They mix the aluminium oxide with molten cryolite. Cryolite is another ionic compound. The molten mixture can be electrolysed at about 850 °C. The large amount of electrical energy that is transferred to the electrolysis cells keeps the mixture molten (Figure 3).

The overall reaction in the electrolysis cell is:

$$\text{aluminium oxide} \xrightarrow{\text{electrolysis}} \text{aluminium} + \text{oxygen}$$
$$2Al_2O_3(l) \rightarrow 4Al(l) + 3O_2(g)$$

Aluminium forms at the negative electrode (cathode).

Oxygen is produced at the positive electrode (anode).

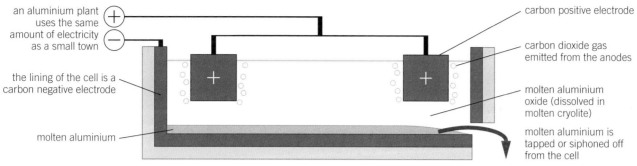

Figure 3 *A cell used in the extraction of aluminium by electrolysis*

Higher

At the cathode (negative electrode):

Each aluminium ion, Al^{3+}, gains three electrons. The ions turn into aluminium atoms. The Al^{3+} ions are reduced (as they gain electrons) to form Al atoms:

$$Al^{3+}(l) + 3e^- \rightarrow Al(l)$$

The aluminium metal formed is molten at the temperature of the cell and collects at the bottom. It is siphoned or tapped off.

At the anode (positive electrode):

Each oxide ion, O^{2-}, loses two electrons. The ions turn into oxygen atoms. The O^{2-} ions are oxidised (as they lose electrons) to form oxygen atoms. These bond in pairs to form molecules of oxygen gas, O_2:

$$2O^{2-}(l) \rightarrow O_2(g) + 4e^-$$

The oxygen reacts with the hot carbon anodes, making carbon dioxide gas:

$$C(s) + O_2(g) \rightarrow CO_2(g)$$

So the carbon anodes gradually burn away and need to be replaced regularly.

Synoptic link

Revisit Figure 2 in Topic C5.3 to see aluminium's position in the reactivity series.

1   a   Explain why aluminium oxide must be molten for electrolysis to take place. [2 marks]
    b   Why is aluminium oxide dissolved in molten cryolite in the extraction of aluminium? [2 marks]

2   Why are the carbon anodes replaced regularly in the industrial electrolysis of aluminium oxide? [2 marks]

Ⓗ 3 a   Write half equations for the changes at each electrode in the electrolysis of molten aluminium oxide. [4 marks]
    b   Explain which ions are oxidised and which ions are reduced in the electrolysis of molten aluminium oxide. [2 marks]

4   a   Explain why the extraction of aluminium requires so much energy. [3 marks]
    b   Suggest why aluminium metal was only discovered in the early 1800s, despite it being the most common metallic element in the Earth's crust. [3 marks]
Ⓗ c   Calculate the maximum mass of aluminium metal that can be extracted from 25.5 tonnes of aluminium oxide. [3 marks]

Key points

- Aluminium oxide, from the ore bauxite, is electrolysed in the extraction of aluminium metal.
- The aluminium oxide is mixed with molten cryolite to lower its melting point, reducing the energy needed to extract the aluminium.
- Aluminium forms at the cathode (negative electrode) and oxygen forms at the anode (positive electrode).
- The carbon anodes are replaced regularly as they gradually burn away as the oxygen reacts with the hot carbon anodes, forming carbon dioxide gas.

C6.4 Electrolysis of aqueous solutions

Learning objectives

After this topic, you should know:

- how to predict the products of the electrolysis of an aqueous solution
- how to investigate the electrolysis of a solution using inert electrodes
- **H** the half equation at each electrode during the electrolysis of an aqueous solution.

The electrolysis of **brine** (concentrated sodium chloride solution) is a very important industrial process. When brine is electrolysed, you get three useful products that are used to make other chemicals:

- chlorine gas is produced at the anode (positive electrode)
- hydrogen gas is produced at the cathode (negative electrode)
- sodium hydroxide solution is also formed.

You can summarise the electrolysis of brine as:

$$\text{sodium chloride solution} \xrightarrow{\text{electrolysis}} \text{hydrogen gas} + \text{chlorine gas} + \text{sodium hydroxide solution}$$

At the anode (+)

The negative chloride ions, Cl^-, are attracted to the positive electrode. When they get there, they each lose one electron. The chloride ions are oxidised, as they lose electrons. The chlorine atoms bond together in pairs and are given off as chlorine gas, Cl_2.

The half equation at the anode is:

$$2Cl^-(aq) \rightarrow Cl_2(g) + 2e^-$$

This can also be written as: $2Cl^-(aq) - 2e^- \rightarrow Cl_2(g)$

Higher

At the cathode (–)

There are H^+ ions in brine, formed when water breaks down:

$$H_2O(l) \rightleftharpoons H^+(aq) + OH^-(aq)$$

These positive hydrogen ions are attracted to the negative electrode. The sodium ions, $Na^+(aq)$, are also attracted to the same electrode. But remember in Topic C6.1, you saw what happens when two ions are attracted to an electrode. It is the less reactive element that gets discharged. In this case, hydrogen ions are discharged and sodium ions stay in solution as aqueous ions.

When the H^+ ions reach the negative electrode, they each gain one electron. The hydrogen ions are reduced, as they each gain an electron. The hydrogen atoms formed bond together in pairs and are given off as hydrogen gas, H_2.

Synoptic link

You will find the tests for gases in Topic C12.3.

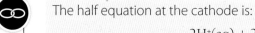

The half equation at the cathode is:

$$2H^+(aq) + 2e^- \rightarrow H_2(g)$$

Higher

The remaining solution

You can test the solution around the cathode (negative electrode) with an acid/base indicator. It shows that the solution is alkaline. This is because you can think of brine as containing aqueous ions of Na^+ and Cl^- (from salt) and H^+ and OH^- (from water). The $Cl^-(aq)$ and $H^+(aq)$ ions are removed during electrolysis. So this leaves a solution containing $Na^+(aq)$ and $OH^-(aq)$ ions, which is a solution of sodium hydroxide, $NaOH(aq)$.

Investigating the electrolysis of a solution

You can now use your knowledge and understanding of electrolysis to investigate the electrolysis of different aqueous solutions using inert electrodes.

Plan your investigation. Think about:

● Which solutions will you test?

● Are there any safety issues with your plan? Are the solutions hazardous? What about the gases that might be given off? How can you reduce any risks?

Predict what products will be formed at each electrode, and explain your choices.

● How will you test and identify any gases formed at the electrodes (Topic 12.3)?

After collecting and recording the data, evaluate your investigation.

If you have time you can investigate whether changing the concentration of some solutions affects the products formed.

Safety: Do not start any practical work until your teacher has checked your plan.

Figure 1 *Bleach is made from chlorine and sodium hydroxide obtained from the electrolysis of a concentrated solution of sodium chloride (brine)*

Figure 2 *The chlorine made when you electrolyse brine in industry is used to kill bacteria in drinking water, and also in some swimming pools*

1 Name the *three* products made when you electrolyse sodium chloride solution (brine). [3 marks]

2 Describe positive tests for the gases named in question 1. [2 marks]

H 3 For the electrolysis of sodium chloride solution (brine), write half equations, including state symbols, for the reactions:
   a at the anode [3 marks]
   b at the cathode. [3 marks]

4 You can also electrolyse *molten* sodium chloride.
   a Compare the products formed with those from the electrolysis of sodium chloride solution. [2 marks]
   b Explain fully any differences. [4 marks]

H 5 A sodium metal manufacturer completely electrolyses 234 tonnes of sodium chloride. Calculate the maximum mass of sodium metal that could be extracted. [2 marks]

Key points

● When you electrolyse sodium chloride solution (brine), you get three products – chlorine gas and hydrogen gas given off at the electrodes, plus sodium hydroxide solution (an alkali) left in solution.

● Hydrogen is produced at the cathode (−), as $H^+(aq)$ ions are discharged from solution in preference to $Na^+(aq)$ ions.

● Chlorine is produced at the anode (+), as $Cl^-(aq)$ ions are discharged from solution in preference to $OH^-(aq)$ ions.

C6 Electrolysis

Summary questions

1 For each of the following statements, state whether it is the anode (+) or cathode (–) that is being referred to.
   a Positive ions move towards this. [1 mark]
   b Oxidation happens here. [1 mark]
   c This is connected to the negative terminal of the power supply. [1 mark]
   d This is connected to the positive terminal of the power supply. [1 mark]
   e Reduction happens here. [1 mark]
   f Negative ions move towards this. [1 mark]

2 a Which of the following ions would move towards the anode (+) and which towards the cathode (–) during electrolysis?

   potassium   oxide
   iodide   magnesium
   calcium   aluminium
   fluoride   bromide   [2 marks]

   H b Write a half equation for the discharge of:
      i magnesium ions [3 marks]
      ii bromide ions [3 marks]

3 The diagram shows the electrolysis of sodium chloride solution in the laboratory.

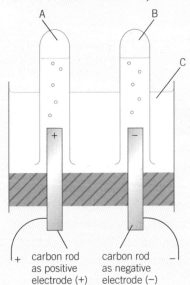

   carbon rod   carbon rod
   as positive   as negative
   electrode (+)   electrode (–)

   a Identify the products A, B, and C on the diagram. [3 marks]
   b Give two uses for substance A. [2 marks]

c What would be the pH of the solution around the cathode? [1 mark]
d How would you carry out a positive test on product B? [1 mark]

H e Write the half equations, including state symbols, for the changes at the anode and cathode. [6 marks]

4 Water can be slightly acidified and broken down into hydrogen and oxygen using electrolysis. The word equation for this reaction is:

   water → hydrogen + oxygen

   a Write a balanced symbol equation for this reaction, including state symbols. [3 marks]
   b In water, a small percentage of molecules ionise (split up). Write a balanced symbol equation, including state symbols, for the ionisation of water molecules. [3 marks]
   H c Write half equations to show what happens at the positive and negative electrodes in the electrolysis of water. [3 marks]
   H d When some water is electrolysed, it produces 0.20 moles of hydrogen gas. Calculate the volume of oxygen gas produced at room temperature and pressure at the same time. (1 mole of any gas occupies $24\,dm^3$ at room temperature and pressure.) [2 marks]
   H e Where does the energy needed to split water into hydrogen and oxygen come from during electrolysis? [1 mark]

H 5 Complete the following half equations:
   a $Li^+ \rightarrow Li$ [1 mark]
   b $Sr^{2+} \rightarrow Sr$ [1 mark]
   c $F^- \rightarrow F_2$ [1 mark]
   d $O^{2-} \rightarrow O_2$ [1 mark]

H 6 *Electrolysis can be thought of as a redox reaction. Discuss this statement by explaining the changes that take place at the anode and cathode during electrolysis, using an example of a molten ionic compound. Write half equations for the reactions occurring at each electrode.* [6 marks]

Practice questions

01 This question is about the electrolysis of lead bromide, $PbBr_2$.
Lead bromide contains Pb^{2+} and Br^- ions.
The apparatus for the electrolysis of lead bromide is shown in **Figure 1**.

Figure 1

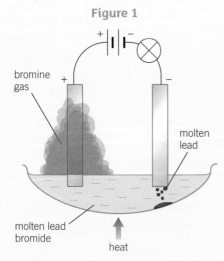

The electrodes were placed in solid lead bromide and the lead bromide was heated. The bulb did not light initially but lit when the lead bromide was completely molten.

01.1 Explain why molten lead bromide conducts electricity. [2 marks]

01.2 Explain why the metal in the wire conducts electricity. [2 marks]

Ⓗ 01.3 Explain the formation of the brown gas at the anode (the positive electrode). Include a half equation in your answer. [4 marks]

Ⓗ 01.4 Explain the formation of the grey droplets at the cathode (the negative electrode). Include a half equation in your answer. [4 marks]

02 This question is about extracting metals.

02.1 Iron can be extracted by reacting iron(III) oxide with carbon.
Aluminium cannot be extracted by reacting aluminium oxide with carbon.
Explain why. [2 marks]
Aluminium is extracted from aluminium oxide by electrolysis. The electrolysis of aluminium oxide Al_2O_3 is shown in **Figure 2**.
Aluminium oxide contains Al^{3+} and O^{2-} ions.

Figure 2

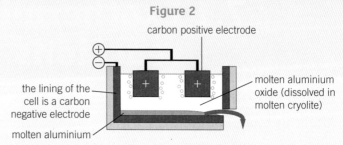

02.2 Why is molten aluminium oxide dissolved in molten cryolite? [2 marks]

02.3 The positive electrodes need frequent replacement. Explain why. [4 marks]

Ⓗ 02.4 Write a half equation to show how aluminium is formed at the negative electrode. [1 mark]

Ⓗ 02.5 Why are aluminium ions reduced? [1 mark]

03 This question is about the electrolysis of sodium chloride solution.
The electrolysis of sodium chloride solution is shown in **Figure 3**.

Figure 3

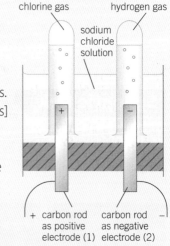

03.1 Give the test and the result for hydrogen gas. [2 marks]
Sodium chloride solution can conduct electricity because the ions can move.
Some electrodes are made of carbon in the form of graphite.

03.2 Explain why electrodes that are made of graphite can conduct electricity. [2 marks]

03.3 The solution contains Na^+ ions and H^+ ions, which are both attracted to the negative electrode.
Explain why hydrogen gas forms instead of sodium metal. [2 marks]

03.4 A few drops of universal indicator were added to the solution. The universal indicator turned blue. Explain why. [2 marks]

Ⓗ 03.5 Complete the half equation for the formation of chlorine gas at the positive electrode.
$$Cl^- \quad \rightarrow \quad Cl_2$$ [2 marks]

Ⓗ 03.6 Complete the half equation for the formation of hydrogen gas at the negative electrode.
$$H^+ \quad \rightarrow \quad H_2$$ [2 marks]

7.1 Exothermic and endothermic reactions

Learning objectives

After this topic, you should know:

- energy cannot be created or destroyed in a chemical reaction
- that energy is transferred to or from the surroundings in chemical reactions, and some examples of these exothermic and endothermic reactions
- how to distinguish between exothermic and endothermic reactions on the basis of the temperature change
- how to carry out an investigation into energy changes in chemical reactions.

Synoptic link

Remind yourself about neutralisation reactions in Topic C5.7.

Figure 1 *When a fuel burns in oxygen, energy is transferred to the surroundings. You usually don't need a thermometer to know that there is a temperature change!*

Have you ever warmed your hands near a fire? If so you will have felt the effects of energy being transferred during a chemical reaction. In fact, whenever chemical reactions take place, energy is always transferred, as chemical bonds in the reactants are broken and new ones are made in the products. However, the total energy remains the same before and after a reaction. So energy cannot be created or destroyed in any chemical reaction.

Many reactions transfer energy from the reacting chemicals to their surroundings. These are called **exothermic** reactions. The energy transferred from the reacting chemicals often heats up the surroundings. As a result of this, you can measure a rise in temperature as the reaction happens.

Other reactions transfer energy from the surroundings to the reacting chemicals. These are called **endothermic** reactions. As they take in energy from their surroundings, these reactions cause a fall in temperature as they happen.

Exothermic reactions

The burning of fuels, such as the combustion of methane gas, is an obvious example of exothermic reactions. When methane (the main gas present in natural gas) burns, it gets oxidised and releases energy to its surroundings.

Neutralisation reactions between acids and alkalis are also exothermic. You can easily measure the rise in temperature using simple apparatus (see the practical Investigating temperature changes).

The products of exothermic reactions have a lower energy content than the reactants. The actual differences in energy are usually expressed in kilojoules per mole (kJ/mol).

For example, for the reaction:

$$CH_4(g) + 2O_2(g) \rightarrow CO_2(g) + 2H_2O(l)$$

the energy transferred to the surroundings = 890 kJ/mol.

Endothermic reactions

Endothermic reactions are much less common than exothermic ones. The reaction between citric acid and sodium hydrogencarbonate is a good example to try in the lab, as it is easy to measure the fall in temperature.

Thermal decomposition reactions are also endothermic. An example is the decomposition of calcium carbonate. When heated, it forms calcium oxide and carbon dioxide. This reaction only takes place if you keep heating the

Figure 2 *All warm-blooded animals rely on exothermic reactions in respiration to keep their body temperatures steady*

calcium carbonate strongly. The calcium carbonate needs to absorb energy from the surroundings (such as the energy provided by a roaring Bunsen flame) to be broken down.

In endothermic reactions the products have a higher energy content than the reactants, so energy is transferred from the surroundings.

For example, for the reaction:

$$CaCO_3(s) \rightarrow CaO(s) + CO_2(g)$$

the energy transferred from the surroundings = 178 kJ/mol.

Figure 3 *When you eat sherbet you can feel an endothermic reaction. Sherbet dissolving in the water in your mouth takes in energy. It provides a slight cooling effect*

Investigating temperature changes

You can use very simple apparatus to investigate the energy changes in reactions involving at least one solution. Often you do not need to use anything more complicated than a poly(styrene) cup and a thermometer.

For example, you can investigate the temperature changes when you mix different combinations of reactants.

● Record the initial temperatures of any solutions and the maximum or minimum temperature reached in the course of the reaction.

To extend yourself, you could consider how the quantities of reactants used might affect the temperature change.

● Make a quantitative prediction and explain your reasoning. This hypothesis could include a sketch graph.

● Test your hypothesis, recording and displaying your data using a suitable table and graph.

● Draw your conclusions.

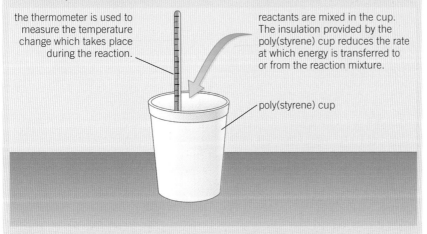

the thermometer is used to measure the temperature change which takes place during the reaction.

reactants are mixed in the cup. The insulation provided by the poly(styrene) cup reduces the rate at which energy is transferred to or from the reaction mixture.

poly(styrene) cup

● Evaluate your investigation, including at least two ways in which you could make the data you collect more accurate.

Safety: Wear eye protection.

Key points

● Energy is conserved in chemical reactions. It is neither created nor destroyed.

● A reaction in which energy is transferred from the reacting substances to their surroundings is called an exothermic reaction.

● A reaction in which energy is transferred to the reacting substances from their surroundings is called an endothermic reaction.

1 a What do you call a reaction that transfers energy to its surroundings? [1 mark]

  b What do you call a reaction that takes in energy transferred from its surroundings? [1 mark]

  c Give *two* examples of:
    i an exothermic reaction [2 marks]
    ii an endothermic reaction. [2 marks]

2 Potassium nitrate dissolving in water is an endothermic process. Explain what you would feel if you held a beaker of water in your hand as you stirred in potassium nitrate. [2 marks]

3 Two solutions are added together and the temperature changes from 19 °C to 27 °C. Explain what you can deduce about the energy transferred between the reaction mixture and its surroundings. [2 marks]

4 The energy required for the thermal decomposition of 16.8 g of magnesium carbonate is 23.4 kJ.

  a Write a balanced symbol equation, including state symbols, for the reaction. [2 marks]

  Ⓗ b Calculate the number of moles of magnesium carbonate broken down. [2 marks]

C7.2 Using energy transfers from reactions

Warming up

Chemical hand and body warmers can be very useful. These products use exothermic reactions to warm you up. People can take hand warmers to places they know will get very cold. For example, spectators at outdoor sporting events in winter can warm their hands. People usually use the chemical body warmers to help ease aches and pains.

Some hand warmers can only be used once. An example of this type makes use of the energy transferred to the surroundings in the oxidation of iron. Iron turns into hydrated iron(III) oxide in an exothermic reaction. The reaction is similar to rusting. Sodium chloride (common salt) is used as a catalyst. This type of hand warmer is disposable. It can be used only once but it lasts for hours.

Other hand warmers can be reused many times. These are based on the formation of crystals from solutions of a salt. The salt used is often sodium ethanoate, $CH_3COO^-Na^+$. A supersaturated solution is prepared by dissolving as much of the salt as possible in hot water. The solution is then allowed to cool.

A small metal disc in the plastic pack is used to start the exothermic change (see Figure 1). When you press this a few times, small particles of metal are scraped off. These 'seed' (or start off) the crystallisation. The crystals spread throughout the solution, transferring energy to the surroundings in an exothermic change. They work for about 30 minutes.

Figure 1 *During the recrystallisation of sodium ethanoate in a hand warmer, energy is transferred from the reaction system to the surroundings, increasing the temperature of the surroundings – your hand*

To reuse the warmer, you simply put the solid pack into boiling water to re-dissolve the crystals. Once it has cooled down, the pack is ready to activate again.

Exothermic reactions are also used in self-heating cans (Figure 2) that make drinks like hot coffee without any external heating device (e.g., a kettle). The reaction used to transfer energy to the food or drink is usually:

calcium oxide + water → calcium hydroxide

You press a button in the base of the can. This breaks a seal and lets the water and calcium oxide mix. Then the exothermic reaction can begin.

Development of the self-heating coffee can took years and cost millions of pounds. Even then, over a third of the can was taken up with the reactants needed to transfer enough energy to the coffee. Also, in some early versions, the temperature of the coffee did not rise high enough in cold conditions.

Cooling down

Endothermic processes can be used to cool things down. For example, chemical cold packs usually contain ammonium nitrate and water. When ammonium nitrate dissolves, it absorbs energy from its surroundings, making them colder.

These cold packs are used as emergency treatment for sports injuries. The decrease in temperature reduces swelling and numbs pain.

The ammonium nitrate and water (sometimes present in a gel) are kept separate in the pack. When squeezed or struck, the bag inside the water pack breaks, releasing ammonium nitrate. The instant cold packs work for about 20 minutes.

Instant cold packs can only be used once, but are ideal where there is no ice available to treat a knock or strain. This type of cold pack is often included in the first aid kit at venues used for amateur sports or outdoor pursuits.

The same endothermic change can also be used to chill cans of drinks.

1 a Give *two* uses of endothermic changes. [2 marks]
   b Which endothermic change is often used in cold packs? [1 mark]
   c The solid used in cold packs is often ammonium nitrate.
      i Give the formula of ammonium nitrate. [1 mark]
      ii State another use of ammonium nitrate. [1 mark]

2 a Which solid is usually used in the base of self-heating coffee cans? [1 mark]
   b Write a balanced symbol equation, including state symbols, for the reaction of water with the solid in part **a**. [3 marks]
   c Why is it essential that the coffee stays out of contact with the solid in part **a**? [1 mark]

3 a Describe the chemical reaction which takes place in a disposable hand warmer. [4 marks]
   b Describe how a reusable hand warmer works. [4 marks]
   c Give an advantage and a disadvantage of each type of hand warmer. [2 marks]
   d Name *one* use of an exothermic reaction in the food industry. [1 mark]

TO HEAT CONTAINER
Turn container UPSIDE DOWN before opening and follow instructions.

STEP 4
HOT SPOT turns from pink to white when beverage is hot (6–8 minutes).

STEP 5
Once hot, shake for 5 to 10 seconds then twist lid to align opening with pull-tab. Open and enjoy.

STEP 3
Wait 5 SECONDS and turn can right side up.

STEP 2
Place container on flat surface. Using thumb, FIRMLY push button DOWNWARD until internal foil seal tears and coloured water drains into the activation chamber.

STEP 1
PULL off tamper-proof metal bottom.

Figure 2 *Development of the self-heating can in the USA took about 10 years*

Figure 3 *Instant cold packs can be applied as soon as an injury occurs to minimise damage*

Key points

- Exothermic changes can be used in hand warmers and self-heating cans. Crystallisation of a supersaturated solution is used in reusable warmers. However, disposable, one-off warmers heat up the surroundings for longer.
- Endothermic changes can be used in instant cold packs for sports injuries.

C7.3 Reaction profiles

Learning objectives

After this topic, you should know:

- how to draw simple reaction profiles for exothermic and endothermic reactions, including the activation energy
- the definition of activation energy
- how to use reaction profiles to identify reactions as exothermic or endothermic
- **H** the energy changes when bonds are broken and when bonds are made.

You can find out more about what is happening in a particular reaction by looking at its reaction profile. These diagrams show the relative amounts of energy contained in the reactants and the products, measured in kilojoules per mole (kJ/mol). A curved line, drawn from reactants to products, shows the course of the reaction. The difference in energy between the reactants and the peak of the curve indicates the energy input required for the reaction to take place.

Exothermic reactions

Figure 1 shows the reaction profile for an exothermic reaction. The products are at a lower energy level than the reactants. This means that when the reactants form the products, energy is transferred to the surroundings.

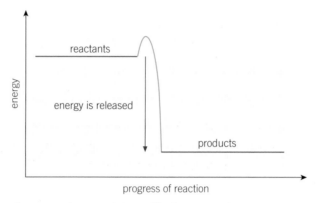

Figure 1 *The reaction profile for an exothermic reaction*

The difference between the energy levels of the reactants and the products is the energy change during the reaction, measured in kJ/mol.

An amount of energy equal to the difference in energy between the products and the reactants is transferred to the surroundings. Therefore, in an exothermic reaction, the surroundings get hotter and their temperature rises.

Endothermic reactions

Figure 2 shows the reaction profile for an endothermic reaction.

Here the products are at a higher energy level than the reactants. As the reactants react to form products, energy is transferred from the surroundings to the reaction mixture. The temperature of the surroundings decreases, because energy is taken in during the reaction. The surroundings get colder.

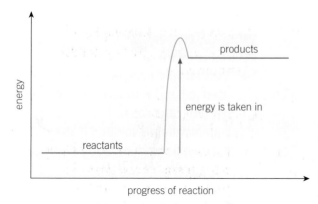

Figure 2 *The reaction profile for an endothermic reaction*

Activation energy

There is a minimum amount of energy needed before colliding particles of reactants have sufficient energy to cause a reaction. This energy needed to start a reaction is called the **activation energy**. This activation energy is shown on reaction profiles. Figure 3 shows the activation energy of an exothermic reaction, and Figure 4 shows that of an endothermic reaction.

If colliding particles of reactants collide with less energy than the activation energy, they will just bounce off each other.

Bond breaking and bond making

Think about what happens as a chemical reaction takes place. First, the chemical bonds between the atoms or ions in the reactants are broken. Then new chemical bonds are formed to make the products.

- Energy has to be ~~supplied~~ absorbed to break chemical bonds. This means that breaking bonds is an *endothermic* process. Energy is taken in from the surroundings.

- However, when new bonds are formed, energy is transferred to the surroundings. Making bonds is an *exothermic* process.

$$H-H \quad + \quad O=O \quad \longrightarrow \quad \overset{H}{\underset{H}{>}}O \quad + \quad \overset{H}{\underset{H}{>}}O$$

Figure 5 *Hydrogen and oxygen react to make water. The bonds between hydrogen atoms and between oxygen atoms have to be broken so that bonds between oxygen atoms and hydrogen atoms in water can be formed*

In reality, the two processes do not happen in sequence – all the bonds do not break and then all the new bonds form. The bond breaking and making processes happen at the same time. However, comparing the energy required to break the bonds with the energy released when the new bonds form gives a good guide to the overall energy change. This allows you to decide if a reaction will be exothermic or endothermic (see Topic C7.4).

1 Draw reaction profiles for the following reactions:
    a $H_2(g) + Cl_2(g) \rightarrow 2HCl(g)$
    For this reaction, the energy transferred to the surroundings is 184 kJ/mol.   [3 marks]
    b $H_2(g) + I_2(g) \rightarrow 2HI(g)$
    For this reaction, the energy taken in from the surroundings is 26.5 kJ/mol.   [3 marks]

Ⓗ 2 Explain why some reactions are exothermic and some are endothermic, in terms of the processes of bond making and bond breaking. ✏   [3 marks]

Ⓗ 3 a Explain why bond breaking is an endothermic process.   [3 marks]
    b Using a diagram like Figure 5, show the bonds being broken and new bonds being made in the reaction between methane, CH_4, and oxygen, which produces carbon dioxide and water.   [2 marks]
    c List the number and type of each bond broken and formed in your answer to part **b**.   [4 marks]

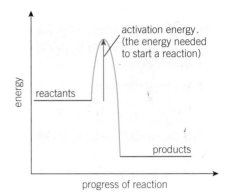

Figure 3 *This reaction profile shows the activation energy for an exothermic reaction*

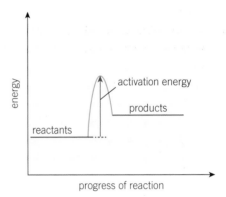

Figure 4 *This reaction profile shows the activation energy for an endothermic reaction*

Synoptic links

For information about how catalysts can increase the rate of a reaction by providing an alternative reaction pathway, which has a lower activation energy, see Topic C8.5.

Study tip

Remember that **B**reaking bonds a**B**sorbs energy, fo**R**ming bonds **R**eleases energy.

Key points

- You can show the relative difference in the energy of reactants and products on reaction profiles.
- Ⓗ Bond breaking is endothermic, whereas bond making is exothermic.

C7.4 Bond energy calculations

Learning objectives

After this topic, you should know:

- how the balance between bond breaking in the reactants and bond making in the products affect the overall energy change of a reaction
- how to calculate the energy transferred in chemical reactions when supplied with bond energies.

Making and breaking bonds

There is always a balance between the energy needed to break bonds and the energy released when new bonds are made in a reaction. This is what decides whether the reaction is endothermic or exothermic.

- In some reactions, the energy released when new bonds are formed (as the products are made) is more than the energy needed to break the bonds in the reactants. These reactions transfer energy to the surroundings. They are *exothermic*.

- In other reactions, the energy needed to break the bonds in the reactants is more than the energy released when new bonds are formed in the products. These reactions transfer energy from the surroundings to the reacting chemicals. They are *endothermic*.

Bond energy

The energy needed to break the bond between two atoms is called the **bond energy** for that bond.

Bond energies are measured in kJ/mol. You can use bond energies to work out the energy change for many chemical reactions. Before you can do this, you need to have a list of the most common bond energies:

Table 1 *Common bond energies*

Bond	Bond energy in kJ/mol		Bond	Bond energy in kJ/mol
C—C	347		H—Cl	432
C—O	358		H—O	464
C—H	413		H—N	391
C—N	286		H—H	436
C—Cl	346		O═O	498
Cl—Cl	243		N≡N	945

To calculate the energy change for a chemical reaction, you need to work out:

1. how much energy is needed to break the chemical bonds in the reactants

2. how much energy is released when the new bonds are formed in the products.

It is very important to remember that the data in the table is the energy required for *breaking* bonds. When you want to know the energy released as these bonds are formed, the amount of energy involved is the same (see Figure 1).

For example, the bond energy for a C—C bond is 347 kJ/mol taken in from the surroundings (an endothermic change). This means that the energy released *forming* a C—C bond is 347 kJ/mol transferred to the surroundings (an exothermic change).

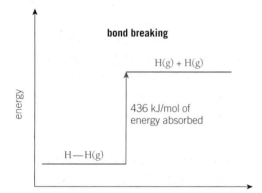

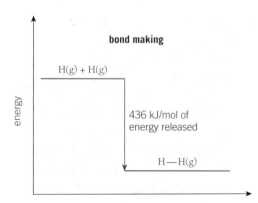

Figure 1 *Breaking and making a particular bond always involves the same amount of energy*

Worked example

Ammonia is made from nitrogen and hydrogen in the Haber process. The balanced symbol equation for this reaction is:

$$N_2(g) + 3H_2(g) \rightleftharpoons 2NH_3(g)$$

Calculate the overall energy change for the forward reaction using bond energies.

Solution

This equation tells you that the bonds in 1 mole of nitrogen molecules and 3 moles of hydrogen molecules need to break in this reaction (see Figure 2).

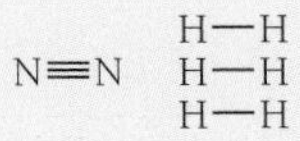

$$N\equiv N \quad \begin{matrix} H-H \\ H-H \\ H-H \end{matrix}$$

Figure 2 *These bonds are broken in the forward reaction*

Nitrogen molecules are held together by a triple bond (written like this: N≡N). This bond is very strong. Using data from the table, its bond energy is 945 kJ/mol.

Hydrogen molecules are held together by a single bond (written like this: H—H). From the table, the bond energy for this bond is 436 kJ/mol.

The energy needed to break 1 mole of N≡N and 3 moles of H—H bonds = 945 + (3 × 436) kJ = 2253 kJ taken in from the surroundings (an endothermic process).

When these atoms form ammonia (NH_3), six new N—H bonds are made since 2 moles of NH_3 are formed (see Figure 3). The bond energy of the N—H bond is 391 kJ/mol.

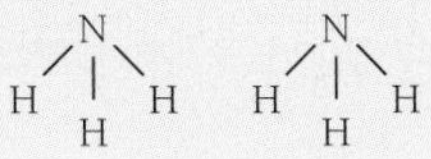

Figure 3 *These bonds are made in the forward reaction*

Energy transferred when 6 moles of N—H bonds are made = 6 × 391 kJ = 2346 kJ (the energy is transferred to the surroundings as this is an exothermic process).

So the *overall* energy change = (2253 kJ − 2346 kJ) = −93 kJ

This is the energy *transferred to the surroundings* in the forward reaction as written above.

The reaction profile in Figure 4 shows the overall energy change for the formation of ammonia.

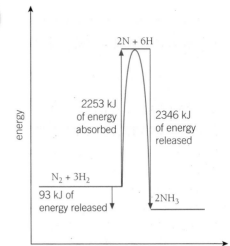

Figure 4 *The formation of ammonia. The energy released, 93 kJ, is from the formation of 2 moles of ammonia, as shown in the balanced equation. So if you wanted to know the energy change for the reaction per mole of ammonia formed, it would release exactly half this, i.e., 46.5 kJ/mol*

Go further

If you study A Level Chemistry, you will do this type of calculation on more complex reactions, and find out why the value for the energy change is an approximation of the true value.

Key points

- In chemical reactions, energy must be supplied to break the bonds between atoms in the reactants.
- When new bonds are formed between atoms in a chemical reaction, energy is released.
- In an exothermic reaction, the energy released when new bonds are formed is greater than the energy absorbed when bonds are broken.
- In an endothermic reaction, the energy released when new bonds are formed is less than the energy absorbed when bonds are broken.
- You can calculate the overall energy change in a chemical reaction using bond energies.

1 If the energy required to break bonds is greater than the energy transferred to the surroundings when bonds are made, will the reaction be exothermic or endothermic? [1 mark]

2 What is meant by the 'bond energy of a chemical bond'? [1 mark]

3 Calculate the energy needed to break all the bonds in 0.0960 g of oxygen gas. [2 marks]

4 Write balanced symbol equations and calculate the energy changes for the following chemical reactions:
   a hydrogen + chlorine → hydrogen chloride [6 marks]
   b hydrogen + oxygen → water [6 marks]
   (Use the bond energies supplied in Table 1.)

C7.5 Chemical cells and batteries

Learning objectives

After this topic, you should know:

- how to interpret data on chemical cells in terms of the relative reactivity of different metals
- how to evaluate the use of chemical cells when given information
- how to plan and carry out an investigation of the voltage produced by simple cells using different metals.

Figure 1 *There are many different types of electrical cells and batteries used in mobile electrical appliances*

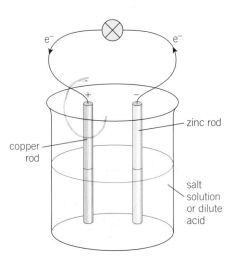

Figure 2 *An electrical cell made from zinc and copper. The electrons flow from the more reactive metal (zinc) to the less reactive metal (copper). So zinc acts as the negative terminal of the cell, providing electrons to the external circuit*

Did you realise that the electrical cells and batteries you use in many mobile appliances rely on the differing reactivity of metals? You have seen in Chapter C5 how metals can be extracted from their ores and how the method used is related to the metal's position in the reactivity series.

You also used this order of reactivity to predict displacement reactions, as a more reactive metal will displace a less reactive metal from its compounds.

For example:

$$Zn(s) + CuSO_4(aq) \rightarrow ZnSO_4(aq) + Cu(s)$$

The sulfate ions do not change in the displacement reaction above. They are spectator ions.

So you can leave them out of the equation and write an ionic equation:

$$Zn(s) + Cu^{2+}(aq) \rightarrow Zn^{2+}(aq) + Cu(s)$$

You can think of this redox reaction as two half equations.

One will represent reduction:

$$Cu^{2+}(aq) + 2e^- \rightarrow Cu(s)$$

The Cu^{2+} ions are reduced to Cu.

The other will be an oxidation reaction:

$$Zn(s) \rightarrow Zn^{2+}(aq) + 2e^-$$

The Zn atoms are oxidised to Zn^{2+} ions.

You can use this difference in reactivity to make electrical cells and batteries. (A battery is made up of two or more cells joined together to increase the voltage available.) If you join the two metals together by a wire and dip them into an electrolyte, such as a salt solution, electrons will flow through the wire from the zinc to the copper.

In the simple cell shown in Figure 2, zinc atoms donate electrons via the connecting wire to the copper(II) ions; so zinc acts as the negative terminal of the cell. The flow of electrons is an electric current. The current will flow in the circuit opposite until one of the reactants, Zn(s) or $Cu^{2+}(aq)$, is used up. You can use the current to light a lamp.

Remember that there is a tendency for any metal atom to give away electrons and form a positive ion. The greater the tendency to form their positive ion, the more reactive the metal is. So the copper(II) ions accept electrons from the zinc atoms and change into copper atoms.

In general:

The greater the difference in reactivity between the two metals used, the higher the voltage produced.

Higher

You can test this out in the following experiment. You can place a voltmeter in the external circuit. The voltage reading will give you a measure of the difference in reactivity between the two metals used in the cell. You can think of this as the difference in the metals' tendency to give away electrons. The larger the voltage, the bigger the difference in their power as reducing agents.

Synoptic links

To revise the reactivity series and how metals are extracted from their ores, see Topic C5.3.

For help identifying spectator ions, see Topic C5.4.

Investigating chemical cells

Use this apparatus to investigate the voltage produced by different metals paired with magnesium ribbon.

You can compare magnesium against zinc, iron, copper, and tin in your electrical cells.

- Record and display your data using a suitable table.
- Then draw your conclusion.
- You can extend your investigation to find out if any other factors affect the voltage produced, besides the two metals used.
- Or you can check predictions of the voltage produced by other pairs of metals using the data in your table.

Safety: Wear eye protection.

The first mass-produced batteries made were called primary cells, and improved versions are still manufactured today. They cannot be recharged. The dry cells, with electrodes made of zinc and carbon, are non-rechargeable, as are the popular, modern alkaline cells that can produce a larger voltage. Once one of the reactants has run out, the cell stops working and should be disposed of in a recycling centre.

Other cells can be recharged and used over and over again. In the recharging process, the battery is connected to a power supply that reverses the reactions that occur at each electrode when the cell is discharging. This regenerates the original reactants.

1 Why is it not possible to make an electrical cell using two electrodes made of zinc metal? [1 mark]

2 An electrical cell is made using the metals iron and zinc as the two electrodes, and a salt solution as the electrolyte.
   a Draw a diagram of the electrical cell set up with a lamp in the external circuit. [3 marks]
   Ⓗ b Name the metal that is reduced in the cell. [1 mark]
   Ⓗ c Explain which metal will act as the negative terminal of the cell. Include two half equations in your answer. 🖉 [6 marks]
3 Using Figure 3, explain two disadvantages of the zinc–carbon cell compared with some other more modern cells. 🖉 [4 marks]

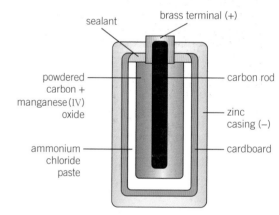

Figure 3 *A zinc–carbon dry cell produces a voltage of about 1.5 V, but cannot be recharged and is prone to leakage if left in an appliance for a long period of time*

Key points

- Metals tend to lose electrons and form positive ions.
- When two metals are dipped in a salt solution and joined by a wire, the more reactive metal will donate electrons to the less reactive metal. This forms a simple electrical cell.
- The greater the difference in reactivity between the two metals, the higher the voltage produced by the cell.

C7.6 Fuel cells

Learning objectives

- how to evaluate the use of hydrogen fuel cells
- **Ⓗ** how to write the half equations for the electrode reactions in the hydrogen fuel cell.

Go further

You will study the latest types of cells and batteries used in cell phones and laptops if you study A Level Chemistry. The performance of lithium ion cells has improved recharging characteristics, and you can now get higher voltages from smaller batteries.

Hydrogen-powered vehicles

Scientists are developing hydrogen as a fuel. It burns well and produces no pollutants:

$$\text{hydrogen} \; + \; \text{oxygen} \; \rightarrow \; \text{water}$$
$$2H_2 \; + \; O_2 \; \rightarrow \; 2H_2O$$

Using hydrogen as a fuel could help reduce the human impact on global warming because the reaction does not produce carbon dioxide. However, there are problems of safety and storage that need to be solved. Supplying the hydrogen to burn in car engines is also an issue. Making hydrogen using electrolysis requires electricity from non-renewable fossil fuels. The power station will still be producing carbon dioxide and using up limited energy resources.

Figure 1 *A small number of hydrogen refuelling stations have been set up to trial the use of hydrogen-powered vehicles*

A more efficient use of the energy from oxidising hydrogen is in a **fuel cell**. These cells are fed with hydrogen and oxygen and produce water. Most of the energy released in the reaction is transferred to electrical energy. This can be used to run a vehicle. However, a constant supply of hydrogen is still needed to run the fuel cell.

Scientists are aware that replacing engines powered by fossil fuels with 'cleaner' energy sources could have great benefits. Therefore they have developed many types of fuel cell and hydrogen-powered engines. The challenge is to match the performance, convenience and price of petrol or diesel cars.

Figure 2 *This London bus runs on fuel cells*

Hydrogen gas is supplied as a fuel to the negative electrode. It diffuses through the graphite electrode and reacts with hydroxide ions to form water and provides a source of electrons to an external circuit

$$2H_2(g) + 4OH^-(aq) \rightarrow 4H_2O(l) + 4e^-$$

Oxygen gas is supplied to the positive electrode. It diffuses through the graphite and reacts to form hydroxide ions, accepting electrons from the external circuit:

$$O_2(g) + 2H_2O(l) + 4e^- \rightarrow 4OH^-(aq)$$

If you add the two electrode reactions together, the electrons and the OH⁻ ions on either side of the half equations cancel out. So you are left with the overall change in the hydrogen fuel cell, that is the oxidation of hydrogen (the fuel):

$$2H_2(g) + O_2(g) \rightarrow 2H_2O(l)$$

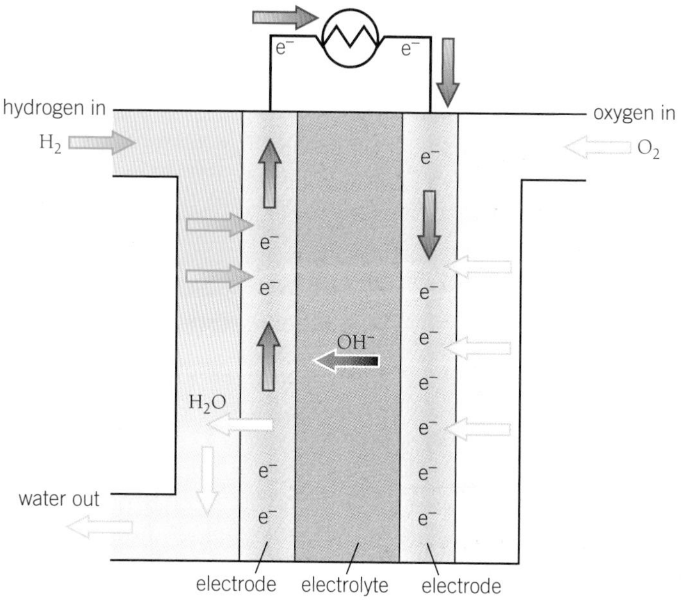

Figure 3 *A hydrogen fuel cell which has an alkaline electrolyte, such as potassium hydroxide solution. Notice that the only waste product is water*

The only waste product is water. So hydrogen fuel cells offer a potential alternative to conventional fossil fuels and to rechargeable cells and batteries.

Advantages of hydrogen fuel cells:
● do not need to be electrically recharged
● no pollutants are produced
● can be a range of sizes for different uses.

Disadvantages of hydrogen fuel cells:
● hydrogen is highly flammable
● hydrogen is sometimes produced for the cell by non-renewable sources
● hydrogen is difficult to store.

1 In the hydrogen fuel cell described above:
   a Which gases are pumped into the fuel cell? [1 mark]
   b What is the waste product of the fuel cell? [1 mark]
   c Write a word equation for the overall reaction in the fuel cell. [1 mark]

Ⓗ 2 Write two half equations that show what happens to the hydrogen and oxygen gases in a hydrogen fuel cell. [4 marks]

3 Imagine that your family decides to buy an electric car that uses batteries that need regular recharging by plugging into an electrical socket in your garage. Why would this not necessarily mean that you had found a way to get around without adding to the carbon dioxide in the air? How could the electric car run without contributing to global warming? ⊘ [5 marks]

Key points

● Much of the world relies on fossil fuels. However, they are non-renewable and they cause pollution.
● Hydrogen is one alternative fuel. It can be burned in combustion engines or used in fuel cells to power vehicles.
● Hydrogen gas is oxidised and provides a source of electrons in the hydrogen fuel cell, in which the only waste product is water.

C7 Energy changes

Summary questions

1 Two solutions are mixed and react in an endothermic reaction. When the reaction has finished, the reaction mixture is allowed to stand until it has returned to its starting temperature.

   a Sketch a graph of temperature (*y*-axis) against time (*x*-axis) to show how the temperature of the reaction mixture changes. [1 mark]

   b Label the graph clearly and explain what is happening wherever you have shown that the temperature is changing. [6 marks]

2 a Draw a reaction profile to show the exothermic reaction between nitric acid and sodium hydroxide, including its activation energy. [3 marks]

   b Draw a reaction profile to show the endothermic change when ammonium nitrate dissolves in water, including its activation energy. [3 marks]

3 When you eat sugar, you break it down to eventually produce water and carbon dioxide.

   a Complete the balanced symbol equation:

$$C_{12}H_{22}O_{11} + O_2 \rightarrow$$ [2 marks]

   b Why must your body *supply* energy in order to break down a sugar molecule? [1 mark]

   c When you break down sugar in your body, energy is released. Explain where this energy comes from in terms of the bonds in molecules. [3 marks]

   d You can get about 1700 kJ of energy by breaking down 100 g of sugar. If a heaped teaspoon contains 5 g of sugar, how much energy does this release when broken down by your body? [1 mark]

H 4 Hydrogen peroxide has the structure H—O—O—H. It decomposes slowly to form water and oxygen:

$$2H_2O_2(aq) \rightarrow 2H_2O(l) + O_2(g)$$

The table shows the bond energies for different types of bond.

Bond	Bond energy in kJ/mol
H—O	464
H—H	436
O—O	144
O=O	498

   a Use the bond energies to calculate the energy change for the decomposition of hydrogen peroxide, as shown in the equation above. [4 marks]

b Explain exothermic and endothermic reactions in terms of bond breaking and bond making. [4 marks]

5 A student carried out an experiment to find out the voltage produced when different combinations of three metals (A, B, and C – not their chemical symbols) were connected in electrical cells.
Two of their results are shown below:

Donator of electrons (attached to negative terminal of the voltmeter)	Acceptor of electrons (attached to positive terminal of the voltmeter)	Voltage in volts
B	A	1.1
C	B	0.5

   a The students also tested metals A and C.

     i Add in the missing row of the table, with your prediction of the voltage produced. [3 marks]

     ii Draw a labelled diagram of the apparatus the student would use to test metals A and C. [3 marks]

   b Put the three metals in order of their reactivity, with the most reactive metal first. [1 mark]

   c Which metal is the least powerful reducing agent? [1 mark]

6 a In an alkaline hydrogen fuel cell, name and give the formula for a suitable electrolyte. [2 marks]

H b i Write a half equation showing the chemical change involving hydrogen gas that provides the source of the electrons to the external circuit in a hydrogen fuel cell. [1 mark]

     ii What type of reaction does the hydrogen undergo at this electrode in the fuel cell? [1 mark]

   c i Name the other gas required to operate a hydrogen fuel cell. [1 mark]

     ii Write the balanced symbol equation, including state symbols, for the overall reaction that takes place in a hydrogen fuel cell. [2 marks]

     iii Another type of fuel cell uses methane gas, CH_4, (the main gas in 'natural gas') instead of hydrogen in a fuel cell. Write the balanced symbol equation, including state symbols, for the overall reaction that takes place in a methane fuel cell. [1 mark]

     iv State two advantages a hydrogen fuel cell has compared with a methane fuel cell. [3 marks]

Practice questions

01 **Figure 1** shows the energy level diagram for the reaction of reaction of ethene and bromine

Figure 1

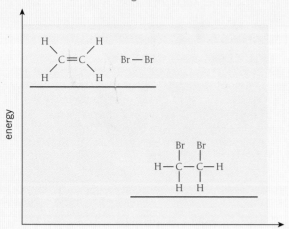

01.1 Copy and complete the diagram and use arrows to label:
- activation energy E_a
- energy released. [3 marks]

01.2 Explain why, in terms of the energy involved in bond breaking and bond making, this reaction is exothermic. [3 marks]

02 A molecule of hydrogen is shown in **Figure 2**.

Figure 2

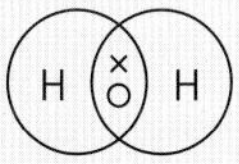

02.1 Describe the attractions in a covalent bond and explain why bond breaking is endothermic. [5 marks]

02.2 Hydrogen reacts with chlorine as shown in the equation.

$$H—H + Cl—Cl \rightarrow 2H—Cl$$

The bond enthalpies are shown in **Table 1**.

Table 1

Bond	Bond energy in kJ/mol
H—H	436
Cl—Cl	243
H—Cl	432

Calculate the energy change for the reaction in kJ/mol. [3 marks]

03 An electrical cell is made when two different metals are connected together in contact with a salt solution. **Figure 3** shows an electrical cell created by copper and metal **A**.

Figure 3

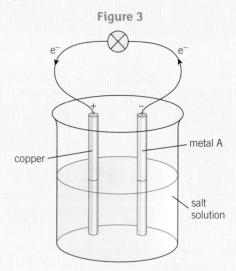

The bigger the difference in the reactivity of the two metals, the larger the voltage produced. The voltages produced with four different metals are shown in **Table 2**.

Table 2

Metal A	Voltage in volts
zinc	1.10
iron	0.79
lead	0.47
silver	−0.46

03.1 Why is potassium unsuitable for metal **A**? [2 marks]

03.2 Give **two** variables that should be controlled in this experiment. [2 marks]

03.3 Chromium is less reactive than zinc but more reactive than iron.
Predict the voltage produced when metal **A** is chromium. [1 mark]

03.4 Suggest why the voltage produced when metal **A** is silver is negative. [2 marks]

03.5 When metal **A** is zinc, Cu^{2+} ions become copper metal and zinc metal becomes Zn^{2+} ions.
Write half equations for the two processes and explain why Cu^{2+} ions are reduced. [4 marks]

3 Rates, equilibrium, and organic chemistry

Chemical reactions can occur at vastly different rates and there are many variables that can be manipulated in order to change their speed. Chemical reactions may also be reversible so conditions will affect the yield of desired product. In industry, chemists and chemical engineers determine the effect of different variables on rate of reaction and yield of product.

The great variety of organic compounds is possible because carbon atoms can form chains and rings linked by C—C bonds. Chemists can modify these organic molecules in many ways to make new and useful materials such as polymers, pharmaceuticals, perfumes, flavourings, dyes, and detergents.

Key questions

- How are reaction rates and reversible reactions affected by changing the conditions?

- How is a range of useful products obtained from crude oil?

- How do functional groups affect the reactions of organic compounds?

- How does the structure of a polymer affect its properties?

Making connections

- You will look at the pollutants from burning fuels from crude oil and their effects in more detail in **Chapter C13 Our atmosphere**.

- The compromise made between the rate of reaction and high yields in industrial processes that use reversible reactions are covered in **Chapter C15 Using our resources**.

- In **Chapter C15 Using our resources** we see the importance of reducing the rate of rusting of iron and steel.

I already know...

the properties of the different states of matter (solid, liquid and gas) in terms of the particle model, including gas pressure.

what catalysts do.

simple techniques for separating mixtures such as distillation.

some examples of combustion and thermal decomposition reactions.

the structure and bonding of some simple molecular substances.

polymers are long molecules made of many repeating groups of atoms.

I will learn...

how to apply the particle model in the collision theory used to explain the effect of changing conditions on the rate of a reaction.

explain how catalysts can affect the rate of a reaction in terms of their effect on the activation energy of the reaction, including reaction profile diagrams.

how fractional distillation is used to separate different fractions from the mixture of hydrocarbons in crude oil.

the products of complete and incomplete combustion of fuels from crude oil, and the use of thermal decomposition in the process of cracking large hydrocarbons into smaller, more useful products.

how to draw the displayed formula of alkanes, alkenes, alcohols, carboxylic acids, and esters.

the different types of bonding between monomers and how this affects the properties of a polymer.

Required Practical

Practical		Topic
5	Concentration and rate of reaction	C8.4

Learning objectives

After this topic, you should know:

- what is meant by the rate of a chemical reaction
- how to collect data on the rate of a chemical reaction
- how to calculate the mean rate of a reaction
- **ⓗ** how to calculate the rate of reaction at a specific time.

The rate of a chemical reaction tells you how fast reactants turn into products. In your body, there are lots of reactions taking place all the time. They happen at the correct rate to supply your cells with what they need, whenever required.

Reaction rate is also very important in the chemical industry. Any industrial process has to make money by producing useful products. This means the right amount of product needed must be made as cheaply as possible. If it takes too long to produce, it will be hard to make a profit when it is sold. The rate of the reaction must be fast enough to make it quickly but safely.

How can you find out the rate of reactions?

Reactions happen at all sorts of different rates. Some are really fast, such as the combustion of chemicals inside a firework exploding. Others are very slow, such as an old piece of iron rusting.

There are two ways you can work out the rate of a chemical reaction. You can find out how quickly:

- the reactants are used up as they make products, or
- the products of the reaction are made.

Here are three techniques you can use to collect this type of data in experiments.

Calculating the rate of reaction at a specific time

You can use a reaction graph to find the rate of the reaction at a given time:

- Draw a tangent to the curve at that time (a straight line that just touches the curve at that point).
- Then construct a right-angled triangle, using the tangent as its longest side (the hypotenuse).

Higher

- Finally, calculate the gradient of the tangent, as shown on the graph below.
- Make sure you include the units of rate, usually g/s, cm³/s or mol/s.

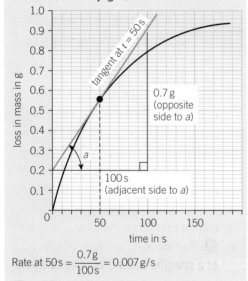

Rate at 50 s $= \dfrac{0.7\,\text{g}}{100\,\text{s}} = 0.007\,\text{g/s}$

(The gradient is the tangent of angle a in the right-angled triangle, i.e. opposite side divided by adjacent side.)

Measuring the decreasing mass of a reaction mixture

You can measure the rate at which the *mass* of a reaction mixture changes if the reaction gives off a gas. As the reaction takes place, the mass of the reaction mixture decreases. You can measure and record the mass at regular time intervals.

Some balances can be attached to a data-logger to monitor the loss in mass continuously.

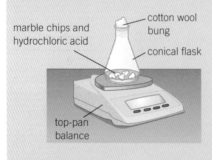

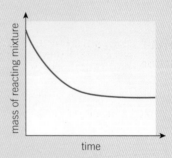

- Why is the cotton wool placed in the neck of the conical flask?
- How would the line on the graph differ if you plot 'loss in mass' on the vertical axis?

Safety: Wear eye protection.

Measuring the increasing volume of gas given off

If a reaction produces a gas, you can use the gas to find out the rate of the reaction. You do this by collecting the gas and measuring the volume given off at time intervals.

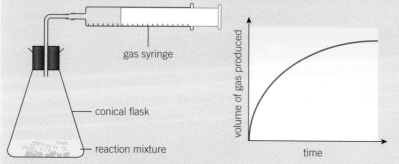

- What are the sources of error when measuring the volume of gas?

Safety: Wear eye protection.

Measuring the decreasing light passing through a solution

Some reactions in solution make a suspension of an insoluble solid (precipitate). This makes the solution go cloudy. You can use this to measure the rate at which the precipitate appears (see Topic C8.4).

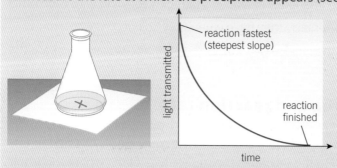

- What are the advantages of using a light sensor rather than the 'disappearing cross' method to monitor precipitation?

You can calculate the mean rate of reaction after a given time using the equation below (see example in Topic C8.2):

$$\text{mean rate of reaction} = \frac{\text{quantity of reactant used}}{\text{time}} \text{ or } \frac{\text{quantity of product formed}}{\text{time}}$$

1 a In the practical, 'Measuring the decreasing mass of a reaction mixture', there were some marble chips left when the fizzing stopped. Why was no more gas given off? [1 mark]

   b How can you tell from the graph when the reaction had finished? [1 mark]

2 Sketch graphs to show the results of measuring:

   a i the mass of products formed in a reaction over time [2 marks]

   ii the mass of reactants remaining in a reaction over time. [2 marks]

   b What does the gradient of the line at any particular time on the graphs in part a tell you about the reaction? [1 mark]

Key points

- You can find out the rate of a chemical reaction by monitoring the amount of reactants used up over time.
- Alternatively, you can find out the rate of reaction by measuring the amount of products made over time.
- The gradient of the line at any given time on the graph drawn from such an experiment tells you the rate of reaction at that time. The steeper the gradient, the faster the reaction.
- Ⓗ To calculate the rate of reaction at a specific time, draw the tangent to the curve, then calculate its gradient.

C8.2 Collision theory and surface area

Learning objectives

After this topic, you should know:

- the factors that can affect the rate of a chemical reaction
- collision theory
- how to use collision theory to explain the effect of surface area on reaction rate.

Synoptic links

For more information about the activation energy of a reaction, look back to Topic C7.3.

You first looked at surface area to volume ratio (in the context of nanoparticles) in Topic C3.11.

Figure 1 *The collision theory is used to explain how reactant particles (atoms, molecules, or ions) react together and why their rate of reaction can vary*

Study tip

Particles collide all the time, but only some collisions lead to reactions.

Increasing the number of collisions in a certain time and the energy of collisions produces faster rates of reaction.

A larger surface area does not result in collisions with more energy, but does increase the frequency of collisions.

In everyday life you often control the rates of chemical reactions without thinking about it. For example, cooking cakes in an oven or adding more detergent to a washing machine. In chemistry you need to know what affects the rate of reactions. You also need to be able to explain why each factor affects the rate of a reaction.

There are four main factors which affect the rate of chemical reactions: temperature, surface area of solids, concentration of solutions or pressure of gases, and the presence of a catalyst.

Reactions can only take place when the particles (atoms, ions, or molecules) of reactants come together. The reacting particles do not only have to bump into each other, but also need to collide with enough energy to cause a reaction to take place. This is known as **collision theory**.

The minimum amount of energy that particles must have before they can react is called the **activation energy**.

So reactions are more likely to happen between reactant particles if you:

- increase the frequency of reacting particles colliding with each other
- increase the energy they have when they collide.

If you increase the chance of particles reacting, you will also increase the rate of the reaction.

Surface area and reaction rate

Imagine lighting a campfire. It is not a good idea to pile large logs together and try to set them alight. You use small pieces of wood to begin with. Doing this increases the surface area of the wood. This results in more wood being exposed to react with oxygen in the air.

When a solid reacts in a solution, the size of the pieces of solid affects the rate of the reaction. The particles inside a large lump of solid are not in contact with the reactant particles in the solution, so they cannot react. The particles inside the solid have to wait for the particles on the surface to react first before they are exposed and have a chance to react.

In smaller lumps, or in a powder, each tiny piece of solid is surrounded by solution. Many more particles of the solid are exposed and able to react at a given time. This means that reactions can take place much more quickly.

You can compare solids with different surface areas quantitatively by looking at their surface area to volume ratio (SA : V). The smaller the size of the pieces of a solid material, the larger its surface area to volume ratio. As the side of a cube decreases in size by a factor of 10, its surface area to volume ratio increases by 10. You get a larger surface area of reactant particles exposed, for the same volume of material. So the larger the SA : V, the faster the reaction.

Worked example 1

A group of students timed how long it took before no more gas was given off from calcium carbonate added to excess dilute hydrochloric acid. They performed three experiments using the same volume and concentration of acid, and 2.50 g of large, medium, and then small marble chips. Here are their results:

Size of marble chips	Time until no more bubbles of gas appeared in s
small	102
medium	188
large	294

Calculate the mean rate of reaction of each size of marble chips used.

Solution

Using the equation in Topic C8.1:

$$\text{mean rate} = \frac{\text{mass of reactant used up (g)}}{\text{time (s)}}$$

Small chips
$$\frac{2.50\,g}{102\,s}$$

Medium chips
$$\frac{2.50\,g}{188\,s}$$

Large chips
$$\frac{2.50\,g}{294\,s}$$

= 0.0245 g/s
(fastest mean rate)

= 0.0133 g/s

= 0.00850 g/s
(slowest mean rate)

Notice that the data are given to 3 significant figures, so the answers are consistent with the data provided.

Worked example 2

On a nanoscale, an individual nanoparticle could have a size of 10 nm. Remember that a nanometre is just 1×10^{-9} m. So what would be the SA : V of a cube of side 10 nm?

Solution

$SA = (10 \times 10^{-9}) \times (10 \times 10^{-9}) \times 6\,m^2$

$\quad\quad = 600 \times 10^{-18}\,m^2$

$V\ = (10 \times 10^{-9})^3\,m^3$

$\quad\quad = 1000 \times 10^{-27}\,m^3$

So the SA : V ratio is:
$600 \times 10^{-18}\,m^2 : 1000 \times 10^{-27}\,m^3$

$\quad\quad = 0.6 \times 10^9/m$

$\quad\quad = 6 \times 10^8/m$ (written in standard form)

So with a surface area to volume ratio of 600 million per metre, it is no wonder nanoparticles are reactive.

1   List the factors that can affect the rate of a chemical reaction.   [4 marks]

2   Draw a diagram to explain why it is easier for an iron nail that has been cut into small pieces to rust than for a whole iron nail to rust.   [2 marks]

3   Explain why the acid in your stomach can help you digest your food more quickly if you chew it well before you swallow.   [2 marks]

4   Explain why the idea of activation energy is an important part of the collision theory used to explain rates of reaction.   [2 marks]

5   In an investigation of the reaction between zinc and dilute sulfuric acid, a student compared the rates of reaction by measuring the time taken for a set volume of hydrogen gas to be given off. The student tested zinc granules, and then zinc pellets of an equal mass. It was found that the granules gave off 25.0 cm³ of hydrogen in 225 s, whereas with the pellets it took 114 s to collect the same volume of gas.
a   Calculate the mean rate of reaction with:
   i   zinc granules   [1 mark]
   ii   zinc pellets.   [1 mark]
b   Draw a conclusion from this data using the collision theory.
   [6 marks]
H c   Calculate the mass of zinc that reacted with the sulfuric acid to give off 2.08×10^{-3} g of hydrogen gas. Show the steps in your calculation.   [3 marks]

Key points

● Particles must collide, with a certain minimum amount of energy, before they can react.

● The minimum amount of energy that particles must have in order to react is called the activation energy of a reaction.

● The rate of a chemical reaction increases if the surface area to volume ratio of any solid reactants is increased. This increases the frequency of collisions between reacting particles.

C8.3 The effect of temperature

Learning objectives

After this topic, you should know:

- how increasing the temperature affects the rate of reactions
- how to use collision theory to explain this effect.

Figure 1 *Lowering the temperature will slow down the reactions that make foods go off*

Figure 2 *Moving faster means it is more likely that you will bump into someone else – and the collision will be harder (more energetic) too*

When you increase the temperature, it increases the rate of reaction. You can use fridges and freezers to reduce the temperature and slow down the rate of reaction. When food goes off it is because of chemical reactions, so reducing the temperature slows down these unwanted reactions.

Collision theory tells you why raising the temperature increases the rate of a reaction. There are two reasons:

- particles collide more often
- particles collide with more energy.

Particles collide more frequently

When you heat up a substance, energy is transferred to its particles. In solutions and in gases, this means that the particles move around faster. When particles move around faster, they also collide more often. Imagine a lot of people walking around in the school playground blindfolded. They may bump into each other occasionally. However, if they start running around, they will bump into each other much more often.

When particles collide more frequently, there are more chances for them to react. This increases the rate of the reaction.

Particles collide with more energy

Particles that are moving around more quickly have more energy. This means that any collisions they have are much more energetic. It is like two people colliding when they are running rather than when they are walking.

When you increase the temperature of a reacting mixture, a higher proportion of the collisions will result in the reaction taking place in any given time. This is because a higher proportion of particles have energy greater than the activation energy.

An increased proportion of particles exceeding the activation energy has a greater effect on rate than the increased frequency of collisions.

Around room temperature, if you increase the temperature of a reaction by 10 °C, the rate of the reaction will roughly double.

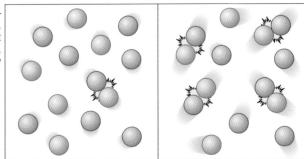

cold – slow movement, less frequent collisions, little energy

hot – fast movement, more frequent collisions, more energy

Figure 3 *More frequent collisions, with more energy – both of these factors lead to the increase in the rate of a chemical reaction that is caused by increasing the temperature*

The effect of temperature on rate of reaction

Alka-Seltzer tablets are a well-known indigestion remedy. They fizz when added to water. The fizzing is caused by carbon dioxide gas, produced by the reaction of sodium hydrogencarbonate and citric acid, both contained in the tablets. These compounds

Figure 4 *When indigestion tablets dissolve in water, sodium hydrogencarbonate and citric acid mix and react to produce carbon dioxide*

can come into contact once the tablet is added to water. By varying the temperature of the water, you can measure its effect on the rate of reaction.

● How could you vary the temperature in the investigation?

● How could you measure the mean rate at different temperatures?

Check your ideas with your teacher before you start your investigation.

● Which variables do you have to control to make this a fair test?

● Why is it difficult to get accurate timings in this investigation?

● How can you improve the **precision**, and hence the accuracy in this case, of any data which has random measurement errors?

Safety: Wear eye protection. Do not raise the temperature above 50 °C.

The results of an investigation like this can be plotted on a graph.

The graph in Figure 5 shows how the time taken for the reaction to finish changes with temperature. The negative slope shows that as the *temperature increases*, the *time* it takes for the reaction to finish *decreases*. However, as the *temperature increases*, the *rate* of the reaction (or 'mean rate of reaction' in this case) also *increases*.

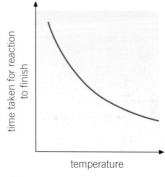

Figure 5 *Example results for an investigation on the effect of temperature on the rate of a reaction*

1 a Why does increasing the temperature increase the rate of a reaction? [2 marks]
   b By how much does a 10 °C rise in temperature increase the reaction rate at room temperature? [1 mark]

2 Look at the experiment in the Practical box above.
   a Why does the tablet fizz in the water? [2 marks]
   b Describe what happens to the time it takes for the reaction to finish as the temperature increases. [1 mark]
   c Explain your answer to part **b** by using collision theory. [3 marks]

3 Water in a pressure cooker boils at a much higher temperature than water in a saucepan because it is under pressure. Explain why food takes longer to cook in a pan than it does in a pressure cooker. [2 marks]

Key points

● Reactions happen more quickly as the temperature increases.

● Increasing the temperature increases the rate of reaction because particles collide more frequently and more energetically. More of the collisions occurring in a given time results in a reaction, because a higher proportion of particles have energy greater than the activation energy.

C8.4 The effect of concentration and pressure

Learning objectives

After this topic, you should know:

- how, and why, increasing the concentration of reactants in solutions affects the rate of reaction
- how, and why, increasing the pressure of reacting gases affects the rate of reaction.

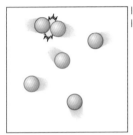
low concentration/
low pressure

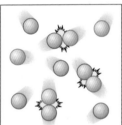
high concentration/
high pressure

Figure 1 *Increasing the concentration of solutions or the pressure of gases both mean that particles are closer together. This increases the frequency of collisions between reactant particles, so the reaction rate increases*

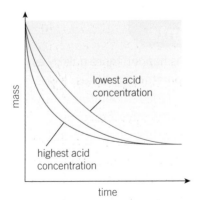

mass

lowest acid concentration

highest acid concentration

time

Figure 2 *How the mass of the reaction mixture of marble chips and hydrochloric acid decreases over time at three different concentrations of acid*

Some of the most beautiful buildings and statues are made of limestone or marble. These have stood for centuries. However, they are now crumbling away at a greater rate than before. This is because both limestone and marble are mainly calcium carbonate. This reacts with acids, leaving the stone soft and crumbly. The rate of this reaction has speeded up because the concentration of acids in rainwater has increased.

Increasing the concentration of reactants in a solution increases the rate of reaction because there are more particles of the reactants moving around in the same volume of solution. The more 'crowded together' the reactant particles are, the more likely it is that they will collide. So the increased frequency of collisions results in a faster reaction.

Increasing the pressure of reacting gases has the same effect. Increased pressure squashes the gas particles more closely together. There are more particles of gas in a given space. This increases the chance that they will collide and react. So increasing the pressure produces more frequent collisions, which will increase the rate of the reaction.

Concentration and rate of reaction

A You can investigate the effect of changing concentration by reacting marble chips with different concentrations of hydrochloric acid:

$$CaCO_3(s) + 2HCl(aq) \rightarrow CaCl_2(aq) + CO_2(g) + H_2O(l)$$

You can find the rate of a reaction by plotting the volume of carbon dioxide gas given off as the reaction progresses over time. You can measure the volume of gas at regular time intervals using the method shown in the Practical box in Topic C8.5. Alternatively, you might choose to time how long it takes to collect a fixed volume of gas using the same apparatus.

- Write a hypothesis explaining your prediction of what will happen. (You might include a sketch graph of your expected results).
- How do you make this a fair test?
- What conclusion can you draw from your results?

B Consider the reaction between sodium thiosulfate and dilute hydrochloric acid:

$$Na_2S_2O_3(aq) + 2HCl(aq) \rightarrow 2NaCl(aq) + SO_2(aq) + S(s) + H_2O(l)$$

Plan an investigation to see how varying the concentration of sodium thiosulfate affects the rate of this reaction. Your timing method should use the increasing cloudiness (turbidity) of the reaction mixture as the reaction proceeds (see the last Practical box in Topic 8.1).

Make sure your teacher checks your plan before any practical work is attempted.

Safety: Wear eye protection.

Higher

Worked example

An investigation was carried out to find how the concentration of dilute hydrochloric acid affected the rate of its reaction with calcium metal. The volume of hydrogen gas given off was monitored over 150 seconds using a gas syringe. One test was carried out using 0.167 g of calcium with an excess of 1.00 mol/dm³ dilute hydrochloric acid, and this was repeated using the same volume of 0.500 mol/dm³ acid, also in excess. The results were plotted on a graph – see the two curves in the graph below.

a Use the results on the graph to find the initial rates of reaction, i.e. at the start when time = 0 seconds.

b Draw a conclusion from part **a**.

Solution

a

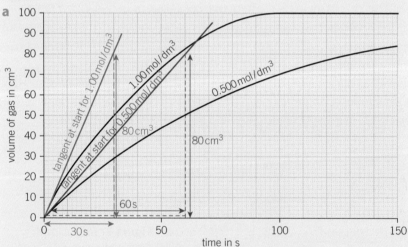

Initial rate for 1.00 mol/dm³ acid (gradient of tangent) at $t = 0$s is: $\dfrac{80\,cm^3}{30\,s} = 2.67\,cm^3/s$

Initial rate for 0.500 mol/dm³ acid (gradient of tangent) at $t = 0$s is: $\dfrac{80\,cm^3}{60\,s} = 1.33\,cm^3/s$

b The rate for 1.00 mol/dm³ dilute hydrochloric acid is twice the rate for the 0.500 mol/dm³ acid. So doubling the concentration doubles the rate of reaction. The rate is *directly proportional* to the concentration for this reaction. This could be because in any given volume of the acids, in the 1.00 mol/dm³ dilute hydrochloric acid there are twice as many $H^+(aq)$ ions as there are in the 0.500 mol/dm³ acid. This makes it twice as likely that collisions will occur between the acidic $H^+(aq)$ ions and the calcium. So in any given time there will be twice as many collisions, resulting in the reaction rate also doubling.

Go further

In A Level Chemistry you will do plenty of experiments and calculations to find out how rates of reactions depend on the concentration of each reactant in a particular reaction – the balanced equation will not give you the answer.

Key points

- Increasing the concentration of reactants in solutions increases the frequency of collisions between particles, and so increases the rate of reaction.
- Increasing the pressure of reacting gases also increases the frequency of collisions, and so increases the rate of reaction.

1 a Explain how you know which line on the graph in Figure 2 shows the fastest reaction.   [2 marks]

   b The electric balance used in the experiment needs a high resolution. What does this mean?   [1 mark]

2 a We can monitor the reaction between marble chips and hydrochloric acid (see Practical box) by measuring the volume of gas given off. Sketch a graph of volume of gas against time for three different concentrations. Label the lines as high, medium, and low concentration.   [3 marks]

   Ⓗ b Explain how you can tell which of the reactants was the limiting reactant (see Topic C4.3).   [4 marks]

3 Acidic cleaners are designed to remove limescale (calcium carbonate) when they are used neat (undiluted). They do not work as well when they are diluted. Explain this using your knowledge of collision theory.   [4 marks]

C8.5 The effect of catalysts

Learning objectives

After this topic, you should know:

- what a catalyst is
- why catalysts are important in industry.

Figure 1 *Catalysts are all around you, in the natural world and in industry. The biological catalysts in living things are called enzymes. These enzymes are large protein molecules that are folded into intricate shapes to accommodate the reactant molecules (called substrates). Enzymes are incredibly efficient catalysts and work at relatively low temperatures, conserving energy in the biotechnology industry*

Figure 2 *The transition metals platinum and palladium are used in the catalytic converters in cars. They are coated onto a honeycombed support to maximise their surface area to volume ratio*

Synoptic link

To remind yourself about reaction profiles and exothermic reactions look back to Topic C7.3.

Sometimes a reaction might only work if you use very high temperatures or pressures. This can cost industry a lot of money. However, you can speed up some reactions and reduce energy costs by using **catalysts**.

A catalyst is a substance that changes the rate of a reaction. However, it is not changed chemically itself at the end of the reaction. A catalyst is not used up in the reaction, so it can be used over and over again.

Different catalysts are needed for different reactions. Many of the catalysts used in industry involve transition metals. Examples include iron, used to make ammonia, and platinum, used to make nitric acid. Catalysts are normally used in the form of powders, pellets, or fine gauzes. This gives them the biggest possible surface area to volume ratio, making them as effective as possible, as the reactions they catalyse often involve gases reacting on their surfaces.

How catalysts work

Catalysts do not increase the frequency of collisions between reactant particles, nor do they make collisions any more energetic. They increase rates of reaction by providing an alternative reaction pathway to the products, with a lower activation energy than the reaction without the catalyst present. So with a catalyst, a higher proportion of the reactant particles have sufficient energy to react. This means that the frequency of *effective* collisions (collisions that result in a reaction) increases and the rate of reaction speeds up.

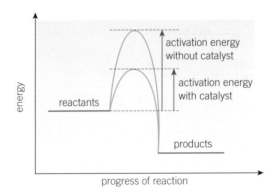

Figure 3 *The reaction profile of an uncatalysed and a catalysed exothermic reaction. The catalyst lowers the activation energy of the reaction*

Advantages of catalysts in industry

Catalysts are often very expensive precious metals. Gold, platinum, and palladium are all costly, but are the most effective catalysts for particular reactions. It is usually cheaper to use a catalyst than to pay for the extra energy needed without one. To get the same rate of reaction without a catalyst would require higher temperatures and/or pressures.

So catalysts save money and help the environment. That is because using high temperatures and pressures often involves burning fossil fuels. So operating at lower temperatures and pressures conserves these non-renewable resources. It also stops more carbon dioxide entering the atmosphere when they are burnt, helping to combat **climate change.**

Not only does a catalyst speed up a reaction, but it also does not get used up in the reaction, so a tiny amount of catalyst can be used to speed up a reaction over and over again.

However, the catalysts used in chemical plants eventually become 'poisoned', so that they do not work any more. This happens because impurities in the reaction mixture combine with the catalyst and stop it working properly.

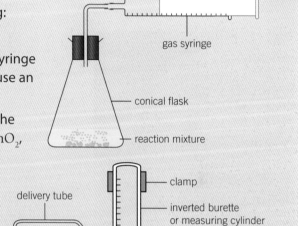

Investigating catalysis

You can investigate the effect of different catalysts on the rate of a reaction, for example hydrogen peroxide solution decomposing:

$$2H_2O_2(aq) \rightarrow 2H_2O(l) + O_2(g)$$

The reaction produces oxygen gas. You can collect this in a gas syringe using the apparatus shown on the right. Alternatively, you can use an inverted measuring cylinder, or burette, filled with water.

You can investigate the effect of many different substances on the rate of this reaction. Examples include manganese(IV) oxide, MnO_2, and potassium iodide, KI.

- State the independent variable in this investigation.

A table of the time taken to produce a certain volume of oxygen can then tell you which catalyst makes the reaction go fastest.

- What type of graph would you use to show the results of your investigation? Why?

Safety: Wear eye protection. If the syringe is glass, try to ensure that the piston does not fall out and break. Manganese(IV) oxide is harmful.

1 a How is a catalyst affected by a chemical reaction that it speeds up? [1 mark]

   b In the reaction shown in the Practical box above, with one of the catalysts tested it took 3 minutes and 12 seconds to collect $50\,cm^3$ of oxygen gas. Calculate the mean rate of the reaction, giving your answer in cm^3/s. [2 marks]

2 Solid catalysts used in chemical processes are often shaped as tiny beads or cylinders with holes through them. Why are they made in these shapes? [1 mark]

3 Suggest why the number of moles of catalyst needed to speed up a chemical reaction is very small compared with the number of moles of reactants. [2 marks]

4 Do some research to find out about four industrial processes that make products using catalysts. Write a word equation for each reaction and name the catalyst used. [4 marks]

5 Evaluate the use of catalysts in the chemical industry. ✪ [6 marks]

Study tip

Catalysts change only the *rate* of reactions. The products do not change. So catalysts do not appear in the balanced equation for a reaction, but can be written above the reaction arrow.

Key points

- A catalyst speeds up the rate of a chemical reaction, but is not used up itself during the reaction. It remains chemically unchanged.
- Different catalysts are needed for different reactions.
- Catalysts are used whenever possible in industry to increase rates of reaction and reduce energy costs.

C8.6 Reversible reactions

Learning objectives

After this topic, you should know:

- what a reversible reaction is
- how you can represent reversible reactions.

Figure 1 *Indicators undergo reversible reactions, changing colour to show you whether solutions are acidic or alkaline. In the top photo, blue litmus paper turns red in an acidic solution, and in the bottom photo, red litmus paper turns blue in an alkaline solution*

In most chemical reactions, the reactants react completely to form the products. You show this by using an arrow pointing *from* the reactants *to* the products:

$$A + B \quad \rightarrow \quad C + D$$
$$\text{reactants} \qquad \text{products}$$

However, in some reactions the products can react together to make the original reactants again. This is called a **reversible reaction**.

A reversible reaction can go in both directions, so two 'half-arrows' are used in the equation. One arrow points in the forwards direction and one in the backwards direction:

$$A + B \rightleftharpoons C + D$$

You still call the substances on the left-hand side of the equation the 'reactants' and those on the right-hand side the 'products'. So it is important that you write down the equation of the reversible reaction you are referring to when you use the words 'reactants' and 'products'. If the equation is written as:

$$C + D \rightleftharpoons A + B$$

the reactants are C and D, and the products are A and B!

Examples of reversible reactions

Have you ever tried to neutralise an alkaline solution with an acid? It is very difficult to get a solution that is exactly neutral. You can use an indicator to tell when just the right amount of acid has been added. An indicator forms compounds that are different colours in acidic solutions and in alkaline solutions.

Litmus is a complex molecule. This can be represented as HLit (where H is hydrogen). HLit is red. If you add alkali, HLit turns into the Lit^- ion by losing an H^+ ion. Lit^- is blue. If you then add more acid, blue Lit^- changes back to red HLit, and so on.

$$HLit(aq) \rightleftharpoons H^+(aq) + Lit^+(aq)$$
$$\text{red litmus} \qquad \text{blue litmus}$$

Other reversible reactions involve salts and their water of crystallisation. For example:

hydrated copper(II) sulfate (blue)	$\rightleftharpoons$	anhydrous copper(II) sulfate (white)	+ water
$CuSO_4 \cdot 5H_2O$	$\rightleftharpoons$	$CuSO_4$	+ $5H_2O$

When you heat ammonium chloride, a reversible reaction takes place.

Heating ammonium chloride
Gently heat a small amount of ammonium chloride (harmful) in a test tube with a loose plug of mineral wool. Use test-tube holders or clamp the test tube at an angle. Make sure you warm the bottom of the tube.

● What do you see inside the test tube? Explain the changes.

Safety: Wear eye protection. Take care if you are asthmatic.

Ammonium chloride breaks down on heating. It forms ammonia gas and hydrogen chloride gas. This is an example of thermal decomposition:

$$\text{ammonium chloride} \xrightarrow{\text{heat}} \text{ammonia} + \text{hydrogen chloride}$$
$$NH_4Cl(s) \longrightarrow NH_3(g) + HCl(g)$$

The two hot gases rise up the test tube. When they cool down near the mouth of the tube, they react with each other (Figure 2). The gases re-form ammonium chloride again. The white ammonium chloride solid forms on the inside of the glass:

$$\text{ammonia} + \text{hydrogen chloride} \xrightarrow{\text{cool}} \text{ammonium chloride}$$
$$NH_3(g) + HCl(g) \longrightarrow NH_4Cl(s)$$

You can show the reversible reaction as:

$$\text{ammonium chloride} \rightleftharpoons \text{ammonia} + \text{hydrogen chloride}$$
$$NH_4Cl(s) \rightleftharpoons NH_3(g) + HCl(g)$$

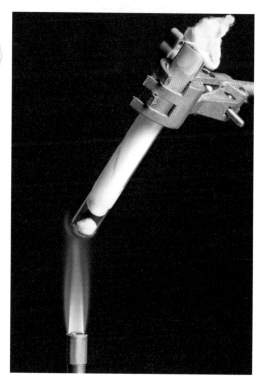

Figure 2 *An example of a reversible reaction – heating ammonium chloride to break it down into ammonia gas and hydrogen chloride gas. When the gases cool down, they recombine to form the white solid ammonium chloride again*

1 What do chemists mean by the term 'reversible chemical reaction'?
[1 mark]

2 Phenolphthalein is an indicator. It is colourless in acid and pure water, but is pink-purple in alkali. In a demonstration, a teacher started with a beaker containing a mixture of water and phenolphthalein. In two other beakers she had different volumes of acid and alkali. The acid and alkali had the same concentration. She then poured the mixture into the beaker containing $2\,cm^3$ of sodium hydroxide solution. Finally, she poured the mixture into a third beaker containing $5\,cm^3$ of hydrochloric acid.
Describe what you would observe happening in the demonstration. [1 mark]

3 You can represent the phenolphthalein indicator in question **2** as HPhe. Assuming it behaves like litmus, write a balanced symbol equation to show its reversible reaction in acid and in alkali. Label the colour of HPhe and Phe⁻ under their formulae in your equation. [1 mark]

4 Thermochromic materials change colours at different temperatures. The change is reversible. Give a use or potential use for thermochromic materials in the home. [1 mark]

5 Find out about two reversible reactions that can be used to test for the presence of water. Give the positive result for each test, including a balanced symbol equation. [4 marks]

Key points

● In a reversible reaction, the products of the reaction can react to make the original reactants.
● You can show a reversible reaction using the $\rightleftharpoons$ sign.

C8.7 Energy and reversible reactions

Learning objective

After this topic, you should know:

- what happens in the energy transfers in reversible reactions.

Synoptic links

To remind yourself about energy transferred in chemical reactions, look back to Topic C7.1.

In Topic C7.1, you saw examples of reactants forming products in exothermic reactions and endothermic reactions. These energy changes are involved in reversible reactions too. An example is given here.

Figure 1 shows a reversible reaction where A and B react to form C and D. The products of this reaction (C and D) can also react to form A and B again.

if the reaction **transfers** energy to the surroundings when it goes in this direction ...

$$A + B \rightleftharpoons C + D$$

... it will **take in** exactly the same amount of energy from the surroundings when it goes in this direction.

Figure 1 *A reversible reaction*

If the reaction between A and B is exothermic, energy will be transferred to the surroundings when the reaction forms C and D.

If C and D then react to make A and B again, the reverse reaction must be endothermic. It will take in exactly the same amount of energy as was transferred when C and D were formed from A and B.

Energy cannot be created or destroyed in a chemical reaction. The amount of energy transferred to the surroundings when the reaction goes in one direction in a reversible reaction must be exactly the same as the energy transferred back in when the reaction goes in the opposite direction.

You can see how this works if you look at what happens when blue copper(II) sulfate crystals are heated. The crystals contain water as part of the lattice formed when the copper(II) sulfate crystallised. The copper sulfate is **hydrated**. Heating the copper(II) sulfate drives off the water from the crystals, producing white **anhydrous** ('without water') copper(II) sulfate. This is an endothermic reaction.

endothermic (in forward reaction)

hydrated copper(II) $\rightleftharpoons$ anhydrous copper(II) + water
sulfate (blue)　　　　　　　sulfate (white)

$$CuSO_4.5H_2O \rightleftharpoons CuSO_4 + 5H_2O$$

exothermic (in reverse reaction)

When you add water to anhydrous copper(II) sulfate, hydrated copper(II) sulfate is formed. The colour change in the reaction, from white to blue, is a useful test for the presence of water. The reaction in this direction is exothermic (Figure 4). In fact, you may see steam rising if you add water dropwise to anhydrous copper(II) sulfate powder.

Figure 2 *Blue hydrated copper(II) sulfate and white anhydrous copper(II) sulfate*

Energy changes in a reversible reaction

Gently heat a few copper(II) sulfate crystals in a test tube (Figure 3). Observe the changes. When the crystals are completely white, allow the tube to cool to room temperature (this takes several minutes). Add two or three drops of water from a dropper and observe the changes. Carefully feel the bottom of the test tube.

- Explain the changes you have observed.

You can repeat this with the same sample of copper(II) sulfate, as it is a reversible reaction, or try with other hydrated crystals, such as cobalt(II) chloride. Some are not so colourful, but the changes are similar.

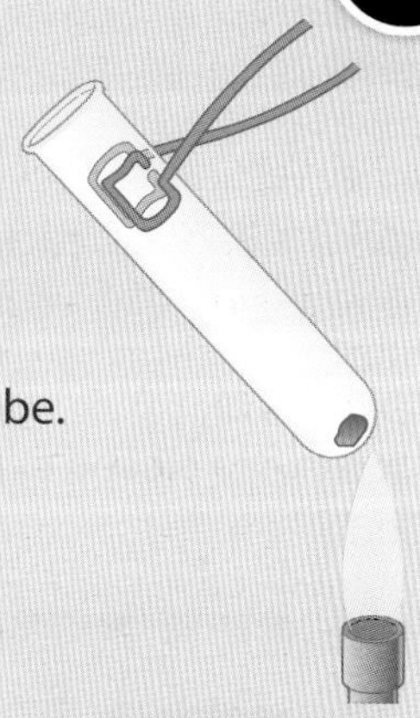

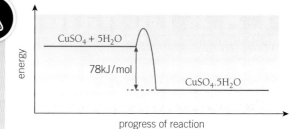

Figure 4 *The reaction profile of the reversible reaction involving anhydrous and hydrated copper(II) sulfate*

Figure 3 *Heating hydrated copper(II) sulfate*

Safety: Wear eye protection. Avoid skin contact with cobalt(II) chloride, which is toxic. Copper salts are harmful.

You can soak filter paper in cobalt(II) chloride solution and allow it to dry in an oven. The blue paper that is produced is called cobalt(II) chloride paper. The paper turns pale pink when water is added to it, and so can be used as an indicator for the presence of water.

1 a How does the enthalpy change for a reversible reaction in one direction compare with the energy change for the reaction in the opposite direction? [1 mark]

  b What can anhydrous copper(II) sulfate be used to test for? [1 mark]

  c Why does blue cobalt(II) chloride turn pink if left out in the open air? [1 mark]

  d When water is added to blue cobalt(II) chloride, is energy transferred to or from the surroundings? [1 mark]

2 A reversible reaction transfers 50 kilojoules (kJ) of energy to the surroundings in the forward reaction. In this reaction W and X react to give Y and Z.

  a Write an equation to show the reversible reaction. [1 mark]

  b State the energy transfer in the reverse reaction. [1 mark]

3 Blue cobalt(II) chloride crystals turn pink when they become damp. The formula for the two forms can be written as $CoCl_2.2H_2O$ and $CoCl_2.6H_2O$.

H a How many moles of water will combine with 1 mole of $CoCl_2.2H_2O$? [1 mark]

  b Write a balanced chemical equation for the reaction, which is reversible. [3 marks]

  c How can pink cobalt(II) chloride crystals be changed back to blue cobalt(II) chloride crystals? [1 mark]

H d Calculate the mass of water lost when 0.50 moles of pink cobalt(II) chloride is turned completely into blue cobalt(II) chloride. [2 marks]

Figure 5 *Blue cobalt(II) chloride paper turns pink when water is added*

Key points

- In reversible reactions, one reaction is exothermic and the other is endothermic.
- In any reversible reaction, the amount of energy transferred to the surroundings when the reaction goes in one direction is exactly equal to the energy transferred back when the reaction goes in the opposite direction.

C8.8 Dynamic equilibrium

Learning objectives

After this topic, you should know:

- how a reversible reaction in a closed system can be 'at equilibrium'
- **H** that the composition of an equilibrium mixture can be altered by changing conditions, such as concentration.

Some reactions are reversible. The products formed can react together to make the original reactants again:

$$A + B \rightleftharpoons C + D$$

So what happens when you start with just the reactants in a reversible reaction in a **closed system**, in which no reactants or products can get in or out?

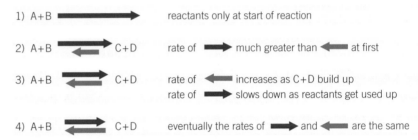

So in a reversible reaction, as the concentration of products builds up, the rate at which they react to re-form reactants increases. As this starts to happen, the rate of the forward reaction is decreasing. That is because the concentration of reactants is decreasing from its original maximum value. Eventually both forward and reverse reactions are happening at the same rate, but in opposite directions.

When this happens, the reactants are making products at the same rate as the products are making reactants. Overall, there is *no change* in the amount of products and reactants. The reaction has reached **equilibrium**.

At equilibrium, the rate of the forward reaction equals the rate of the reverse reaction.

As the forward and reverse reactions are continuously taking place (unnoticed because they occur at the atomic level), a state of 'dynamic' equilibrium has been reached.

Figure 1 *The situation at equilibrium is just like running up an escalator that is going down – if you run up as fast as the escalator goes down, you will get nowhere!*

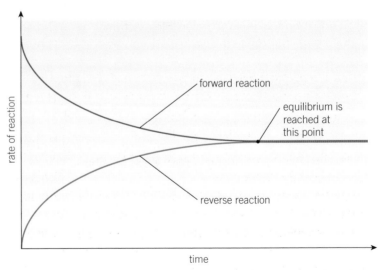

Figure 2 *In a reversible reaction at equilibrium, the rate of the forward reaction is the same as the rate of the reverse reaction*

Affecting the composition of an equilibrium mixture

One example of a reversible reaction is the reaction between iodine monochloride, ICl, and chlorine gas. Iodine monochloride is a brown liquid, while chlorine is a yellowish green gas. These substances can be reacted together to make yellow crystals of iodine trichloride, ICl_3.

When there is plenty of chlorine gas, the forward reaction makes iodine trichloride crystals, which are quite stable. However, if the concentration of chlorine gas is lowered, the rate of the forward reaction decreases and the rate of the reverse reaction increases. This starts turning more iodine trichloride back to iodine monochloride and chlorine, until equilibrium is established again.

You can change the relative amounts of the reactants and products in a reacting mixture at equilibrium by changing the conditions. This is an application of **Le Châtelier's Principle**. Henry Louis Le Châtelier was a French chemist who observed equilibrium mixtures. He noticed that whenever a change in conditions is introduced to a system at equilibrium, the position of equilibrium shifts so as to cancel out the change. The change in conditions can be changes in concentration, pressure or temperature.

This principle is very important in the chemical industry. In a process with a reversible reaction, industrial chemists need to find the conditions that give as much product as possible, in as short a time as possible.

However, there are always other economic and safety factors to consider when chemists manipulate reversible reactions in industry.

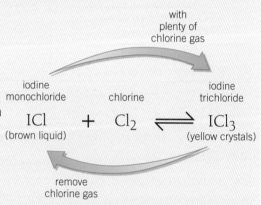

Figure 3 *This reacting mixture can be changed by adding or removing chlorine from the mixture*

Synoptic links

For more information about economic and safety issues associated with reversible reactions in industry, see Topic C15.6 and Topic C15.8.

1 How does the rate of the reverse reaction compare with the rate of the forward reaction in a reaction at equilibrium? [1 mark]

2 What can you say about the concentration of reactants and products in a reversible reaction at equilibrium? [1 mark]

3 Explain why chemists describe chemical equilibrium as dynamic as opposed to static. [2 marks]

H 4 An equilibrium mixture is set up in a closed system with iodine monochloride, chlorine gas, and iodine trichloride. In order to make more iodine trichloride, would you pump more chlorine gas into the mixture or remove chlorine gas? Explain your answer using Le Châtelier's Principle. [2 marks]

Key points

- In a reversible reaction, the products of the reaction can react to re-form the original reactants.
- In a closed system, the rate of the forward and reverse reactions is equal at equilibrium.
- Changing the reaction conditions can change the amounts of products and reactants in a reaction mixture at equilibrium.

C8.9 Altering conditions

Pressure and equilibrium

You saw in Topic C8.8 how changing concentration can affect a reversible reaction at equilibrium. In general, the position of equilibrium shifts as if trying to cancel out any change in conditions. Think about increasing the concentration of a reactant. This will cause the position of equilibrium to shift to the right, in favour of the products, in order to reduce the concentration of that reactant. It opposes the change that is introduced.

If a reversible reaction involves changing numbers of gas molecules, altering the pressure can also affect the equilibrium mixture. In many reversible reactions, there are more molecules of gas on one side of the equation than on the other. By changing the pressure at which the reaction is carried out, you can change the amount of products that are made.

Look at the table below:

If the forward reaction produces *more* molecules of gas …	If the forward reaction produces *fewer* molecules of gas …
… an increase in pressure decreases the amount of products formed.	… an increase in pressure increases the amount of products formed.
… a decrease in pressure increases the amount of products formed.	… a decrease in pressure decreases the amount of products formed.

You can show this with the following reversible reaction:

$$2NO_2(g) \rightleftharpoons N_2O_4(g)$$
$$\text{brown gas} \qquad \text{pale yellow gas}$$

In this reaction you can see from the balanced symbol equation that there are two molecules of gas on the left-hand side of the equilibrium equation and one molecule of gas on the right-hand side.

Imagine that you increase the pressure in the reaction vessel. The position of equilibrium will shift to reduce the pressure. It will move in favour of the reaction that produces fewer gas molecules. In this case that is to the right, in favour of the forward reaction. So more N_2O_4 gas will be made. The colour of the gaseous mixture will get lighter (Figure 1).

Note that pressure changes do not affect all gaseous reversible reactions at equilibrium. When there are *equal* numbers of molecules of gas on both sides of the balanced equation, changing the pressure has no effect on the composition of the equilibrium mixture. However, increasing the pressure will speed up both the forward and reverse reactions by the same amount.

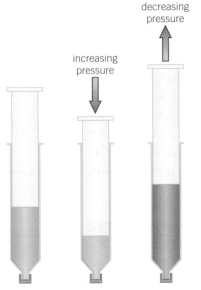

decreasing pressure

increasing pressure

Figure 1 *The effect of changing the pressure on $2NO_2(g) \rightleftharpoons N_2O_4(g)$*

Temperature and equilibrium

When you have a closed system, no substances are added or taken away from the reaction mixture. In a closed system, the relative amounts of the reactants and products in a reversible reaction at equilibrium depend on the temperature.

By changing the temperature, you can plan to get more of the products and less of the reactants. Look at the table below:

If the forward reaction is exothermic ...	If the forward reaction is endothermic ...
... an increase in temperature decreases the amount of products formed.	... an increase in temperature increases the amount of products formed.
... a decrease in temperature increases the amount of products formed.	... a decrease in temperature decreases the amount of products formed.

You can show this by looking at the reaction involving $NO_2(g)$ and $N_2O_4(g)$ again. The forward reaction is exothermic, so the reverse reaction is endothermic.

$$\underset{\text{endothermic}}{\overset{\text{exothermic}}{2NO_2(g) \rightleftharpoons N_2O_4(g)}}$$

If the temperature is increased, the equilibrium shifts as if to try to reduce the temperature. The reaction that is endothermic (taking in energy from the surroundings) will cool it down. So in this case, the reverse reaction is favoured and more NO_2 is formed (Figure 2).

| more NO_2 formed | heat up ← | $2NO_2(g) \rightleftharpoons N_2O_4(g)$ | cool down → | more N_2O_4 formed |

(favours endothermic reaction) / (favours exothermic reaction)

Figure 2 *The effect of changing the temperature on $2NO_2(g) \rightleftharpoons N_2O_4(g)$*

1 How does increasing the pressure affect the amount of products formed in a reversible reaction that produces a larger volume (more molecules) of gas at equilibrium? [1 mark]

2 Look at the reversible reaction below:
$$H_2O(g) + C(s) \rightleftharpoons CO(g) + H_2(g)$$
The forward reaction is endothermic.
Describe how the amount of hydrogen gas formed will change if the temperature is increased. [1 mark]

3 a In Figure 1, how will increasing the pressure affect the colour of a mixture of NO_2 and N_2O_4 gases? Explain your answer. [2 marks]
  b In Figure 2, what will happen to the colour of a mixture of NO_2 and N_2O_4 gases if you increase the temperature? Explain your answer. [2 marks]

4 Explain what effect increasing the pressure would have on the equilibrium mixture below:
$$H_2(g) + I_2(g) \rightleftharpoons 2HI(g)$$ [3 marks]

Go further

If you go on to study A Level Chemistry you can do calculations to find where the position of equilibrium lies – does it favour the reactants or the products? Equilibrium constants will tell you the answer.

Observing equilibrium

Using a gas syringe, your teacher can show an equilibrium mixture of NO_2 and N_2O_4 gases.

You can see the effect of changing temperature by using warm water and ice.

Pushing and pulling the plunger of the syringe can change the pressure.

● Explain your observations.

Safety: Take care if you are asthmatic. You and your teacher should be wearing chemical splash-proof eye protection. The gases are 'very toxic' and must be prepared in a fume cupboard by your teacher.

Key points

● Pressure can affect reversible reactions involving gases at equilibrium. Increasing the pressure favours the reaction that forms fewer molecules of gas. Decreasing the pressure favours the reaction that forms the greater number of molecules of gas.

● You can change the relative amount of products formed at equilibrium, by changing the temperature at which you carry out a reversible reaction.

● Increasing the temperature favours the endothermic reaction. Decreasing the temperature favours the exothermic reaction.

C8 Rates and equilibrium

Summary questions

1 Two students investigated the reaction of some marble chips with dilute nitric acid.

Time in minutes	Investigation A: mass of gas produced in g	Investigation B: mass of gas produced in g
0.0	0.00	0.00
0.5	0.56	0.28
1.0	0.73	0.36
1.5	0.80	0.39
2.0	0.82	0.41
2.5	0.82	0.41

a Name the gas produced in the reaction. Describe the positive test for this gas. [1 mark]

b i Marble chips contain calcium carbonate. Write a word equation for the reaction investigated. [1 mark]

ii Write a balanced symbol equation, including state symbols, for the reaction. [3 marks]

c The students were investigating the effect of concentration on rate of reaction. Suggest the method the students used to collect the data for their table. ⊘ [6 marks]

d Plot a graph of these results, with time on the *x*-axis. [4 marks]

Ⓗ e i Use your graph to calculate the rate of reaction at 30 seconds in both investigations, and give your answers in units of g/min and in mol/s. [2 marks]

ii Compare the rate of reaction in Investigation B with the rate of reaction in Investigation A. [1 mark]

f Compare the final mass of gas produced in Investigation B with that produced in Investigation A. [1 mark]

g From the results, what can you say about the initial concentration of the acids in Investigations A and B? [1 mark]

h Use the collision theory to explain the data obtained. [4 marks]

2 A pair of students are studying the effect of surface area on rate of reaction. They test the reaction of zinc metal with dilute hydrochloric acid.

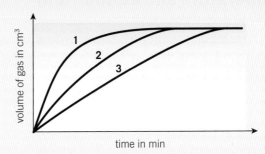

a What would the students investigating the reaction need to do to make it a *fair test*? [3 marks]

b Which gas was given off in the reaction? How could you test for this gas? [2 marks]

c Write a balanced symbol equation, including state symbols, for the reaction under investigation. [2 marks]

d Which line (1, 2, or 3) shows results for the zinc with the largest surface area to volume ratio? [1 mark]

e What size of pieces of zinc would react most slowly? [1 mark]

f The students doing the experiments also tried reacting the same mass of *powdered* zinc with the acid. What would their results look like on the graph? [2 marks]

g Use the collision theory to explain the results in this investigation. [3 marks]

Ⓗ h The students had reacted zinc with excess dilute hydrochloric acid. If they had added 0.13 g of zinc, calculate the total volume of hydrogen gas collected at r.t.p. when the reaction had finished. (1 mole of any gas at r.t.p. occupies a volume of 24 dm³.) [3 marks]

3 a Define the word catalyst. [2 marks]

b Hydrogen peroxide, H_2O_2, solution decomposes to form water and oxygen gas. Write a balanced symbol equation, including state symbols, for this reaction. [3 marks]

c The reaction is catalysed by some metal oxides. You are provided with oxides of copper, lead, manganese, and iron. Describe how you can test which is the best catalyst for the decomposition of hydrogen peroxide, collecting quantitative (measured) data to support your conclusion. ⊘ [5 marks]

Ⓗ d In one test, it was found that 40 cm³ of oxygen gas was collected in 16 s. Calculate the mean rate of reaction in that test, expressing your answer in cm³/s and mol/s (1 mole of any gas at room temperature and pressure occupies a volume of 24 dm³.) [3 marks]

e How can you get a pure dry sample of the metal oxides from the mixture left at the end of the reaction? [3 marks]

Practice questions

01 This question is about the reactions of sodium thiosulfate.

A student investigated the rate of reaction between sodium thiosulfate and hydrochloric acid.

The reaction was set up in a conical flask as shown in **Figure 1**.

The student recorded the time from mixing the sodium thiosulfate and hydrochloric acid until the mixture became so cloudy that the cross below the flask was no longer visible.

Figure 1

The equation for the reaction is

$Na_2S_2O_3(aq) + 2HCl(aq) \rightarrow 2NaCl(aq) + H_2O(l) + SO_2(g) + S(s)$

01.1 Which product caused to mixture to become cloudy? [1 mark]

01.2 The student changed the concentration of sodium thiosulfate. The student did each reaction at the same temperature

Give one other variable the student should control. [1 mark]

01.3 The rate of reaction increased as the concentration of sodium thiosulfate increased. Explain why. [2 marks]

01.4 The rate of reaction would also increase as the temperature increased. Explain why. [4 marks]

02 A student investigated the rate of reaction between calcium carbonate chips and dilute hydrochloric acid. The apparatus is shown in **Figure 2**.

Figure 2

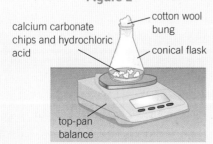

calcium carbonate chips and hydrochloric acid

cotton wool bung

conical flask

top-pan balance

The equation for the reaction is

$CaCO_3(s) + 2HCl(aq) \rightarrow CaCl_2(aq) + H_2O(l) + CO_2(g)$

The mass decreased during the reaction. The student's results are shown in **Figure 3**.

Figure 3

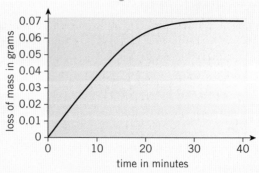

02.1 Use the equation to explain why the mass decreased during the reaction. [2 marks]

02.2 Suggest why the student used a cotton wool plug? [1 mark]

02.3 The student used an excess of calcium carbonate chips.

Explain why the rate of reaction slows down as the reaction proceeds. [2 marks]

02.4 The same mass of calcium carbonate was used in another experiment. Suggest one reason why the rate of reaction was higher. Explain your answer. [3 marks]

Ⓗ 03 This question is about reversible reactions.

03.1 Consider the two reversible reactions.

Reaction 1 $C_2H_4(g) + H_2O(g) \rightleftharpoons C_2H_5OH(g)$
exothermic

Reaction 2 $N_2O_4(g) \rightleftharpoons 2NO_2(g)$
endothermic

Use the equations to fill in the gaps in **Table 1**.

Table 1

	Effect of increasing pressure on rate	Effect of increasing pressure on yield	Effect of increasing temperature on yield
Reaction 1	increases		
Reaction 2	increases		

[4 marks]

03.2 Explain why increasing the pressure increases the rate of a reaction involving gases. [2 marks]

The equation for the industrial manufacture of ammonia is:

$N_2(g) + 3H_2(g) \rightleftharpoons 2NH_3(g)$

The reaction is exothermic in the forwards direction.

03.3 Explain why high pressure is used for the process. [3 marks]

Learning objectives

After this topic, you should know:

- what crude oil is made up of
- what alkanes are
- how to represent alkanes by their chemical formula or displayed formula
- the names and formulae of the first four alkanes.

Figure 1 *The price of nearly everything you buy is affected by oil prices. The cost of the fuels used to move goods to the shops affects how much you have to pay for them*

Synoptic links

For more information about the Earth's finite (non-renewable) resources, see Topic C14.1.

So far, some of the 21st century's most important chemicals come from the organic carbon compounds found in crude oil. These chemicals play a major part in our lives. The massive variety of natural and synthetic carbon compounds occur due to the ability of carbon atoms to form families of similar compounds. The carbon atoms can bond to each other to form chains and rings that form the 'skeletons' of organic molecules. Organic compounds are used as fuels to run cars, to warm homes, and to generate electricity.

Fuels are important because they keep you warm and on the move. So when oil prices rise, it affects everyone. Countries that produce crude oil can affect the whole world economy by the price they charge for their oil.

Crude oil

Crude oil is a finite resource found in rocks. It was formed over millions of years from the remains of tiny, ancient sea animals and plants, mainly plankton, that were buried in mud. Over time, layer upon layer of rock was laid down on top, creating the conditions (high pressure and temperature, in the absence of oxygen) to make crude oil.

The crude oil formed is a dark, smelly liquid. It is a **mixture** of many different carbon compounds. A mixture contains two or more elements or compounds that are not chemically combined together. Nearly all of the compounds in crude oil are compounds containing only hydrogen and carbon atoms. These compounds are called **hydrocarbons**.

Distillation of crude oil

Mixtures of liquids can be separated using distillation. This can be done in the lab on a small scale. Your teacher will heat the crude oil mixture so that it boils. The different fractions vaporise between different ranges of temperature. The vapours can be collected by cooling and condensing them.

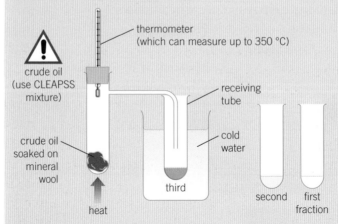

Figure 2 *The distillation of crude oil in the lab*

Crude oil straight from the ground is not much use. There are too many substances in it, all with different boiling points. Before crude oil can be used, it must be separated into different substances with similar boiling points. These are known as **fractions**. Because the properties of substances do not change when they are mixed (as opposed to reacted), the mixtures of substances in crude oil can be separated in the lab by **distillation**. Distillation separates liquids with different boiling points.

Alkanes

Most of the hydrocarbons in crude oil are **alkanes**. You can see some examples of alkane molecules in Figure 3. Notice how all their names end in '-ane'.

The first part of the name of each alkane tells you how many carbon atoms are in its molecules (Table 1).

You describe alkanes as **saturated hydrocarbons**. All the carbon–carbon bonds are single covalent bonds. This means that they contain as many hydrogen atoms as possible in each molecule. No more hydrogen atoms can be added.

The formulae of the first four alkane molecules are given below:

$$CH_4 \text{ (methane)}$$

$$C_2H_6 \text{ (ethane)}$$

$$C_3H_8 \text{ (propane)}$$

$$C_4H_{10} \text{ (butane)}$$

Can you see a pattern in the formulae of the alkanes? You can write the **general formula** for alkane molecules like this:

$$C_nH_{(2n+2)}$$

This means that 'for every n carbon atoms there are $(2n + 2)$ hydrogen atoms' in an alkane molecule. For example, if an alkane molecule contains 12 carbon atoms, its formula will be $C_{12}H_{26}$.

1 a What is crude oil? [1 mark]
   b Why is oil so important to society? [1 mark]
   c Why can you separate crude oil using distillation? [1 mark]
2 Crude oil is drilled from beneath the ground or seabed. Why is this crude oil not very useful as a product itself? [2 marks]
3 a Write the general formula of the alkanes. [1 mark]
   b Write the formulae of the alkanes which have six to ten carbon atoms. Then find out their names. [5 marks]
4 a Draw the displayed formula of octane, whose molecules have eight carbon atoms. [2 marks]
   b How many hydrogen atoms are there in an alkane molecule which has 22 carbon atoms? [1 mark]
   c How many carbon atoms are there in an alkane molecule which has 32 hydrogen atoms? [1 mark]
   d Why are alkanes described as saturated hydrocarbons? [2 marks]

Synoptic links

To remind yourself about covalent bonding, look back at Topic C3.5.

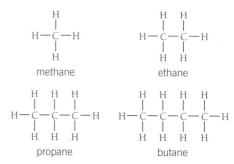

Figure 3 *You can represent alkanes like this, showing all of the atoms in the molecule. These are called displayed formulae. The line drawn between two atoms in a molecule represents a single covalent bond*

Table 1 *Prefixes*

Prefix (start of name)	Number of carbon atoms
meth-	1
eth-	2
prop-	3
but-	4

Key points

- Crude oil is a mixture of many different compounds.
- Most of the compounds in crude oil are hydrocarbons – they contain only hydrogen and carbon atoms.
- Alkanes are saturated hydrocarbons. They contain as many hydrogen atoms as possible in their molecules.
- The general formula of an alkane is: $C_nH_{(2n+2)}$

C9.2 Fractional distillation of oil

Learning objectives

After this topic, you should know:

- how the volatility, viscosity, and flammability of hydrocarbons are affected by the size of their molecules
- how to separate crude oil into fractions
- how to explain the separation of crude oil by fractional distillation
- how the fractions from crude oil are used.

The properties of hydrocarbons

There is a great variety of hydrocarbon molecules. Some are quite small, with relatively few carbon atoms in short chains. These short-chain molecules make up the hydrocarbons that tend to be most useful. These hydrocarbons make good fuels as they ignite easily and burn well, with less smoky flames than hydrocarbons made up of larger molecules. They are described as very **flammable** (Figure 1). Other hydrocarbons have lots of carbon atoms in their long-chain molecules, and may have branches (side-chains) or form rings.

Figure 1 shows the trends in the properties of hydrocarbons.

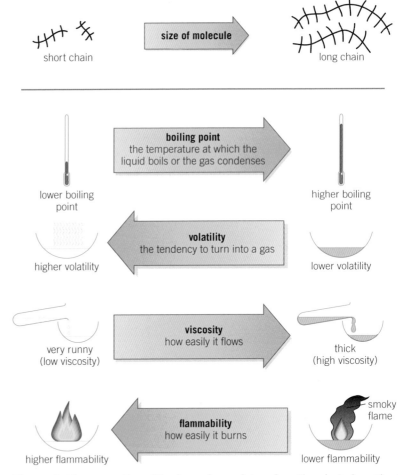

Figure 1 *The properties of hydrocarbons depend on the chain length of their molecules*

Synoptic links

To remind yourself about the process of fractional distillation, look back to Topic C1.4. To revise intermolecular forces, refer to Topic C3.6.

Fractional distillation of crude oil

Crude oil is separated into hydrocarbons with similar boiling points, called fractions. This process is called **fractional distillation**. Each hydrocarbon fraction contains molecules with similar numbers of carbon atoms. Each of these fractions boils at a different temperature range because of the different sizes of the molecules in it.

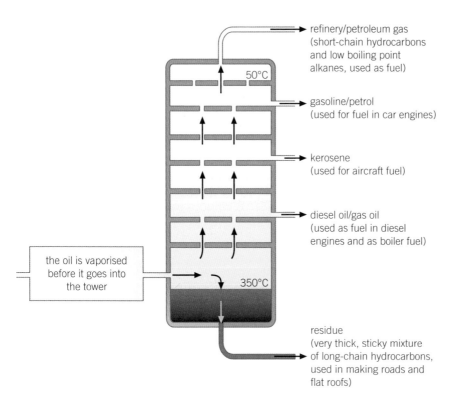

refinery/petroleum gas
(short-chain hydrocarbons
and low boiling point
alkanes, used as fuel)

50°C

gasoline/petrol
(used for fuel in car engines)

kerosene
(used for aircraft fuel)

diesel oil/gas oil
(used as fuel in diesel
engines and as boiler fuel)

the oil is vaporised
before it goes into
the tower

350°C

residue
(very thick, sticky mixture
of long-chain hydrocarbons,
used in making roads and
flat roofs)

Figure 2 *The boiling point of a hydrocarbon depends on the size of its molecules. Differences in boiling points can be used in fractional distillation to separate the hydrocarbons in crude oil into fractions*

Crude oil is heated and fed in near the bottom of a tall tower (called a fractionating column) as hot vapour. The column is kept very hot at the bottom and much cooler at the top, so the temperature decreases going up the column. The gases move up the column and the hydrocarbons condense when they reach the temperature of their boiling points. The different fractions are collected as liquids at different levels. The fractions are collected from the column in a continuous process.

Hydrocarbons with the smallest molecules have the lowest boiling points. They are piped out of the cooler top of the column as gases. At the bottom of the column, the fractions have high boiling points. They cool to form very thick liquids or solids at room temperature.

Once collected, the fractions need more processing before they can be used.

1 a How does the size of a hydrocarbon molecule affect:
   i   the boiling point   [1 mark]
   ii the volatility   [1 mark]
   iii the viscosity of a hydrocarbon?   [1 mark]
   b A hydrocarbon catches fire very easily. Is it likely to have molecules with long hydrocarbon chains or short ones? Will it give off lots of black smoke when it burns?   [1 mark]

2 Make a table to summarise how the properties of hydrocarbons depend on the size of their molecules.   [4 marks]

3 Explain the steps involved in the fractional distillation of crude oil. ✏   [6 marks]

Comparing fractions

Your teacher compares the viscosity (thickness) and flammability of some fractions (mixtures of hydrocarbons with similar boiling points) from crude oil.

Figure 3 *An oil refinery at night*

Key points

● Crude oil is separated into fractions using fractional distillation.
● The properties of each fraction depend on the size of the hydrocarbon molecules in it.
● Lighter fractions make better fuels as they ignite more easily and burn well, with cleaner (less smoky) flames.

C9.3 Burning hydrocarbon fuels

Learning objectives

After this topic, you should know:

- the products formed when you burn hydrocarbon fuels in a good supply of air
- how to test for the products of complete combustion of a hydrocarbon
- why carbon monoxide gas is also given off when incomplete combustion takes place
- how to write balanced equations for the complete combustion of hydrocarbons with a given formula.

Figure 1 *On a winter's day you can often see the water produced when the hydrocarbons in petrol or diesel burn, as the steam formed in combustion of the fuel condenses when it cools down*

Synoptic links

To revise the tests for water and the reactions they are based on, look back to Topic C8.7.

You will look at testing for gases in more detail in Topic C12.3.

Complete combustion

The lighter fractions from crude oil are very useful as fuels. When hydrocarbons burn in plenty of air they transfer lots of energy to the surroundings.

For example, you might have seen orange gas cylinders used in mobile heaters and for gas cookers in areas without mains gas. These cylinders contain propane, one of the petroleum gases that come from the top of the fractionating column in oil refineries. When propane gas burns:

$$\text{propane} + \text{oxygen} \rightarrow \text{carbon dioxide} + \text{water}$$
$$C_3H_8(g) + 5O_2(g) \rightarrow 3CO_2(g) + 4H_2O(g)$$

The carbon and hydrogen in the fuel are **oxidised** completely when they burn like this. Remember that one definition of oxidation means adding oxygen in a chemical reaction.

The products of the complete combustion of a hydrocarbon are carbon dioxide and water.

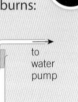

Products of complete combustion

You can test the products given off when a hydrocarbon burns:

small luminous Bunsen flame (airhole closed)

natural gas

ice bath

blue cobalt chloride paper

limewater

to water pump

- What happens to the limewater? Which gas is given off?
- What happens in the U-tube? Which substance is present?

Safety: Wear eye protection. Handle cobalt chloride papers as little as possible.

As well as using blue cobalt chloride paper to test for the water formed in the complete combustion of a hydrocarbon, you can also use white anhydrous copper sulfate. On contact with water, the white powder turns blue.

Incomplete combustion

All fossil fuels – oil, coal, and natural gas – produce carbon dioxide and water when they burn in plenty of air. However, when there is not enough oxygen, for example, inside an engine, there is incomplete combustion. Instead of all the carbon in the fuel turning into carbon dioxide, carbon monoxide gas, CO, is also formed. Carbon monoxide is a toxic gas. It is colourless and odourless. Your red blood cells pick up this gas and carry it around in your blood instead of oxygen.

Higher

Worked example: Combustion of butane

A typical small cylinder of butane, C_4H_{10}, used for camping contains 5.000 kg of liquefied butane.

a Write a balanced symbol equation for the complete combustion of butane.

b If all the gas was burnt in a plentiful supply of air, what would be the maximum mass of carbon dioxide released into the atmosphere?

Solution

a $2C_4H_{10}(g) + 13O_2(g) \rightarrow 8CO_2(g) + 10H_2O(g)$

b In the cylinder we have 5.000 kg of butane, i.e., 5000 g of the gas.
Using A_r of C = 12 and H = 1, the M_r of $C_4H_{10} = (4 \times 12) + (10 \times 1) = 58$.
So knowing that 1 mole of butane = 58 g, in 5000 g we have

$\frac{5000}{58}$ moles of butane (from number of moles = $\frac{mass}{M_r}$).

The balanced equation tells us that 2 moles of butane gives off 8 moles of CO_2, so 1 mole of butane gives off 4 moles of CO_2 and $\frac{5000}{58}$ moles of butane gives off ($\frac{5000}{58}$) × 4 moles of CO_2.

1 mole of CO_2 has a mass of $[12 + (2 \times 16)]\,g = 44\,g$

so, using mass = no. of moles × M_r:

($\frac{5000}{58}$) × 4 moles of CO_2 will have a mass of:

$[(\frac{5000}{58}) \times 4] \times 44\,g$

= **15 170 g (or 15.17 kg)**

The answer is given to 4 significant figures in line with the mass of butane provided in the question.

Figure 2 *Compressed butane gas is sold in blue cylinders and is used to fuel camping stoves*

Synoptic links

To find out more about the effects of the pollution caused by burning hydrocarbon fuels, see Topic C13.5.

1 a Name the products of the complete combustion of a hydrocarbon. [2 marks]

b Describe a positive test for each of the products of combustion. [1 mark]

2 a Natural gas is mainly methane, CH_4. Write a balanced symbol equation, including state symbols, for its complete combustion. [3 marks]

b When natural gas burns in a faulty gas heater, it can produce carbon monoxide (and water). Write a balanced symbol equation, including state symbols, to show just this reaction. [3 marks]

c Describe why victims of carbon monoxide poisoning are unaware that carbon monoxide is building up in the air they breathe. [2 marks]

H 3 A cylinder of propane, C_3H_8, contains 16.00 kg of liquefied gas. Using the balanced chemical equation, calculate the minimum mass of oxygen gas needed for complete combustion of the gas in the cylinder. [4 marks]

Key points

- When hydrocarbon fuels are burned in plenty of air, the carbon and hydrogen in the fuel are completely oxidised. They produce carbon dioxide and water.
- You can test the gases formed in complete combustion of a hydrocarbon: the carbon dioxide turns limewater cloudy, and the water turns blue cobalt chloride paper pink (or white anhydrous copper sulfate blue).
- Incomplete combustion of a hydrocarbon produces carbon monoxide (a toxic gas) as one of its products.

C9.4 Cracking hydrocarbons

Learning objectives

After this topic, you should know:

- how and why larger, less useful hydrocarbon molecules are cracked to form smaller ones
- examples to illustrate the usefulness of cracking and how modern life depends on the uses of hydrocarbons
- what alkenes are and how they differ from alkanes.

Figure 1 *In an oil refinery, huge crackers like this are used to break down large hydrocarbon molecules into smaller, more useful ones. The petrochemical industry source the chemicals used to make products such as solvents, lubricants, polymers, and detergents, as well as the fuels that modern life depends on*

Why crack hydrocarbons?

Some of the heavier fractions from the fractional distillation of crude oil are not in high demand. The hydrocarbons in them are made up of large molecules. They are thick liquids or solids with high boiling points. They are difficult to vaporise and do not burn easily – so they are poor fuels, although they do have their uses (see Topic C9.2). Yet the main demand from crude oil is for fuels and starting materials (feedstock) for the chemical industry. Fortunately, the larger, less useful hydrocarbon molecules can be broken down into smaller, more useful ones in a process we call **cracking**.

The process takes place at an oil refinery in steel vessels called crackers. In the cracker, a heavy fraction distilled from crude oil is heated to vaporise the hydrocarbons. The vapour is then either:

- passed over a hot catalyst, or
- mixed with steam and heated to a very high temperature.

The hydrocarbons are cracked as **thermal decomposition** reactions take place. The large molecules split apart to form smaller, more useful ones.

An example of cracking

Decane, $C_{10}H_{22}$, is a medium-sized alkane molecule. When it is heated to 500 °C with a catalyst, it breaks down. One of the molecules produced is pentane, C_5H_{12}, which is used in petrol.

$$
\begin{array}{c}
\text{H}\quad\text{H}\quad\text{H}\quad\text{H}\quad\text{H}\\
|\quad\;|\quad\;|\quad\;|\quad\;|\\
\text{H}-\text{C}-\text{C}-\text{C}-\text{C}-\text{C}-\text{H}\\
|\quad\;|\quad\;|\quad\;|\quad\;|\\
\text{H}\quad\text{H}\quad\text{H}\quad\text{H}\quad\text{H}
\end{array}
$$
pentane (displayed formula)

Propene and ethene are also made, which the chemical industry can use to produce polymers and other chemicals, such as solvents:

$$C_{10}H_{22} \xrightarrow{\;500\,°C + catalyst\;} C_5H_{12} + C_3H_6 + C_2H_4$$
decane pentane propene ethene

This cracking reaction is an example of thermal decomposition.

Notice how cracking produces different types of molecules. One of the molecules is pentane. The first part of its name tells us that it has five carbon atoms (*pent-*). The last part of its name (*-ane*) shows that it is an alkane. Like all other alkanes, pentane is a saturated hydrocarbon. Its molecules have as much hydrogen as possible in them.

The other molecules in this reaction have names that end slightly differently. They end in *-ene*. We call this type of molecule an **alkene**. The different ending tells us that these molecules are unsaturated. Unsaturated compounds contain at least one **double bond** between their carbon atoms. In Figure 2, you can see that these alkenes have one C=C double covalent bond in their molecules. As carbon atoms form four covalent bonds, it means that the unstaturated alkene molecules have two fewer hydrogen atoms in their molecules than the saturated alkane molecules with the same number of carbon atoms.

The experiment outlined in Figure 3 shows how cracking can be done in the lab.

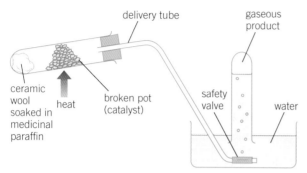

Figure 3 *Apparatus for cracking medicinal paraffin*

Tests on the gaseous products from the experiment above show that alkenes:

● burn in air (but not as well as equivalent small alkanes which are used as fuels)

● react with bromine water, which is orange in colour, decolourising it.

Alkenes are generally more reactive than alkanes. The reaction with bromine water is used as a test to see if an organic compound is unsaturated (like the alkenes with their C══C double bond).

A positive test for an unsaturated hydrocarbon is that it turns orange bromine water colourless.

The alkanes do not react with bromine water, so you can use this test to distinguish between an alkene and an alkane.

1 **a** Why is cracking so important? [1 mark]
  b How can large hydrocarbon molecules be cracked in an oil refinery? [2 marks]

2 Cracking a hydrocarbon makes two new hydrocarbons, A and B. When bromine water is added to A, nothing happens. Bromine water added to B turns from an orange solution to colourless.
  a i Which hydrocarbon, A or B, is unsaturated? [1 mark]
   ii Define an unsaturated hydrocarbon. [1 mark]
   iii Name the type of unsaturated hydrocarbon formed in cracking. [1 mark]
  b i Which hydrocarbon, A or B, is used as a fuel? [1 mark]
   ii Name the type of saturated hydrocarbon formed in cracking. [1 mark]
  c What type of reaction is cracking an example of? [1 mark]

3 Dodecane (an alkane with 12 carbon atoms) can be cracked into octane (with eight carbon atoms) and ethene. Write a balanced symbol equation for this reaction. [3 marks]

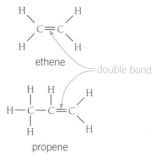

Figure 2 *A molecule of ethene, C_2H_4, and a molecule of propene, C_3H_6. These are both alkenes – each molecule has a carbon–carbon double bond in it*

Synoptic links

To find out more about the reaction between an alkene and bromine water, see Topic C10.1.

Go further

Although the alkanes are less reactive than the alkenes, they do undergo other reactions besides combustion. In A Level Chemistry, you will learn about their reactions with halogens and why they happen in sunlight but not in darkness.

Key points

● Large hydrocarbon molecules can be broken up into smaller molecules by passing the vapours over a hot catalyst, or by mixing them with steam and heating them to a very high temperature.

● Cracking produces saturated hydrocarbons, used as fuels, and unsaturated hydrocarbons (called alkenes).

● Alkenes (and other unsaturated compounds containing carbon–carbon double bonds) react with orange bromine water, turning it colourless.

C9 Crude oil and fuels

Summary questions

1 a i Copy and complete this general formula for the alkanes: $C_nH_?$ [1 mark]

ii What is the formula of the alkane with 18 carbon atoms? [1 mark]

iii What is the formula of the alkane with 18 hydrogen atoms? [1 mark]

b Look at the boiling points in this table below:

Alkane	Number of carbon atoms	Boiling point in °C
methane	1	−161.0
ethane	2	−88.0
propane	3	−42.0
butane	4	−0.5
pentane	5	
hexane	6	69.0

Draw a graph of the alkanes' boiling points (vertical axis) against the number of carbon atoms (horizontal axis). [3 marks]

c What is the general pattern you see from your graph? [1 mark]

d Use your graph to predict the boiling point of pentane. [1 mark]

2 a The alkanes are all saturated hydrocarbons.

i Define a hydrocarbon. [1 mark]

ii What does saturated mean when describing an alkane? [1 mark]

b i Give the name and formula of this alkane:

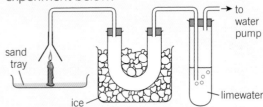

[1 mark]

ii What do the letters represent in this displayed formula? [1 mark]

iii What do the lines between the letters represent? [1 mark]

c The alkane above is used in portable gas heaters.

i Write a word equation to show its complete combustion. [1 mark]

ii Write a balanced symbol equation, including state symbols, for its complete combustion. [3 marks]

d There must be good ventilation in the room when using a portable gas heater. Name the toxic gas that will be produced if there is a poor supply of air in to the heater. [1 mark]

3 One alkane, A, has a boiling point of 344 °C and another, B, has a boiling point of 126 °C.

a Which one will be collected nearer the top of a fractionating column in an oil refinery? Explain your choice. [2 marks]

b Which one will be the better fuel? Explain your choice. [2 marks]

c State the differences you would expect between A and B in terms of their:

i viscosity [1 mark] **ii** volatility. [1 mark]

4 Two students are testing the products formed when the hydrocarbons in wax burn. They set up the experiment below:

a Why did they put ice around the U-tube? [1 mark]

b How can they test for the substance formed in the U-tube? [1 mark]

c Explain what happens to the limewater. [2 marks]

d There is a small amount of carbon dioxide in the air. How can they show that the carbon dioxide they test for is not just the result of the carbon dioxide in the air? [1 mark]

e One of the hydrocarbons found in candle wax is pentacosane, a straight-chain alkane containing 25 carbon atoms.

i Write the chemical formula of pentacosane? [1 mark]

ii Write a word equation for its complete combustion. [1 mark]

iii Write a balanced symbol equation, including state symbols, for the complete combustion. [3 marks]

5 Look at the reaction below:

$$C_{15}H_{32} \rightarrow C_8H_{18} + C_3H_6 + 2C_2H_4$$

a This is a reaction that can take place in an oil refinery. What do we call this type of thermal decomposition reaction? [1 mark]

b State two ways in which the reaction could be carried out using gaseous $C_{15}H_{32}$. [2 marks]

c Using the equation, write the chemical formulae of:

i any alkanes [2 marks]

ii any alkenes. [2 marks]

d In an oil refinery, why is this reaction carried out? [2 marks]

Practice questions

01 The displayed formulae of five hydrocarbons are shown in **Figure 1**.

Figure 1

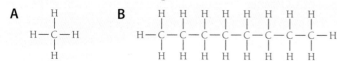

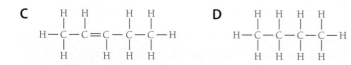

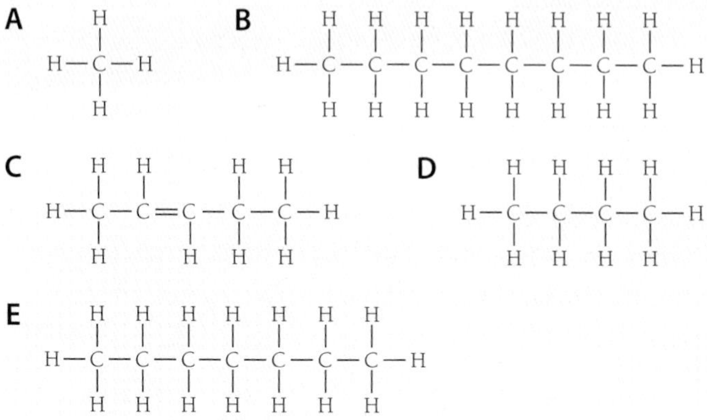

E

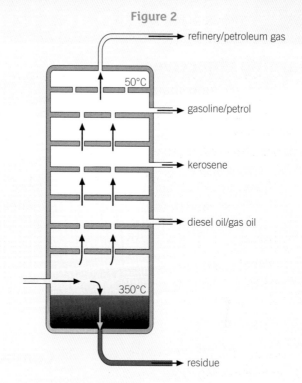

Figure 2

01.1 Define hydrocarbon. [2 marks]

Choose which hydrocarbon **A**, **B**, **C**, **D**, or **E** is the answer to questions **01.2** to **01.6**.

01.2 Which hydrocarbon has the lowest boiling point? [1 mark]

01.3 Which hydrocarbon has the general formula C_nH_{2n}? [1 mark]

01.4 Which hydrocarbon can be cracked to form **D** and C_3H_6 as the only products? [1 mark]

01.5 Which hydrocarbon decolourises bromine water? [1 mark]

01.6 Which hydrocarbon has a relative formula mass M_r of 58? [1 mark]

01.7 The fractional distillation of crude oil produces too much $C_{14}H_{30}$.
Describe how $C_{14}H_{30}$ can be cracked to produce shorter chain hydrocarbons. [2 marks]

02 Crude oil is a mixture of many useful substances. Crude oil is separated into fractions by fractional distillation as shown in **Figure 2**.

02.1 Describe and explain how crude oil is separated into fractions by fractional distillation.
Use **Figure 2** to help you answer this question.
[4 marks]

02.2 The fractional distillation of crude oil produces more kerosene than required.
Kerosene can be cracked.
Complete the equation for the cracking of $C_{12}H_{26}$ to produce C_7H_{16} and two different alkenes.
$C_{12}H_{26} \rightarrow C_7H_{16} + \ldots\ldots\ldots + \ldots\ldots\ldots$
[2 marks]

02.3 The equation shows the incomplete combustion of $C_{12}H_{26}$.
$C_{12}H_{26} + \ldots\ldots O_2 \rightarrow 9CO + 3CO_2 + 13H_2O$
Complete the balancing of the equation. [1 mark]

02.4 Explain why the incomplete combustion of $C_{12}H_{26}$ is dangerous. [3 marks]

03 The equation shows the cracking of $C_{18}H_{38}$ to produce two liquid products.
$C_{18}H_{38} \rightarrow C_{13}H_{28} + C_5H_{10}$

03.1 Explain why $C_{18}H_{38}$ needs to be heated to produce $C_{13}H_{28}$ and C_5H_{10}. [2 marks]

03.2 The two liquid products were poured into separate test tubes.
Describe a chemical test that would show which tube contained C_5H_{10}.
Explain your answer. [3 marks]

03.3 Explain why $C_{18}H_{38}$ has a higher boiling point than $C_{13}H_{28}$. [3 marks]

Ⓗ 04 1 mole of alkene C_xH_{2x} was fully burnt in oxygen.
The products were analysed. 264 g of CO_2 and 108 g of H_2O were produced.
Use the information to balance the equation and to work out the identity of C_xH_{2x}.
$C_xH_{2x} + \ldots\ldots\ldots O_2 \rightarrow \ldots\ldots\ldots CO_2 + \ldots\ldots\ldots H_2O$
[4 marks]

Learning objectives

After this topic, you should know:

- the names of the alkenes ethene, propene, butene, and pentene
- how alkenes react with oxygen in air
- how to draw displayed structural formulae of the first four members of the alkenes, and the products of their addition reactions with:
  - hydrogen
  - water (steam)
  - chlorine, bromine, or iodine.

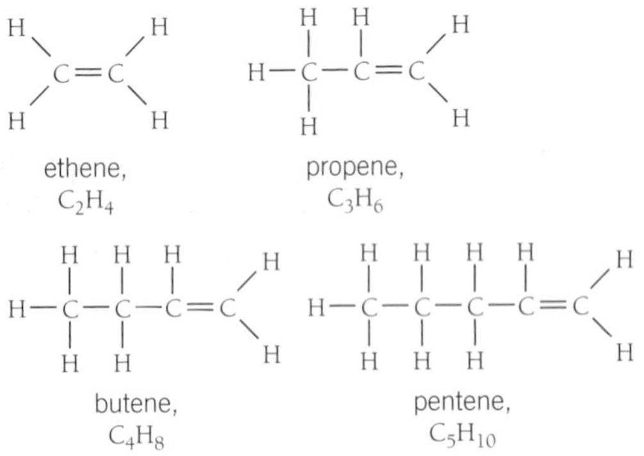

ethene,
C_2H_4

propene,
C_3H_6

butene,
C_4H_8

pentene,
C_5H_{10}

Figure 1 *The first four members of the homologous series of alkenes. The lines between atoms represent covalent bonds. Notice the carbon–carbon double bond (C═C) in each alkene molecule. This type of 2D drawing to show the structure of a molecule is called a 'displayed formula'. Note that the C═C bond could also be located between carbon atoms in the middle of a butene or pentene molecule*

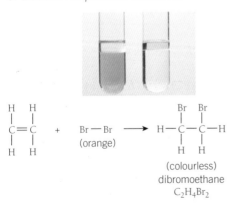

Figure 2 *An alkene will decolourise orange bromine water because the dibromoethane formed is colourless*

The substances that form the basis of all living things, including ourselves, are organic compounds. Organic molecules all contain carbon atoms. The carbon atoms tend to form the 'backbone' of organic molecules. The alkenes you met in Chapter C9 all have a C═C group of atoms in their molecules. The C═C grouping is an example of a **functional group**. A functional group gives a 'family' of organic compounds their characteristic reactions.

A 'family' of organic compounds with the same functional group is called a **homologous series**. Successive members of a homologous series differ from each other by an extra carbon atom that has two hydrogen atoms bonded to it. Look at the first four members of the homologous series of alkenes in Figure 1.

The pattern you can see in the formula of the alkenes shows that their *general formula* is:

$$C_nH_{2n}$$

Combustion of the alkenes

The experiment in Topic C9.4 produces ethene. You can test the ethene gas using a lighted spill. Alkenes do burn but with a smokier, yellow (luminous) flame compared with alkanes because there is incomplete combustion. For the same reason, they also release less energy per mole in combustion than alkanes. So the alkanes are used as fuels, whereas the alkenes are not. During complete combustion, you get:

ethene + oxygen → carbon dioxide + water

$$C_2H_4 + 3O_2 → 2CO_2 + 2H_2O$$

Addition reactions of the alkenes

It is the carbon–carbon double bond, C═C, that makes the alkenes far more reactive than the alkanes. Molecules that react with alkene molecules 'add' to the alkene across their double bond. In general:

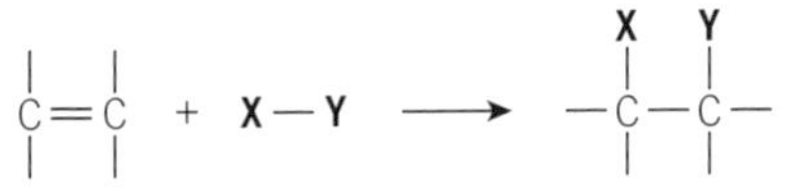

Reaction with halogens

When you test ethene with bromine water, as in Topic C9.4, it reacts. Ethene takes the orange colour out of the reaction mixture and leaves it colourless. The orange colour in bromine water comes from $Br_2(aq)$ molecules, which are converted into a colourless product, called dibromoethane, when they react with ethene:

ethene + bromine → dibromoethane

$$C_2H_4 + Br_2 → C_2H_4Br_2$$

Figure 2 shows the same equation, using fully displayed structural formulae.

Once you know the reaction of one member of a homologous series, you know the reactions of the whole series. For example, propene reacts with bromine as shown in Figure 3.

Reaction with hydrogen

The alkenes are unsaturated compounds. Because of the C=C bond, their molecules do not contain the maximum number of hydrogen atoms possible in a hydrocarbon molecule. But the addition of a hydrogen molecule, H_2, across the double bond will form a saturated hydrocarbon – an alkane. The reaction takes place at 60 °C in the presence of a nickel catalyst. For example:

$$\text{pentene + hydrogen} \xrightarrow{\text{catalyst}} \text{pentane}$$
$$C_5H_{10} + H_2 \longrightarrow C_5H_{12}$$

This reaction is used to add hydrogen across the C=C bonds in unsaturated oils to straighten their molecules and increase their melting points. For example, the right amount of hydrogen is added to make a margarine that can be spread easily straight from the fridge.

Reaction with water (steam)

Ethanol (an alcohol) for industrial use as a fuel or solvent can be made from ethene gas. Ethene is the main by-product produced in cracking. Ethene gas can react with steam to make ethanol.

$$\text{ethene + steam} \underset{\text{catalyst}}{\rightleftharpoons} \text{ethanol}$$
$$C_2H_4 + H_2O \rightleftharpoons C_2H_5OH$$

The reaction requires energy to heat the gases and to generate a high pressure. The reaction is reversible, so ethanol can break down back into ethene and steam. So unreacted ethene and steam are recycled over the catalyst. (You can see the displayed formula of ethanol in Topic C10.2.)

1 The alkene pentene contains five carbon atoms.
  a Write its chemical formula. [1 mark]
  b Draw the displayed formula of pentene, with its double bond at the end of the carbon chain, showing all its bonds. [1 mark]
  c Nonene is an alkene with nine carbon atoms. Write its chemical formula. [1 mark]

2 Draw a displayed formula, showing all the bonds, in the product formed when:
  a propene reacts with iodine [1 mark]
  b ethene reacts with steam [1 mark]
  c butene, with its double bond in the middle of its molecule, reacts with chlorine. [1 mark]

3 a Name the product formed when butene reacts with hydrogen in the presence of a nickel catalyst. [1 mark]
  b i Write the balanced equation for the reaction between propene and steam in the presence of concentrated phosphoric acid catalyst. [1 mark]
    ii Draw the displayed formula of two possible products formed in this reaction, showing all the bonds. [2 marks]
  c Give the balanced symbol equation for the complete combustion of pentene. [1 mark]

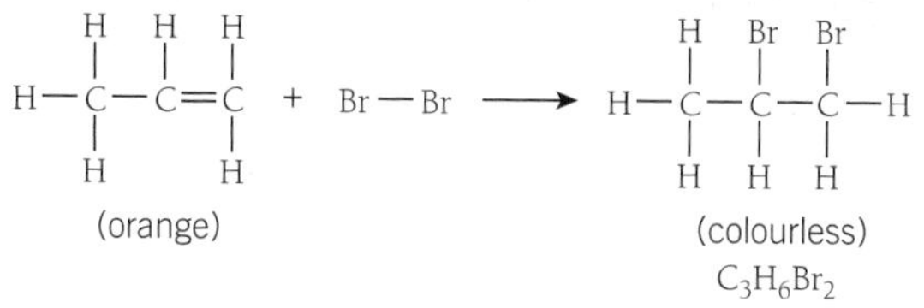

Figure 3 *The alkenes also react in the same way with the other halogens, chlorine and iodine. The reaction of propene with chlorine produces $C_3H_6Cl_2$, and with iodine it produces $C_3H_6I_2$*

Synoptic links

For more information on chemical equilibrium, look back to Topic C8.8.

Go further

You can find out more about the nature of the double bond in alkenes and how it affects their reactivity through studying reaction mechanisms in A Level Chemistry. The mechanism explains a reaction step by step, which will help you make sense of the reactions of different functional groups.

Key points

- The general formula of the alkenes, containing one C=C bond, is C_nH_{2n}.
- Complete combustion of an alkene forms carbon dioxide and water.
- Alkenes react with halogens, hydrogen, and water (steam) by adding atoms across the C=C bond, forming a saturated molecule.

C10.2 Structures of alcohols, carboxylic acids and esters

Learning objectives

After this topic, you should know:

- the names and formulae of the first four members of the alcohols and carboxylic acids, as well as the ester, ethyl ethanoate
- how to represent the structures of their molecules using displayed formulae.

You have already met some organic compounds in Chapter C9, so you will probably recall how to draw molecules of alkanes and alkenes using their displayed formula (Figure 1).

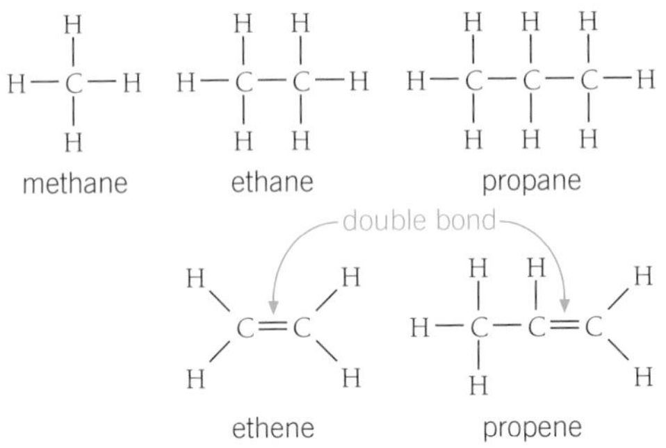

Figure 1 *The displayed formulae of the smallest alkane and alkene molecules*

Both alkanes and alkenes are made of only carbon and hydrogen atoms. However, there are many more homologous series ('families') of organic compounds that also contain other types of atom.

In the rest of this chapter, you will learn about some organic compounds made up of carbon, hydrogen, and oxygen. The three homologous series you will look at are called alcohols, carboxylic acids, and esters. You will consider their characteristic structures first.

Alcohols

You have also met a member of the alcohol 'family' before. This was ethanol, made by reacting ethene with steam (Topic C10.1). So what is an alcohol?

Imagine removing an H atom from an alkane molecule and replacing it with an –OH group. This would give an alcohol molecule. The –OH group of atoms is another example of a functional group. You need to know the first four members of the homologous series of alcohols (Figure 2).

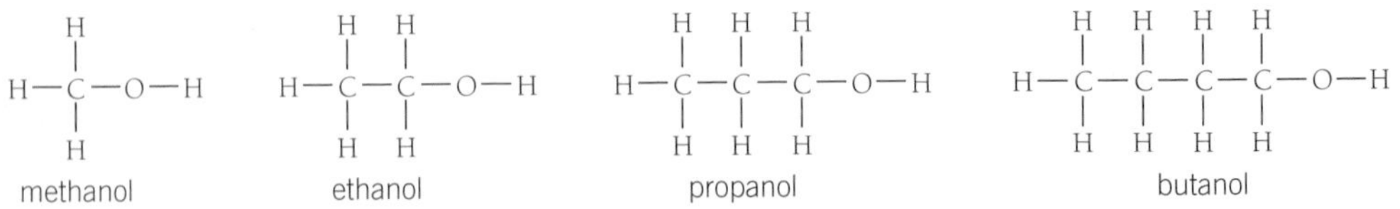

Study tip

When you draw displayed formulae, make sure you show all the bonds, including those in the functional group (as lines between atoms) and all the atoms (as their chemical symbols).

Figure 2 *The structures (displayed formulae) and names of the first four alcohol molecules. You name alcohols from the alkane with the same number of carbon atoms. Just take the '-ane' from the end of the alkane's name and replace it with '-anol'*

If you were asked for the formula of ethanol, you might count the atoms in a molecule and write C_2H_6O.

This is correct, but chemists can give more information in a *structural formula*. This does not show all the bonds, as in a displayed formula, but shows what is bonded to each carbon atom. So for ethanol, a chemist will often show its formula as CH_3CH_2OH (often shortened to C_2H_5OH). Can you see how this relates to the structure of ethanol shown in Figure 2?

Carboxylic acids

You will certainly know of one carboxylic acid. Ethanoic acid is the main acid in vinegar. All carboxylic acids contain the –COOH functional group. Look at the first four members of the homologous series of carboxylic acids in Figure 3:

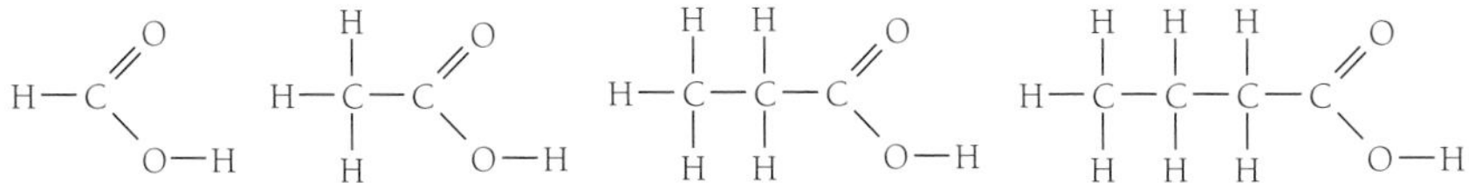

methanoic acid   ethanoic acid   propanoic acid   butanoic acid

Figure 3 *The structures and names of the first four carboxylic acid molecules*

In Figure 3 the structural formula of each of the carboxylic acids is shown as $HCOOH$, CH_3COOH, CH_3CH_2COOH (often shortened to C_2H_5COOH), and $CH_3CH_2CH_2COOH$ (also written as C_3H_7COOH).

Esters

Esters are closely related to carboxylic acids. If you replace the H atom in the –COOH group by a hydrocarbon (alkyl) group, such as $-CH_3$ or $-C_2H_5$, you get an ester. Here is the ester called ethyl ethanoate:

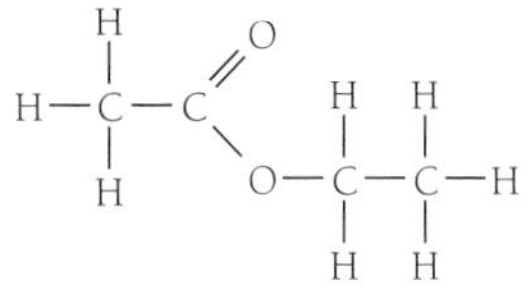

ethyl ethanoate

Figure 5 *The structure of the ester ethyl ethanoate*

An ester's structural formula always contains the –COO– functional group. The structural formula of ethyl ethanoate is $CH_3COOCH_2CH_3$ (or shortened to $CH_3COOC_2H_5$). You will look at the carboxylic acid and ester homologous series in more detail in Topic C10.4.

1 Which homologous series do the following compounds belong to?
   a $CH_3COOCH_2CH_2CH_3$ [1 mark]
   b $CH_3CH_2CH_2CH_2CH_2CH_2OH$ [1 mark]
   c $CH_3CH_2CH_2CH_2COOH$ [1 mark]

2 Name the following compounds:
   a $CH_3CH_2CH_2OH$ [1 mark]
   b $CH_3COOC_2H_5$ [1 mark]
   c $HCOOH$ [1 mark]

3 Draw a displayed formula, showing all the bonds, in:
   a ethanol   [1 mark]
   b ethyl ethanoate   [1 mark]
   c butanoic acid.   [1 mark]

Go further

Looking at butanol in Figure 2, can you see another position of the –OH group on the carbon chain that will result in a different displayed formula, with the same molecular formula of C_4H_9OH? The two compounds are called 'isomers'. To distinguish them, the one shown in Figure 2 is called butan-1-ol and the other butan-2-ol. You will learn more about naming organic compounds and isomers in A Level Chemistry.

Figure 4 *Vinegar contains less than 10% ethanoic acid in an aqueous solution, but the acid provides the characteristic sharp taste and smell*

Key points

- The homologous series of alcohols contains the –OH functional group.
- The homologous series of carboxylic acids contains the –COOH functional group.
- The homologous series of esters contains the –COO– functional group.

C10.3 Reactions and uses of alcohols

Learning objectives

After this topic, you should know:

- how to write balanced chemical equations for the combustion of alcohols
- what is formed in the reaction of alcohols with sodium and when they are oxidised
- some uses of alcohols.

Synoptic links

For more information about depleting the Earth's reserves of non-renewable resources, see Topic C14.1 and Topic C14.5.

Figure 1 *Ethanol is the main solvent in many perfumes, but a key ingredient in some perfumes is octanol. It vaporates more slowly than ethanol and so holds the perfume on the skin for longer*

Uses of alcohols

Alcohols, especially ethanol, are commonly used in everyday products. Ethanol is the main alcohol in alcoholic drinks. It is made by fermenting sugars from plant material with yeast and is becoming an important alternative fuel to petrol and diesel. Ethanol made by **fermentation** is termed a biofuel. Can you see why?

$$\text{glucose} \xrightarrow{\text{yeast}} \text{ethanol} + \text{carbon dioxide}$$
$$C_6H_{12}O_6(aq) \rightarrow 2C_2H_5OH(aq) + 2CO_2(g)$$

However, ethanol can also be made in industry from reacting ethene (obtained by cracking heavy fractions from crude oil) and steam in the presence of a catalyst. This method uses up some of the diminishing supplies of crude oil, a non-renewable resource.

Alcohols dissolve many of the same substances as water. In fact, the alcohols with smaller molecules mix very well with water, giving neutral solutions. The alcohols can also dissolve many other organic compounds. This property makes them useful as solvents. For example, you can remove ink stains from permanent marker pens using methylated spirits.

Methylated spirits ('meths') is mainly ethanol, but has the more toxic methanol mixed with it. It also has a purple dye and other substances added to make it unpleasant to drink. Alcohols are also used as solvents in products such as perfumes, aftershaves, and mouthwashes.

Reactions of alcohols
Combustion

The use of ethanol (and also methanol) as fuels shows that the alcohols are flammable. Ethanol is used in spirit burners and can be used as a biofuel in cars. It burns with a 'clean' blue flame, according to the reaction in the following equation.

$$\text{ethanol} + \text{oxygen} \rightarrow \text{carbon dioxide} + \text{water}$$
$$C_2H_5OH + 3O_2 \rightarrow 2CO_2 + 3H_2O$$

Reaction with sodium

The alcohols react in a similar way to water when sodium metal is added. For example, with ethanol, the sodium effervesces (gives off bubbles of gas), producing hydrogen gas, and gets smaller and smaller as it forms a solution of sodium ethoxide in the ethanol. Their reactions are not as vigorous as the reaction you observe between sodium and water. For example, with ethanol:

$$\text{sodium} + \text{ethanol} \rightarrow \text{sodium ethoxide} + \text{hydrogen}$$
$$2Na + 2C_2H_5OH \rightarrow 2C_2H_5ONa + H_2$$

(You do not need to know a balanced symbol equation for this reaction by heart.) If sodium ethoxide, or any other sodium alkoxide, is dissolved in water, you get a strongly alkaline solution.

Oxidation

Combustion is one way to oxidise an alcohol. However, when you use chemical oxidising agents, such as potassium dichromate(VI), you get different products. An alcohol is oxidised to a carboxylic acid when boiled with acidified potassium dichromate(VI) solution. So ethanol can be oxidised to ethanoic acid:

$$\text{ethanol} + \frac{\text{oxygen atoms from}}{\text{oxidising agent}} \rightarrow \text{ethanoic acid} + \text{water}$$

$$C_2H_5OH + \quad 2[O] \quad \rightarrow CH_3COOH + H_2O$$

(You do not need to know a balanced symbol equation for this reaction by heart.)

The same reaction takes place if ethanol is left exposed to air. Microbes in the air produce ethanoic acid from the ethanol. That is why bottles of beer or wine taste and smell like vinegar when they are left open for too long.

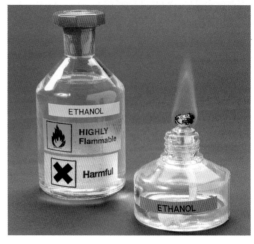

Figure 2 *Alcohols are flammable. They produce carbon dioxide and water in their combustion reactions*

Comparing the reactions of alcohols

a Ignite and observe the flame in three spirit burners – one containing methanol, one ethanol, and the other propanol. Compare the three combustion reactions.

b Watch your teacher add a small piece of sodium metal to each of the alcohols. Compare the reactions. Which gas is given off?

c Watch your teacher heat some of each alcohol with acidified potassium dichromate(VI) solution. What do you see happen in each reaction?

Safety: Ethanol is highly flammable and harmful. Methanol is highly flammable and toxic. Propanol is highly flammable and an irritant. Do not handle the alcohols near naked flames. Wear chemical splash-proof eye protection.

1 Give the name and structural formula of the main alcohol in alcoholic drinks. [2 marks]

2 List the main uses of alcohols. [3 marks]

3 a Write a word equation for the reaction between sodium metal and butanol. [1 mark]
   b Name and give the structural formula of the acidic product formed when butanol, $CH_3CH_2CH_2CH_2OH$, is boiled with acidified potassium dichromate(VI), a chemical oxidising agent. [2 marks]
   c Write a word equation for the complete combustion of butanol. [1 mark]

4 Write a balanced symbol equation for the complete combustion of methanol. [1 mark]

5 Methanol, ethanol, and propanol are all liquids at room temperature. Plan an investigation to see which alcohol – methanol, ethanol, or propanol – transfers most energy to the surroundings per gram when it burns. 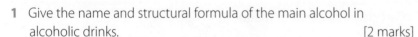 [5 marks]

Key points

- Alcohols are used as solvents and fuels, and ethanol is the main alcohol in alcoholic drinks.
- Alcohols burn in air, forming carbon dioxide and water.
- Alcohols react with sodium metal to form a solution of sodium alkoxide, and hydrogen gas is given off.
- Ethanol can be oxidised to ethanoic acid, either by chemical oxidising agents or by the action of microbes in the air. Ethanoic acid is the main acid in vinegar.

C10.4 Carboxylic acids and esters

Learning objectives

After this topic, you should know:

- how to recognise carboxylic acids from their properties
- why carboxylic acids are described as weak acids
- how to make esters.

Synoptic links

Look back to Chapter C5 to see the typical reactions of acids.

You have already learned about the structure of carboxylic acids in Topic C10.2. The most well-known carboxylic acid is ethanoic acid, which is the acid you use every time when you put vinegar on your food.

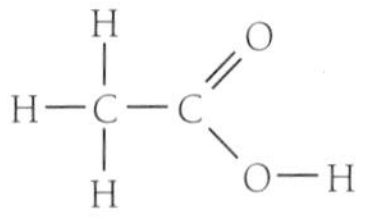

ethanoic acid

Figure 1 *Ethanoic acid, CH_3COOH, is the main acid in vinegar. Its old name was 'acetic acid'. Carboxylic acids are also used to make polyester fibres*

Carboxylic acids, as their name suggests, form acidic solutions when they dissolve in water. You can look at some of their reactions in the Practical box below.

Carboxylic acids have the typical reactions of all acids. For example, reaction with a metal carbonate forms a salt, water, and carbon dioxide. The equation for part **b** in the Practical below is similar to this:

ethanoic   +   sodium   →   sodium   + water + carbon
acid   carbonate   ethanoate   dioxide

$$2CH_3COOH(aq) + Na_2CO_3(s) \rightarrow 2CH_3COONa(aq) + H_2O(l) + CO_2(g)$$

(You do not need to know a balanced symbol equation for this reaction by heart.)

Properties of ethanoic acid solution

Write down and explain your observations for tests **a–d** on an aqueous solution of ethanoic acid.

a   Take the pH of the solution.

b   Add a little potassium carbonate to the acid.

c   Add a small piece of magnesium ribbon to the acid.

d   Test the conductivity of the solution in a small beaker using a simple circuit of a cell, two carbon electrodes, and a bulb. If possible, using the apparatus shown in Topic C6.4, collect the gas given off from the negative electrode in a test tube. Then test it with a lighted spill.

Safety: Wear eye protection.

Figure 2 *Carboxylic acids produce carbon dioxide gas when a metal carbonate is added*

Higher

Why are carboxylic acids called 'weak acids'?

Carbon dioxide gas is given off more slowly when a metal carbonate reacts with a carboxylic acid, as compared with hydrochloric acid of the same concentration. That is because the carboxylic acids are *weak* acids, as opposed to *strong* acids, such as sulfuric acid or hydrochloric acid. Remember that acids must dissolve in water before they show their acidic properties. That is because in water all acids ionise (split up). Their molecules split up to form $H^+(aq)$ ions and negative ions (such as the ethanoate ions, $CH_3COO^-(aq)$ from ethanoic acid). Strong acids ionise completely, whereas weak acids do not. In carboxylic acids, most of the molecules stay as they are (as CH_3COOH molecules in ethanoic acid), and only a small proportion will ionise in their solutions. A reversible reaction takes place, in which the position of equilibrium lies well over to the left in the equation shown below:

$$CH_3COOH(aq) \underset{water}{\rightleftharpoons} CH_3COO^-(aq) + H^+(aq)$$

ethanoic acid   ethanoate ions   hydrogen ions
(high proportion)   (low proportion)

So, a carboxylic acid solution has a higher pH (meaning a lower concentration of $H^+(aq)$ ions) than a strong acid, resulting in a slower reaction with a metal carbonate.

Making esters

Carboxylic acids also react with alcohols to make esters. Water is also formed in this reversible reaction. An acid, usually sulfuric acid, is used as a catalyst. For example:

$$\text{ethanoic acid} + \text{ethanol} \underset{\text{sulfuric acid catalyst}}{\rightleftharpoons} \text{ethyl ethanoate} + \text{water}$$
$$CH_3COOH + C_2H_5OH \rightleftharpoons CH_3COOC_2H_5 + H_2O$$

In general:

$$\textbf{carboxylic acid} + \textbf{alcohol} \underset{\textbf{acid catalyst}}{\rightleftharpoons} \textbf{ester} + \textbf{water}$$

The esters formed have distinctive smells. They are volatile (evaporate easily). Many smell sweet and fruity. This makes them ideal to use in perfumes and food flavourings.

1 a Write a word equation, including the catalyst, to show the reversible reaction between ethanoic acid and ethanol. [1 mark]
  b Esters are volatile compounds. What does this mean? [1 mark]
  c Give two uses of esters. [2 marks]

2 a i Which gas is made when propanoic acid reacts with potassium carbonate? [1 mark]
   ii Name the salt formed. [1 mark]
  b How could the reaction in part a help to distinguish between three test tubes half full of colourless liquids, one containing propanoic acid solution, one containing dilute hydrochloric acid of the same concentration as the propanoic acid solution, and the last containing a solution of propanol in water? Explain your reasoning. [6 marks]

Ⓗ 3 Explain in detail why methanoic acid is described as a weak acid. [5 marks]

Synoptic links

To revise the differences between strong and weak acids, revisit Topic C5.8.

Making esters

Your teacher will show you how different esters are made using carboxylic acids and alcohols.

After the acid has been neutralised with sodium hydrogencarbonate gently waft the gas towards you and carefully smell the test tubes containing the different esters formed.

● Write word equations for each reaction.

Key points

● Solutions of carboxylic acids have a pH value of less than 7. Carbonates gently fizz in solutions of carboxylic acids, releasing carbon dioxide gas.
● Aqueous solutions of carboxylic acids, which are weak acids, have a higher pH value than solutions of strong acids of the same concentration.
● Esters are made by reacting a carboxylic acid with an alcohol, in the presence of a strong acid catalyst.
● Esters are volatile, fragrant compounds used in flavourings and perfumes.

Summary questions

1 Look at the three organic molecules A, B, and C below:

A

B

C

Answer the following questions about A, B, and C.

a Which one is a carboxylic acid? [1 mark]

b Which one is an alcohol? [1 mark]

c Which homologous series of organic compounds does B belong to? [1 mark]

d Which of the compounds can be represented as the structural formula $CH_3CH_2COOCH_2CH_2CH_3$? [1 mark]

e Using a structural formula as shown in part **d**, give the structural formulae of the other two compounds. [2 marks]

2 a i Describe what you would *see* happen if a small piece of sodium metal was dropped into a beaker containing some ethanol. [2 marks]

   ii Name the gas given off in this reaction. [1 mark]

   iii Name the sodium compound formed in the reaction. [1 mark]

   iv The sodium in the compound formed is present as ions. Give the formula of a sodium ion. [1 mark]

   v Using your knowledge of the periodic table and trends in reactivity, suggest how ethanol's reaction with lithium metal would differ from its reaction with sodium. [1 mark]

b Write a word equation for the reaction of sodium with propanol. [1 mark]

3 Copy and complete the following equations:

a pentene + ... → pentane [1 mark]

b ethene + steam → ...... [1 mark]

c propanol + acidified ...... → propanoic acid [1 mark]

d [1 mark]

e $C_4H_9OH + ...O_2 → ...CO_2 + ...H_2O$ [1 mark]

4 a Draw the displayed formula of methanoic acid, showing all the atoms and bonds. [1 mark]

b i Some potassium carbonate powder is dropped into a test tube of a solution of methanoic acid. How would you positively identify the gas given off? Include a labelled diagram of the apparatus you could use. [1 mark]

   ii Write a word equation for the reaction. [1 mark]

H c i Which aqueous solution of the following would you expect to have the highest pH value, if they all had the same concentration?

hydrochloric acid   **sulfuric acid**

methanoic acid   **nitric acid**

ethanol [1 mark]

   ii Which of the five solutions listed above would have the second highest pH value? Explain your choice. [4 marks]

5 a What is the name of this compound?

[1 mark]

b Name the carboxylic acid and alcohol used to make the compound in part **a** and give one use for each. [2 marks]

c Write a word equation showing how this compound can be made. Include the name of the catalyst. [2 marks]

d The compound shown in part **a** is volatile. What does this mean? [1 mark]

e Give a use of the homologous series of compounds to which the compound in part **a** belongs. [1 mark]

6 Describe how you could distinguish between samples of propanol, propanoic acid, and ethyl ethanoate using simple experimental tests. Give the results of any tests suggested. [5 marks]

H 7 You are given a 0.50 mol/dm³ solution of nitric acid, and a 0.50 mol/dm³ solution of methanoic acid.

a What does 'a 0.50 mol/dm³ solution' mean? [1 mark]

b What can you predict about the relative pH values of the two solutions? [1 mark]

c Explain in detail your answer to part **b**, including equations to show the ionisation of nitric acid and methanoic acid in aqueous solution. [5 marks]

d Express the concentration of the 0.50 mol/dm³ methanoic acid in g/dm³. [2 marks]

Practice questions

01 The displayed formula of five organic compounds is shown in **Figure 1**.

Figure 1

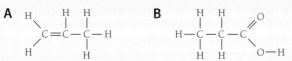

A
H₂C=C–C–H structure

B
H–C–C–C structure with O and O–H

C
H–C–C–C–O–H structure

D
H–C–C–C–H structure

E
H–C–C–C–H structure with O

Use **Figure 1** to answer questions **01.1** to **01.5**.
Choose which organic compound **A**, **B**, **C**, **D**, or **E**:

01.1 is a saturated hydrocarbon [1 mark]

01.2 is an alkene [1 mark]

01.3 has the same molecular formula as organic compound **E** [1 mark]

01.4 will react with sodium carbonate to produce carbon dioxide gas [1 mark]

01.5 will react with **B** in the presence of an acid catalyst to form: [1 mark]

H–C–C–C with O–C–C–C–H structure

02 Ethanol has the displayed formula shown in **Figure 2**.

Figure 2

H–C–C–O–H structure

02.1 Outline how ethanol can be made by fermentation, and how ethanol can be made from ethene. [4 marks]

02.2 Ethanol can react to form ethanoic acid. The displayed formula of ethanoic acid is shown in **Figure 3**.

Figure 3

H–C–C with O and O–H structure

What type of reaction is the reaction of ethanol to form ethanoic acid? [1 mark]

02.3 Ethanol and ethanoic acid are both clear colourless liquids.
A student has two test tubes. One test tube contains ethanol and the other test tube contains ethanoic acid.
Describe a chemical test and the result of the test to show which test tube contains ethanoic acid. [2 marks]

03 This question is about propene.
The displayed formula of propene is shown in **Figure 4**.

Figure 4

H₂C=C–C–H structure

03.1 Copy and complete the dot and cross diagram to show the bonds between the carbon atoms in propene. Show the electrons as *x*. [2 marks]

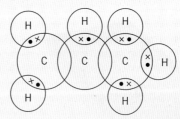

03.2 Draw the displayed formula of the product formed when propene reacts with hydrogen gas in the presence of a nickel catalyst. [1 mark]

03.3 Draw the displayed formula when propene reacts with steam in the presence of a concentrated phosphoric acid catalyst. [1 mark]

03.4 Propene also reacts with hydrogen bromide HBr. Deduce the displayed formula of the product formed. [1 mark]

03.5 The reactions in questions **03.2** to **03.4** are all addition reactions.
Explain why propene undergoes addition reactions. [3 marks]

Learning objectives

After this topic, you should know:

- how to recognise addition polymers and monomers from their displayed formulae
- how to draw diagrams to represent the formation of a polymer from a given alkene monomer
- how to relate the repeating unit of a polymer to its monomer.

Synoptic links

To remind yourself of the products obtained from crude oil, look back to Topic C9.2 and Topic C9.4.

Figure 2 *Polymers produced from compounds derived from crude oil are all around you, and are part of your everyday life*

The fractional distillation of crude oil and cracking in an oil refinery produces a large range of hydrocarbons. These are very important to your way of life. Crude oil-based products are all around you. It is difficult to imagine life without them. Hydrocarbons are the main fuels used to power vehicles. Then there are the chemicals made from crude oil. These are used to make a wide variety of things, ranging from cosmetics to explosives. One of the most important ways that chemicals from crude oil are used is to make **polymers**.

Polymers

The plastics you use in everyday life are made up of huge molecules made from lots of small molecules joined together. These small molecules are called **monomers**. The huge molecules they make are called polymers (*mono* means 'one' and *poly* means 'many'). By using different monomers, you can make various types of polymers that have very different properties.

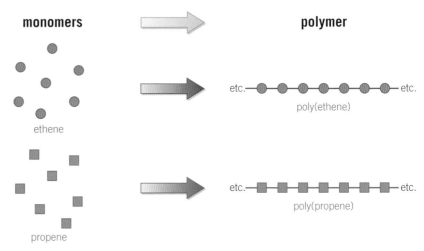

Figure 1 *Polymers are made from many smaller molecules called monomers*

Ethene, C_2H_4, is the smallest unsaturated hydrocarbon molecule. It can be used to make a polymer known as poly(ethene) or polythene. Poly(ethene) is a really useful plastic as it is strong and easy to shape. It is transparent, except when you add colouring material to it. 'Plastic' carrier bags, some drinks bottles, dustbins, washing-up bowls, and cling film are all examples of the uses of poly(ethene):

ethene monomers → poly(ethene).

Propene, C_3H_6, is another alkene. It can be used to make a polymer called poly(propene). Poly(propene) forms a very strong, tough plastic. You can use it to make many things, including carpets, milk crates, and ropes.

propene monomers → poly(propene).

How do monomers join together?

When alkene molecules join together, the double covalent bond between two carbon atoms in each molecule 'opens up'. It is replaced by a single carbon–carbon covalent bond between the two carbon atoms. In this way, thousands of molecules join together, end to end. The polymer chains they form are made up of a 'backbone' of thousands of carbon atoms. This type of reaction is called **addition polymerisation** and the polymer formed is known as an addition polymer. You can show what happens when the addition polymer poly(ethene) is made using the displayed formulae of a few of the ethene molecules involved in the reaction:

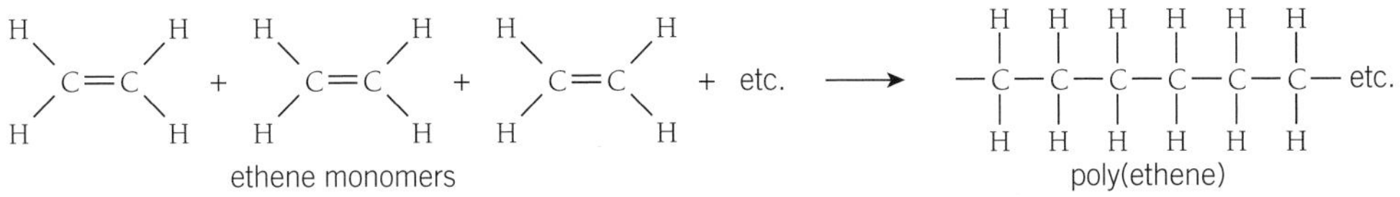

You can also write this more simply as a chemical equation as follows:

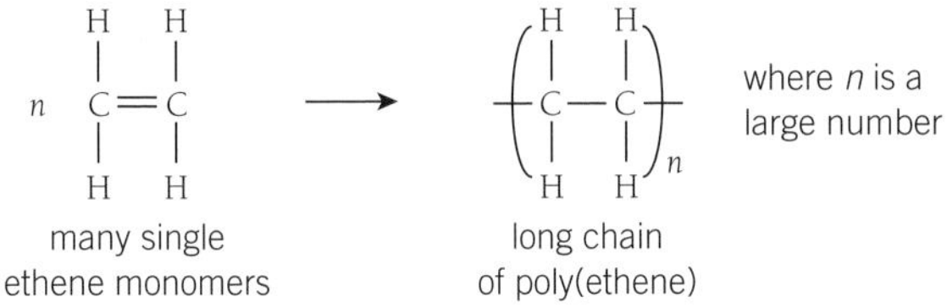

The long chain molecules in polymers are made up of repeated patterns of atoms. In poly(ethene), the repeating unit is shown bracketed in the chemical equation above. In addition polymers, the repeating unit of the polymer has the same displayed formula as one of its monomers, but the C=C double bond in the monomer is changed to a single bond, with a single bond sticking out at each end.

Notice that in addition polymerisation there is only one product formed in the reaction, just like in all the other addition reactions of the alkenes.

The repeating unit of poly(ethene)

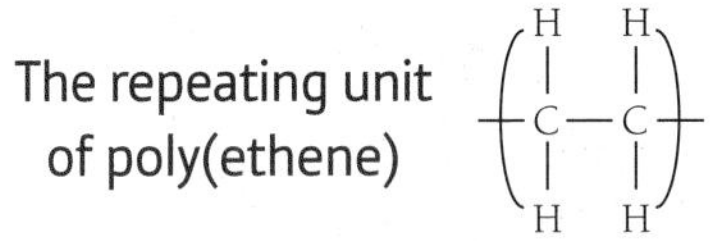

Figure 3 *The repeating unit of poly(ethene)*

1 a State the definition of a monomer and of a polymer. [2 marks]
   b Name the type of reaction that takes place between a large number of monomers. [1 mark]
   c Write a chemical equation to represent the formation of poly(ethene) from ethene. [2 marks]
   d Give two uses of poly(ethene). [2 marks]
2 a Draw the displayed formula of a propene molecule, showing all its bonds. [1 mark]
   b Draw an equation using displayed formulae to show how three propene molecules join together to form part of the chain in poly(propene). [2 marks]
   c State *two* uses of poly(propene). [2 marks]
   d Explain the polymerisation reaction in part **b**. [3 marks]
3 a State and explain the percentage atom economy (see Topic C4.5) in the polymerisation of butene to form poly(butene). [3 marks]
   b Assuming the C=C bond is in the middle of each butene monomer, draw the repeating unit of poly(butene). [1 mark]

C11.2 Condensation polymerisation

Learning objectives

After this topic, you should know:

- the basic principles of condensation polymerisation, including:
  - the functional groups in the monomers
  - the repeating units in the polymers
- how polyesters are formed.

Figure 1 *This cotton/polyester fabric is durable and does not crease as easily as pure cotton fabric*

Synoptic link

To remind yourself how an ester is made, look back at Topic C10.4.

You are probably wearing something made with the polymers commonly known as polyester or nylon. Even if an article of clothing is made mainly of cotton, incorporating polyester fibres will make it more hard-wearing and stops it wrinkling as much. Nylon fabrics are strong and lightweight. They are easy to care for, as they are non-absorbent and dry quickly, with few wrinkles when washed. Both polyester and nylon are examples of condensation polymers, as opposed to the addition polymers introduced in Topic C11.1.

The difference between addition and condensation polymers

As well as addition polymerisation, in which the monomers all contain a $C═C$ bond, chemists can also make polymers from another type of reaction. This is called condensation polymerisation. In addition polymerisation, there is only one product formed in the reaction, the polymer, whereas in condensation polymerisation there are two different products. The main product is the polymer, but you usually also get a small molecule given off. The small molecule is most commonly either water, H_2O, or hydrogen chloride, HCl. To summarise:

$$\text{addition polymerisation} \rightarrow \text{the addition polymer}$$

$$\text{condensation polymerisation} \rightarrow \text{the condensation polymer} + \text{a small molecule}$$

In the plastics industry, the monomers used for addition polymerisation are often all the same alkene. However, in condensation polymerisation there are often two different monomers used. One monomer will have a certain functional group at both ends of its molecule. The other monomer will have a different functional group at its ends. The important thing is that the functional groups on the two different monomers must react together. This is how the long polymer chains are formed.

Forming a polyester

Think back to the how an ester is made. An alcohol (with the –OH functional group) and a carboxylic acid (with the –COOH functional group) react together to give an ester plus water. So to make a polyester chain, you would start with one alcohol monomer with an –OH group at each end, and another carboxylic acid monomer with a –COOH group at each end.

For example, here is ethanediol and a general way of representing any diol (an alcohol containing two –OH groups):

$$HO — CH_2 — CH_2 — OH \quad \textbf{or} \quad HO —\boxed{}— OH$$

and here is hexanedioic acid and a general way of representing any dicarboxylic acid (a carboxylic acid containing two –COOH groups):

$$HOOC — CH_2 — CH_2 — CH_2 — CH_2 — COOH \quad \textbf{or} \quad HOOC —\boxed{}— COOH$$

The monomers link together as they polymerise by 'ester links', and a water molecule is given off as each link is made in the reaction:

$$n \, HO-CH_2-CH_2-OH \; + \; n \, HOOC-CH_2-CH_2-CH_2-CH_2-COOH$$

a diol
(ethanediol)

a dicarboxylic acid
(hexanedioic acid)

$$\left(CH_2-CH_2-OOC-CH_2-CH_2-CH_2-CH_2-COO\right)_n \; + \; 2nH_2O$$

a polyester

water

Study tip

Remember the general equation to make a polyester is:

a diol + a dicarboxylic acid
↓
a polyester + water

or generally represented as:

$$n \, HO-\boxed{}-OH + n \, HOOC-\boxed{}-COOH \longrightarrow \left(\boxed{}-OOC-\boxed{}-COO\right)_n + 2nH_2O$$

When you draw the displayed formula of the polyester for the equation of the repeating unit, imagine the smallest piece of the polymer chain that could be drawn over and over again, end to end, to show a complete section of the polymer chain. Then draw this part in brackets, as in the equation above.

Making a condensation polymer

Making nylon

Put a thin layer of monomer A into the bottom of a very small beaker.

Carefully pour a layer of monomer B on top of this. Gently draw a thread out of the beaker using a pair of tweezers.

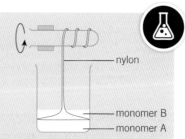

nylon

monomer B
monomer A

Wind it around a test-tube, as shown in the diagram.

Safety: Wear gloves and eye protection. Do not touch the nylon formed. Carry out in a fume cupboard or well-ventilated lab.

The fumes given off as the two different monomers react together to form nylon are hydrogen chloride, HCl gas. This is an example of the small molecule that is always given off in a condensation polymerisation. The monomers have reactive parts at both ends of their molecules so they join together, end-to-end, to make a long nylon polymer chain.

Figure 2 *Nylon has a high tensile strength, as there are strong intermolecular forces between its polymer chains*

1 What is the main difference between an addition and a condensation polymerisation? [2 marks]

2 Explain how a condensation polymer is formed. [5 marks]

3 Draw the displayed formula of the repeating unit of the polyester formed when ethanediol and hexanedioic acid are the monomers. [2 marks]

4 a Complete the following representation of the polymerisation used to make polyesters: [2 marks]

$$n \, HO-\!\!\!\!\bigcirc\!\!\!\!-OH \; + \; n \, HOOC-\!\!\!\!\diamondsuit\!\!\!\!-COOH \longrightarrow \; ..... \; + \; .....$$

b If the oval shape in the incomplete equation above represents $-CH_2-CH_2-CH_2-$ and the diamond shape $-CH_2-$, give the formula that represents the polyester formed. [1 mark]

Key points

- Condensation polymerisation usually involves a small molecule released in the reaction, as the polymer forms.
- The monomers used to make the simplest condensation polymers are usually two different monomers, with two of the same functional groups on each monomer.
- Polyesters are formed from the condensation polymerisation of a diol and a dicarboxylic acid, with H_2O given off in the reaction.

C11.3 Natural polymers

Learning objectives

After this topic, you should know:

- sugars can undergo polymerisation in living things to make polymers, such as starch and cellulose
- proteins are polymers of amino acids
- **H** how amino acids react together
- **H** the formation of polypeptides and proteins by condensation polymerisation.

Figure 1 *Glucose can provide you with a quick energy boost*

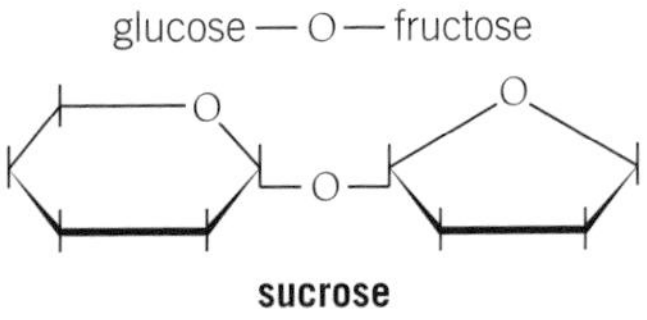

glucose — O — fructose

sucrose

Figure 3 *This is a simplified model of a sucrose molecule*

Synoptic link

To remind yourself of the displayed formula of sucrose (a disaccharide), look back to Topic C3.6.

The polymers you looked at in the first half of this chapter are the synthetic polymers commonly used in everyday life. However, there are also naturally occurring polymers found in all living things. You will look at the polymers that make up starch, cellulose, and proteins in this topic, and those that make up DNA in Topic C11.4.

Making polysaccharides from sugars

You will already know about the carbohydrates in your food and the energy source they provide for you. In terms of their chemistry, carbohydrates are compounds made up of molecules containing carbon, hydrogen, and oxygen atoms. They often have a general formula of $C_x(H_2O)_y$, and are made up of one or more types of sugar molecules.

The most commonly known sugar is the glucose, $C_6H_{12}O_6$, made during photosynthesis in plants. Figure 2 shows its structure in various ways.

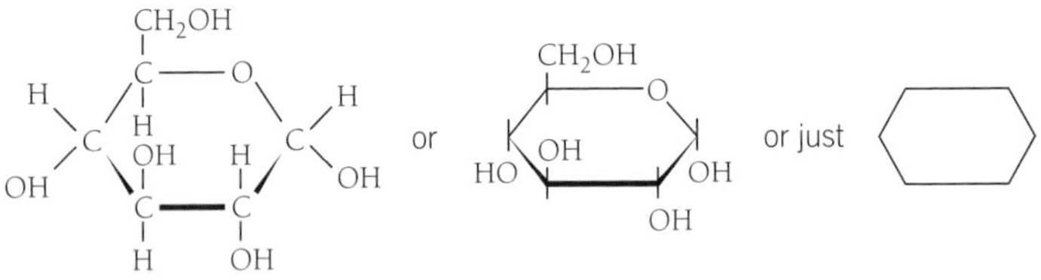

Figure 2 *The structure of glucose shown by its displayed formula and other models that simplify its structure*

Glucose is called a monosaccharide (made of one sugar unit), as is another sugar, fructose, a low-calorie alternative to glucose used to sweeten food. Whereas glucose molecules are made up of a six-membered ring structure, fructose has five-membered rings. Monosaccharides can bond together to make larger molecules. For example, sucrose is made from a glucose and a fructose molecule bonded together (via a condensation reaction in which H_2O is lost when making the link). Its structure is shown in Figure 3.

The monosaccharide sugars can also act as the monomers to make polymers, called polysaccharides. They can be made up of thousands of sugar monomers. Figure 4 shows simplified structures of starch and cellulose, both made up from glucose monomers joined in condensation polymerisation.

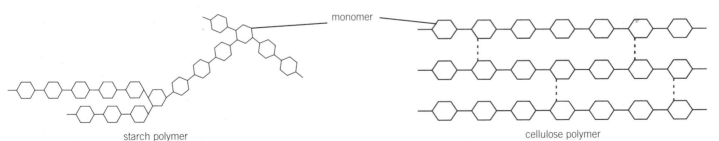

starch polymer

cellulose polymer

Figure 4 *Starch is made up from up to 1500 glucose monomers, arranged in branched chains. Cellulose is made up from around 10 000 glucose monomers, in straighter chains than starch. This means that they can line up neatly next to each other, forming stronger intermolecular forces between the polymer chains*

glucose monomers → starch polymers + water

glucose monomers → cellulose polymers + water

Plants use the starch they make from glucose as energy stores, and the cellulose they make is used to give the plant its structure. **Proteins** are also natural polymers.

variety of amino acid monomers → protein polymers + water

Go further

If you study A Level Chemistry, you can find out how you can analyse the amino acids present in proteins.

Making polypeptides and proteins from amino acids

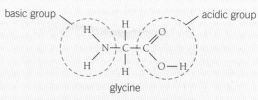

Higher

The monomers of proteins are called amino acids. The amino acids have two functional groups, one basic (the amine group, $-NH_2$) and one acidic (the carboxylic acid group, $-COOH$). The simplest amino acid is called glycine (Figure 5).

Figure 5 *The displayed formula of glycine. Its structural formula is H_2NCH_2COOH*

As you know, acids and bases react together, so with amino acids you have potential monomers with reactive groups at each end. Figure 6 shows how three glycine molecules can react together in a condensation reaction.

Many more glycine molecules can link together to form a polypeptide molecule. You can show this as:

polymerisation

$$nH_2NCH_2COOH \longrightarrow -(HNCH_2CO)_n + nH_2O$$

There are about 20 different amino acids that join together (polymerise) in a great variety of sequences to make the more than 10 000 proteins in your body. Each of the different amino acid monomers is given a three-letter abbreviation when listing their sequence in a protein.

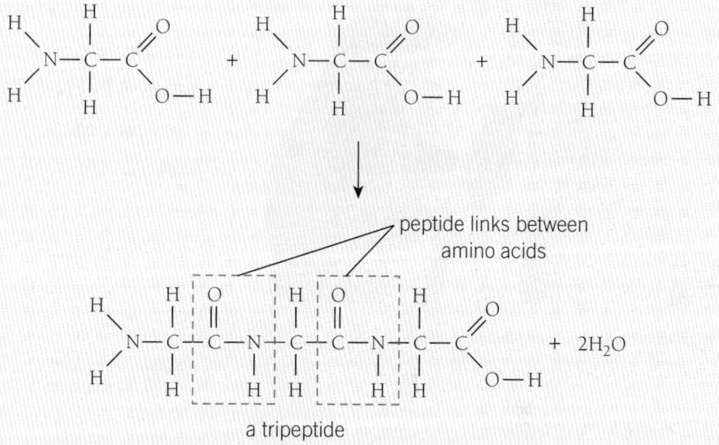

Figure 6 *Glycine forming a tripepide molecule in a condensation reaction*

1 a Name the two simple sugars that join together to make sucrose. [2 marks]

b Name and give the chemical formula of the monomer that makes up the polymers starch and cellulose. [2 marks]

c Describe two differences between the structures of a starch polymer and a cellulose polymer. [2 marks]

Ⓗ 2 a Name the simplest amino acid and draw its displayed formula. [2 marks]

b Another amino acid, called alanine, has one of the H atoms bonded to the central carbon atom of the molecule you drew in part **a** replaced by a $-CH_3$ group. Draw its displayed formula. [1 mark]

c i Draw the dipeptide molecule formed when the two amino acids in this question react together. [1 mark]

ii Name the other product formed in this reaction. [1 mark]

iii What is this type of reaction called? [1 mark]

Ⓗ 3 Explain in detail how a protein is formed from amino acids. ⊘ [6 marks]

Key points

- Simple carbohydrates (monosaccharides) polymerise to make polymers such as starch and cellulose.
- Proteins are polymers made from different amino acid monomers.
- Ⓗ Amino acids have an acidic and a basic functional group in the same molecule.
- Ⓗ Amino acids react together during condensation polymerisation to make polypeptides and proteins made of long sequences of different monomers.

C11.4 DNA

Learning objectives

After this topic, you should know:

- the basic structure of the monomers (nucleotides) used to make DNA
- the way the monomers are arranged in DNA.

Figure 1 *The unique code in DNA enables forensic scientists to identify suspects from a crime scene. They use a technique called 'DNA fingerprinting'*

You will be aware of 'DNA' from studying genetics in Biology or from the DNA testing used in forensic science. **DNA (deoxyribonucleic acid)** is another important natural polymer that is essential for life. It enables living things to develop and function. DNA's structure contains a genetic code that determines the different amino acid sequences of every protein in your body. The genetic code can be copied to make protein molecules with exactly the same sequence of amino acids.

The nucleotide monomers of DNA

DNA is made by the condensation polymerisation of repeating units of monomers called **nucleotides**. So DNA is known as a polynucleotide:

millions of nucleotides ⟶ DNA (a polynucleotide) + water

The structure of DNA

The DNA molecule consists of a double helix made up of two long polymer strands of nucleotides. The two strands run in opposite directions to each other. They are held in place by the intermolecular forces down the length of each polymer strand. There are four different nucleotide monomers that can react with each other to form DNA polymers.

Figure 2 *The double helix of DNA*

Flattening out the two spiralling polymer strands helps you the see the structure of the DNA molecule:

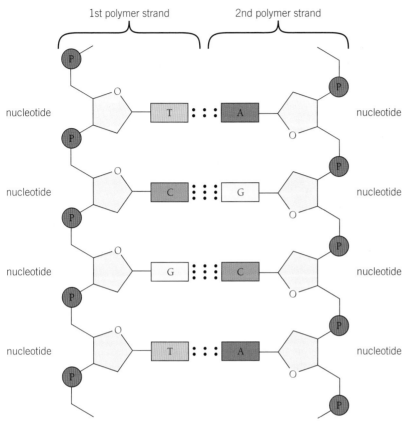

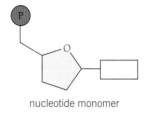

nucleotide monomer

Figure 3 *A simplified (flattened) structure of part of the DNA molecule, showing how the nucleotides bond to each other in each strand (via covalent bonds) and how the strands are linked to each other by intermolecular forces*

You do not need to remember how to draw this structure, but it helps you to see the 'backbone' of each polymer strand. It also shows the four different nucleotides clearly.

1   a   How many different monomers are found in the polymer strands of DNA?   [1 mark]
    b   How many polymer strands make up a molecule of DNA?   [1 mark]

2   a   What do the letters 'DNA' stand for?   [1 mark]
    b   Describe the shape of a DNA molecule.   [1 mark]
    c   i   What type of reaction occurs when the monomers of DNA react together?   [1 mark]
        ii Which small molecule is lost during this reaction?   [1 mark]
        iii Name the type of monomers that form DNA.   [1 mark]

3   Use the information in Figure 3 to answer these questions.
    a   What do the dots between the two sections of the DNA polymer chains represent, and why are they important?   [2 marks]
    b   The rectangular boxes in Figure 3 represent bases in the monomers that make DNA. Write the letters used to represent the bases and find out their names.   [3 marks]
    c   Part of each monomer that makes up DNA is derived from a sugar. Draw how the sugar is represented in Figure 3.   [1 mark]

Go further

You can find out how knowledge of the structure of DNA helps to produce drugs that help fight cancer if you go on to study A Level Chemistry.

Key points

● DNA (deoxyribonucleic acid) is made up from monomers called nucleotides.
● The nucleotides are based on the sugar deoxyribose, bonded to a phosphate group and a base. There are four possible bases that bond to the sugar.
● A DNA molecule consists of two polymer strands (with sugars bonded to phosphate groups) intertwined into a double helix.

C11 Polymers

Summary questions

1 Propene is a hydrocarbon molecule containing three carbon atoms and six hydrogen atoms.
Propene molecules will react together to form long chains.
a What is this type of reaction called? [1 mark]
b Name the product of the reaction. [1 mark]
c What is the general name given to the many propene molecules that react together? [1 mark]
d State the most obvious difference between the reactants and the product. [2 marks]

2 Write a symbol equation to represent the formation of poly(ethene) from ethene. [2 marks]

3 Look at part of a polymer chain below:

$$-\overset{\overset{\displaystyle CH_3}{|}}{\underset{\underset{\displaystyle H}{|}}{C}}-\overset{\overset{\displaystyle H}{|}}{\underset{\underset{\displaystyle H}{|}}{C}}-\overset{\overset{\displaystyle CH_3}{|}}{\underset{\underset{\displaystyle H}{|}}{C}}-\overset{\overset{\displaystyle H}{|}}{\underset{\underset{\displaystyle H}{|}}{C}}-\overset{\overset{\displaystyle CH_3}{|}}{\underset{\underset{\displaystyle H}{|}}{C}}-\overset{\overset{\displaystyle H}{|}}{\underset{\underset{\displaystyle H}{|}}{C}}-$$

a Name the polymer above. [1 mark]
b Give the displayed formula of the monomer that made the polymer shown in part a, showing all its atoms and bonds. [1 mark]

H 4 A polyester is made from the monomers ethanediol and propanedioic acid.
a Draw the displayed formulae of the two monomers. [2 marks]
b i Name the type of polymerisation reaction that takes place between ethanediol and propanedioic acid. [1 mark]
ii Name the small molecule produced in the reaction. [1 mark]
c Draw the repeating unit of the polyester formed. [2 marks]

5 The following diagram shows part of the polymer chain of poly(chloroethene), also known as PVC.

$$-\overset{\overset{\displaystyle H}{|}}{\underset{\underset{\displaystyle H}{|}}{C}}-\overset{\overset{\displaystyle H}{|}}{\underset{\underset{\displaystyle Cl}{|}}{C}}-\overset{\overset{\displaystyle H}{|}}{\underset{\underset{\displaystyle H}{|}}{C}}-\overset{\overset{\displaystyle H}{|}}{\underset{\underset{\displaystyle Cl}{|}}{C}}-\overset{\overset{\displaystyle H}{|}}{\underset{\underset{\displaystyle H}{|}}{C}}-\overset{\overset{\displaystyle H}{|}}{\underset{\underset{\displaystyle Cl}{|}}{C}}-\overset{\overset{\displaystyle H}{|}}{\underset{\underset{\displaystyle H}{|}}{C}}-\overset{\overset{\displaystyle H}{|}}{\underset{\underset{\displaystyle Cl}{|}}{C}}-\overset{\overset{\displaystyle H}{|}}{\underset{\underset{\displaystyle H}{|}}{C}}-\overset{\overset{\displaystyle H}{|}}{\underset{\underset{\displaystyle Cl}{|}}{C}}-\overset{\overset{\displaystyle H}{|}}{\underset{\underset{\displaystyle H}{|}}{C}}-\overset{\overset{\displaystyle H}{|}}{\underset{\underset{\displaystyle Cl}{|}}{C}}-\overset{\overset{\displaystyle H}{|}}{\underset{\underset{\displaystyle H}{|}}{C}}-\overset{\overset{\displaystyle H}{|}}{\underset{\underset{\displaystyle Cl}{|}}{C}}-$$

a Draw the displayed formula of the monomer from which PVC is made. [1 mark]

b i State the full name of the type of reaction in which PVC is made. [1 mark]
ii What is the percentage atom economy of this reaction? [1 mark]
c Another polymer called poly(tetrafluoroethene) is used in the lining of non-stick pans. The chemical formula of its monomer is C_2F_4. Draw an equation using displayed formulae to represent the formation of poly(tetrafluoroethene) from its monomers. [2 marks]

H 6 Look at the displayed formula below:

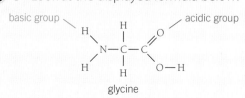

glycine

a State the common name of this compound. [1 mark]
b This, along with other closely related compounds, is the monomer that make proteins. State the name of this class of compound. [1 mark]
c Part of the molecule shown above is acidic and part is basic. State the general name of the functional group that is:
i acidic [1 mark] ii basic. [1 mark]
d The molecule above can undergo condensation polymerisation to form a polypeptide. Draw the repeating unit of this polypeptide. [1 mark]

7 Here is a simplified model used to represent the monomers used to make DNA molecules in the body.

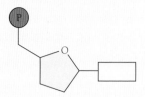

a State the general name of this type of monomer. [1 mark]
b Write the formula of the other product formed in the polymerisation of these monomers. [1 mark]
c i What is the shape of a DNA molecule? [1 mark]
ii How many polymer chains are found in a DNA molecule? [1 mark]
iii How many different types of monomer make the DNA molecule? [1 mark]
d Name two natural polymers made from glucose monomers. [2 marks]

Practice questions

01 **Figure 1** shows how three ethene molecules join together to form a very short section of a polymer

Figure 1

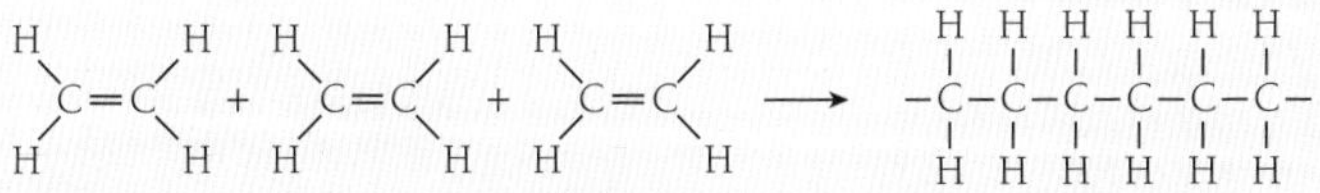

01.1 Give the correct chemical name for the polymer produced in **Figure 1**. [2 marks]

01.2 Ethene, C_2H_4, can be produced by cracking. Describe how ethene can be produced from decane, $C_{10}H_{22}$.
Include an equation in your answer.
You should refer to the structure shown in Figure 1 in your answer. [3 marks]

01.3 The polymer in **Figure 1** is unreactive. Explain why. You should refer to the structure shown **Figure 1** in your answer. [2 marks]

01.4 **Figure 2** shows an equation for the formation of poly(chloroethene).

Figure 2

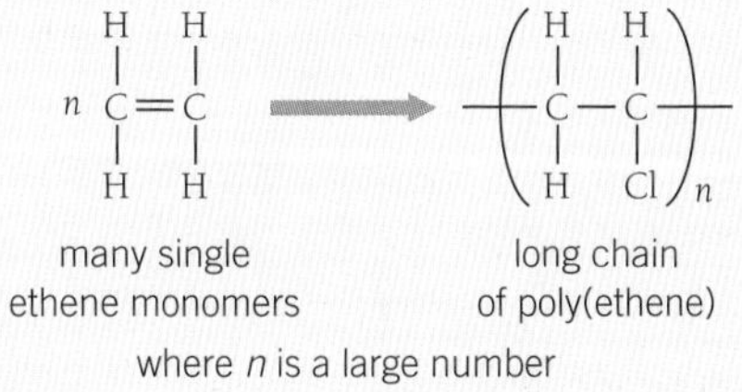

many single ethene monomers

long chain of poly(ethene)

where *n* is a large number

Use the information in **Figure 2** to complete the equations to show the formation of polymers. [5 marks]

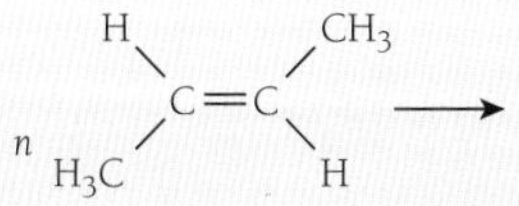

❶ 02 This question is about condensation polymers.

02.1 Some condensation polymers occur naturally. **Table 1** shows the name of a naturally occurring monomer and a naturally occurring polymer. Complete **Table 1**.

Table 1

Name of monomer	Name of polymer
amino acid	
	starch

[2 marks]

02.2 An amino acid called alanine is shown in **Figure 3**.

Figure 3

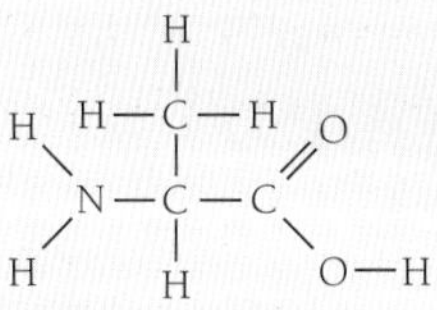

Circle the carboxylic acid functional group in **Figure 3**. [1 mark]

02.3 Copy and complete the structure of the compound formed when two molecules of alanine join together. [3 marks]

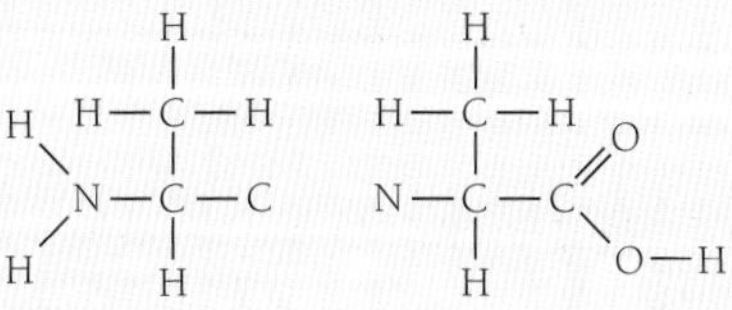

02.4 **Figure 4** shows the repeating unit of a condensation polymer called a polyester. **Figure 5** shows five different organic compounds.

Figure 4

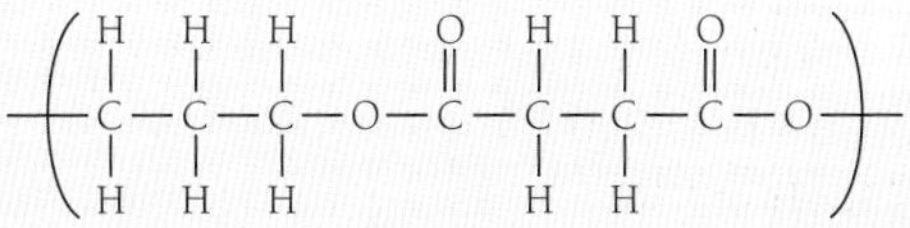

Figure 5

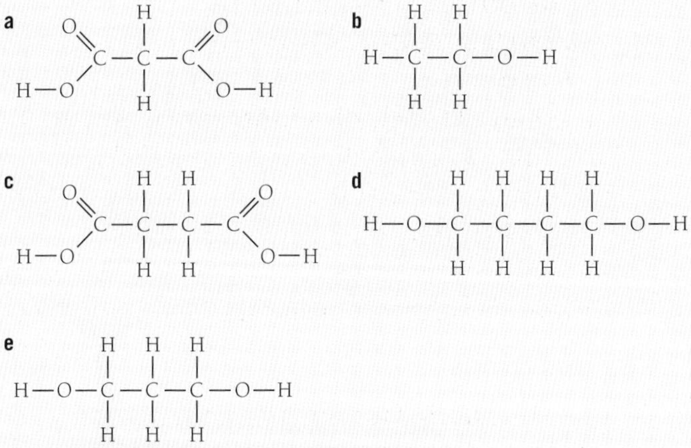

Which two compounds would be the monomers for the polyester in **Figure 4**? [2 marks]

02.5 Which other compound is produced when a polyester is formed from these monomers? [1 mark]

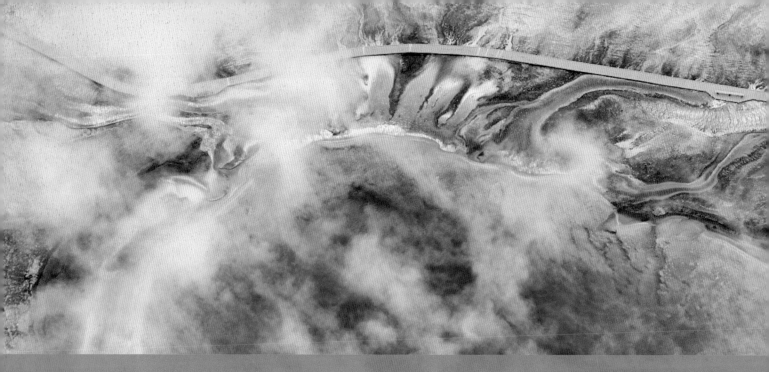

4 Analysis and the Earth's resources

Analytical chemists have developed many tests to detect specific chemicals. These tests are based on reactions that produce a gas with distinctive properties, or a colour change or an insoluble solid that appears as a precipitate. Instrumental analysis provides fast, sensitive, and accurate results, as used by forensic scientists and anti-doping scientists.

The Earth's atmosphere is dynamic and always changing. Some of these changes are man-made and sometimes part of natural cycles. Scientists and engineers are trying to solve the problems caused by increased levels of air pollutants. Industries use the Earth's natural resources to manufacture useful products. In order to operate sustainably, chemists seek to minimise the use of limited resources, the use of energy, waste produced, and environmental impact.

Key questions

- How can we use chemical tests to identify unknown substances?

- What are the advantages and disadvantages of using instrumental methods of analysis?

- How is human activity affecting the Earth's atmosphere?

- How are we seeking to make sustainable use of the Earth's limited resources?

Making connections

- The analysis of chromatograms to identify substances in **C12 Chemical analysis** is based on carrying out paper chromatography from **C1 Atomic structure**.

- Understanding the properties of polymers in **C15 Using our resources** relies on the work done in **C11 Polymers**.

- The atmospheric pollution discussed in **C13 Our atmosphere** builds on the work in **C9 Crude oil and fuels**.

I already know...

about the difference between pure substances and mixtures and how to identify some pure substances.

the composition of the atmosphere.

the production of carbon dioxide by human activity and its impact on climate.

about the Earth as a source of limited resources and the efficacy of recycling.

some properties of ceramics, polymers, and composites.

the use of carbon in obtaining metals from metal oxides.

I will learn...

a wider range of chemical tests to identify unknown gases and ions and why instrumental analysis is used in many applications.

how the atmosphere developed over the Earth's history before arriving at its current composition.

how climate change is caused by increasing levels of greenhouse gases and how this issue needs to be addressed.

how to analyse data on our diminishing finite resources, including order of magnitude estimations, and carry out Life Cycle Assessments to judge the impact of making new materials.

how to explain the properties of ceramics, polymers and composites in terms of their chemical structures.

about the use of biological methods to extract some metals, such as copper, from low grade deposits of metal ores.

Required Practicals

Practical		Topic
6	Finding R_f values	C12.2
7	Identifying unknown ionic compounds	C12.5
8	Purifying and testing water	C14.2

Learning objectives

After this topic, you should know:

- how to use melting point data to distinguish pure from impure substances
- how to identify examples of useful mixtures called formulations, given appropriate information.

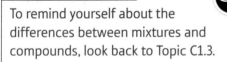

Figure 1 *This mineral water might be described as pure as a mountain stream, but it would not be called a pure substance by a chemist*

Synoptic link

To remind yourself about the differences between mixtures and compounds, look back to Topic C1.3.

What is meant by purity?

When you talk about something being pure in everyday life, often you are not referring to its chemical purity. For example, you often hear of orange juice or mineral water advertised as pure, but look at the label on a bottle of mineral water (such as the one in Figure 1).

In advertising a product, pure is taken to mean 'has had nothing added to it', and that it is in its natural state. For example, in pure orange juice means from freshly squeezed oranges.

However, to a chemist:

A pure substance is one that is made up of just one substance. That substance can be either an element or a compound.

Analysing pure substances and mixtures

You can use boiling points and melting points to identify pure substances.

Do you remember a test for water? For example, it turns white anhydrous copper sulfate blue. But that only tells you that water is present. It does not tell you if the water is pure or not. The test for *pure* water is that its melting point is exactly 0 °C, and its boiling point is exactly 100 °C.

The melting and boiling points of an element or a compound are called its fixed points.

You can use melting points or boiling points to identify substances because pure substances have characteristic, specific temperatures at which they melt and boil. These fixed points can be looked up in databooks or databases stored on computers.

The melting point and boiling point of a mixture will vary, depending on the composition of the mixture. A mixture does not have a sharp melting point or boiling point. It changes state over a range of temperatures. This difference between pure substances and mixtures can be used to distinguish if an unknown sample is a pure substance or a mixture of substances. So doing an experiment to find a melting point (or melting range) is a quick and easy test of a compound's purity.

Impurities tend to lower the melting point of a substance and raise its boiling point. The size of the difference from the fixed point of a pure substance depends

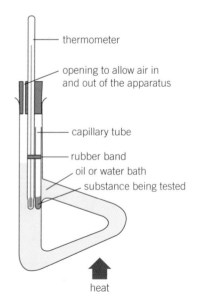

Figure 2 *This simple apparatus can be used to determine the melting point of a solid in powdered or crystal form*

on the amount of any impurities mixed with it. The purer the compound is, the narrower the melting point range. For example, the melting point range of a purified sample of caffeine is 234–237 °C. However, crude caffeine extracted from tea melts in the range 180–220 °C, showing it is an impure sample.

Formulations

A formulation is a mixture that has been designed to produce a useful product. Many consumer products are made up of complex mixtures. For example, medicinal drugs are formulations. They will often only contain between 5% and 10% of the active drug, which is the specific compound that affects the body to relieve symptoms or cure an illness. If taken in tablet form, they can also contain colorants, sweeteners, smooth coatings to aid swallowing, fillers, and other compounds to aid their dissolving at the most effective place in the digestive tract.

Paints are also common formulations. In general, paints will contain:

- a pigment, to provide colour
- a binder, to help the paint attach itself to an object and to form a protective film when dry
- a solvent, to help the pigment and binder spread well during painting by thinning them out.

Other formulations are found in the range of cleaning agents used in the home. For example, washing-up liquids generally contain:

- a surfactant, the actual detergent that removes the grease
- water, to thin out the mixture so it can squirt more easily from the bottle
- colouring and fragrance additives, to improve the appeal of the product to potential customers
- rinse agent, to help water drain off crockery.

Fuels, alloys, fertilisers, pesticides, cosmetics, and food products are all examples of formulations.

1 Explain the difference between the use of the term pure in advertising and its use in chemistry. [2 marks]

2 A white powder was placed in the melting point apparatus shown in Figure 2. The oil in the apparatus had a high boiling point. The chemist carrying out the test noted that the white powder started to melt at 158 °C and finished melting at 169 °C.
   a What was the melting range of the white powder? [1 mark]
   b What does this information tell us about the white powder? [1 mark]
   c Why was oil, and not water, used in the apparatus? [1 mark]

3 An insecticide formulation contains a very powerful toxic substance to kill the insects that feed on crops.
   Suggest two reasons why the major part of this pesticide formulation is made up of a solvent. One reason should benefit the environment and the other should help the farmer. 🖊 [2 marks]

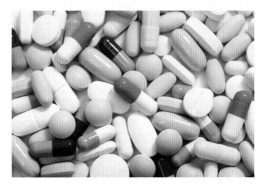

Figure 3 *Scientists have to test the effects that the other compounds in a formulation will have on the active drug in medicines*

Synoptic links

Nanoparticles are becoming an increasingly important component in many formulations, such as cosmetics and paints. To remind yourself about these tiny particles, look back to Topic C3.11 and Topic C3.12.

Key points

- Pure substances can be compounds or elements, but they contain only one substance. An impure substance is a mixture of two or more different elements or compounds.
- Pure elements and compounds melt and boil at specific temperatures, and these fixed points can be used to identify them.
- Melting point and boiling point data can be used to distinguish pure substances (specific fixed points) from mixtures (that melt or boil over a range of temperatures).
- Formulations are useful mixtures, made up in definite proportions, designed to give a product the best properties possible to carry out its function.

C12.2 Analysing chromatograms

Learning objectives

After this topic, you should know:

- how chromatography can be used to distinguish pure substances from impure substances
- how paper chromatography separates mixtures
- how to interpret chromatograms
- how to determine R_f values from chromatograms.

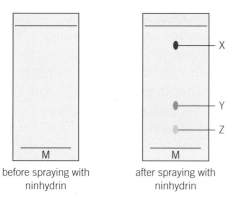

before spraying with ninhydrin ・ after spraying with ninhydrin

Figure 1 *A chromatogram produced by a mixture of amino acids. The spots are made by different amino acids forming coloured compounds with the locating agent*

Synoptic link

To remind yourself about how to set up a paper chromatogram, look back to Topic C1.4. You first met amino acids in Topic C11.3.

Figure 2 *Black ink can be separated out into its different colours on a chromatogram*

Scientists have many *instruments* that they can use to identify unknown compounds. Many of these are more sensitive, automated versions of the techniques, such as paper chromatography, that you use in school labs. For example, chromatography can be used to separate and identify mixtures of amino acids. The amino acids are colourless, but appear as purple spots on the paper when sprayed with a locating agent and dried (Figure 1).

You will have tried paper chromatography before, and probably used it to separate dyes in inks or food colourings (Figure 2).

Chromatography always involves a mobile phase and a stationary phase. The mobile phase moves through the stationary phase, carrying the components of the mixture under investigation with it. Each component in the mixture will have a different attraction for the mobile phase and the stationary phase. A substance with stronger forces of attraction between itself and the mobile phase than between itself and the stationary phase will be carried a greater distance in a given time. A substance with a stronger force of attraction to the stationary phase will not travel as far in the same time.

So in paper chromatography the mobile phase is the solvent chosen, and the stationary phase can be thought of as the paper. In Figure 1, amino acid X from the mixture M has the strongest attraction to the solvent, and amino acid Z has the strongest attraction to the paper.

Given an unknown organic solution, chromatography can usually tell you if it is a single compound or a mixture. If the unknown sample is a mixture of compounds, there will probably be more than one spot formed on the chromatogram. On the other hand, a single spot indicates the possibility of a pure substance.

Identifying unknown substances using chromatography

Once the compounds in a mixture have been separated using chromatography, they can be identified. You can compare spots on the chromatogram with others obtained from known substances (Figure 3).

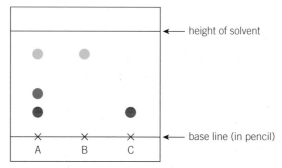

Figure 3 *This chromatogram shows that A is a mixture of three substances, B and C plus one other unknown substance*

Following the chromatography in Figure 3, mixture A still has one substance left unknown. A scientist making the chromatogram often does not know which pure compounds to include in their experiment to make a positive identification. It is also not practical to store actual chromatograms or their images, even on a computer. To make valid comparisons, every variable that affects a chromatogram would need to be exactly the same in all the chromatograms.

It is far more effective to measure data taken from any chromatogram of the unknown sample, then match it against a database. So the data are presented as an R_f (**retention factor**) value. This is a ratio, calculated by dividing the distance a spot travels up the paper (measured to the centre of the spot) by the distance the solvent front travels:

$$R_f = \frac{\textbf{distance moved by substance}}{\textbf{distance moved by solvent}}$$

As the number generated in the calculation is a ratio, it does not matter how long you run your chromatography experiment or what quantities you use. For comparisons against an R_f database to be valid, you just have to ensure that the solvent and the temperature used are the same as those quoted in the database or databook. Figure 4 shows how to get the measurements to calculate R_f values.

Finding R_f values

Using a capillary tube, pencil, **pipette**, water, boiling tube, and a narrow strip of chromatography paper, find out the R_f values of the different dyes in the mixture of food colourings provided.

Present your evidence clearly. Include your dried chromatogram, calculations, and an evaluation.

Worked example

Find the R_f value of compounds A and B using the chromatogram below.

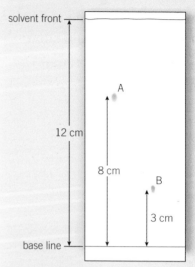

Figure 4 *The R_f value of an unknown substance, in a particular solvent at a given temperature, can be compared with values in a database to identify the substance*

Solution

The R_f value of A $= \dfrac{8}{12} = 0.67$

The R_f value of B $= \dfrac{3}{12} = 0.25$

1 Describe how you can positively identify the dyes mixed to make a food colouring from its chromatogram. [5 marks]

2 What is the R_f value of this substance X?

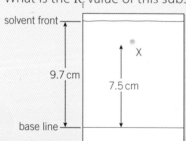

[1 mark]

3 The R_f values of two substances, Y and Z, were taken from a chromatogram run in 50% water and 50% ethanol solvent at 20 °C. The R_f value of Y was 0.54 and that Z was 0.79.
What can you deduce about Y and Z from these values, and how could you use them to identify Y and Z? [3 marks]

4 In order to positively identify a compound from a chromatogram, explain why the solvent and temperature must be the same as those used to generate the R_f values in a database. [4 marks]

Key points

- Scientists can analyse unknown substances in solution by using paper chromatography.
- R_f values can be measured and matched against databases to identify specific substances.
- $R_f = \dfrac{\text{distance moved by substance}}{\text{distance moved by solvent}}$

C12.3 Testing for gases

Learning objectives

After this topic, you should know:

- the tests and the positive results for the gases:
  - hydrogen
  - oxygen
  - carbon dioxide
  - chlorine.

Many of the reactions you will study in chemistry give off gases as a product. So chemists have devised quick and easy tests to identify different gases.

Test for hydrogen

The reaction between zinc and dilute acid is a convenient way to make some hydrogen gas to test:

$$zinc + sulfuric\ acid \rightarrow zinc\ sulfate + hydrogen$$
$$Zn(s) + H_2SO_4(aq) \rightarrow ZnSO_4(aq) + H_2(g)$$

If you want the gas to be produced more quickly, a few crystals of copper(II) sulfate can be added, or magnesium can be used instead of zinc.

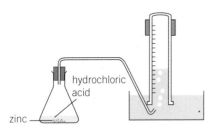

Figure 1 *Collecting hydrogen over water*

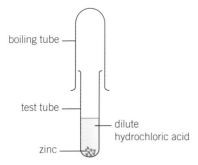

Figure 2 *Collecting hydrogen by upward delivery (downward displacement of air)*

> **Testing for hydrogen gas**
>
> Collect a test tube of hydrogen gas, using either of the sets of apparatus shown in Figures 1 and 2.
>
> - Record your observations when you hold a lighted splint at the open end of the test tube of hydrogen gas.
>
> - Explain your observations.
>
> - What do the methods of collecting hydrogen gas tell you about its properties?
>
> **Safety:** Wear eye protection. Hydrogen gas is flammable.

Positive test for hydrogen: a lighted splint 'pops'.

Test for oxygen

A convenient way to make some oxygen gas to test is the decomposition of hydrogen peroxide solution, with a little manganese(IV) oxide added as a catalyst.

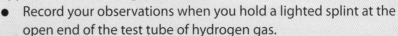

$$hydrogen\ peroxide \xrightarrow{\text{manganese(IV) oxide}} water + oxygen$$
$$2H_2O_2(aq) \rightarrow 2H_2O(l) + O_2(g)$$

> **Testing for oxygen gas**
>
> Collect 15 cm³ hydrogen peroxide solution in a small conical flask.
>
> Add a small amount of manganese(IV) oxide from the end of a spatula.
>
> Insert a glowing splint (made by blowing out a lighted splint) in the mouth of the flask.
>
> - Record an explain your observations.
>
> **Safety:** Wear eye protection.

Positive for oxygen gas: a glowing splint relights.

Synoptic links

To revise how you can use the gases given off in reactions to monitor rates of reaction, look back to Topic C8.1.

To remind yourself about the reaction of metals with dilute acids, look back to Topic C5.1 and Topic C5.4.

Study tip

When asked how to identify a given gas, always give the test **and** its result.

Test for carbon dioxide

You can make carbon dioxide gas to test by reacting marble chips (calcium carbonate) and dilute hydrochloric acid:

calcium carbonate + hydrochloric acid → calcium chloride + water + carbon dioxide

$$CaCO_3(s) + 2HCl(aq) \rightarrow CaCl_2(aq) + H_2O(g) + CO_2(g)$$

Testing for carbon dioxide gas

Bubble carbon dioxide gas through limewater (calcium hydroxide solution), using the apparatus shown below:

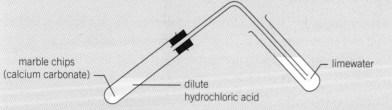

marble chips (calcium carbonate)

dilute hydrochloric acid

limewater

- Record your observations.
- Explain your observations.

Safety: Wear eye protection.

Positive test for carbon dioxide gas: limewater turns milky (cloudy white).

Test for chlorine

Chlorine is a toxic gas, so care must be taken when working with this gas. Your teacher will show you the test for chlorine gas (see the box at the top of this page).

Positive for chlorine gas: damp blue litmus paper turns white (as it gets bleached).

1 Explain why hydrogen gas 'pops' when a lighted splint is applied. [2 marks]

2 **a** Write a word equation for the reaction of magnesium carbonate plus dilute sulfuric acid, which produces carbon dioxide gas. [1 mark]

  b Look at the apparatus in the Practical box 'Testing for carbon dioxide gas'. Carbon dioxide gas is denser than air. Suggest an alternative way to collect the gas from the reaction in part **a** and to test it with limewater. [3 marks]

3 During the electrolysis of a chloride solution, a student predicted that a mixture of chlorine and oxygen gases would be given off from the anode. Suggest how you could test this prediction. [6 marks]

4 **a i** Write a balanced symbol equation, including state symbols, to describe the reaction between magnesium and dilute hydrochloric acid, to produce hydrogen gas. [3 marks]

  (H) ii Write an ionic equation, including state symbols, for the reaction in part **i**. [3 marks]

  (H) b In the reaction in part **a**, which species is reduced and which is oxidised? Explain your answer. [4 marks]

Testing for chlorine gas

Your teacher will carefully add concentrated hydrochloric acid (corrosive) to a spatula of moistened potassium manganate(VII) crystals in a boiling tube held in a rack inside a fume cupboard.

A piece of damp blue litmus paper can be held in the mouth of the boiling tube.

- Record and explain your observations.

Synoptic link

You might be asked to test the chlorine gas given off during the electrolysis of a chloride solution, as in Topic C6.4.

Key points

- Hydrogen gas burns rapidly with a 'pop' when you apply a lighted splint.
- Oxygen gas relights a glowing splint.
- Carbon dioxide gas turns limewater milky (cloudy).
- Chlorine gas bleaches damp blue litmus paper white.

C12.4 Tests for positive ions

Learning objectives

After this topic, you should know:

- how to identify positive ions by:
  - the flame tests for the ions of Li, Na, K, Ca, and Cu
  - the precipitates formed in the reactions that produce insoluble hydroxides.

Metal ion	Flame colour
lithium, Li^+	crimson
sodium, Na^+	yellow
potassium, K^+	lilac
calcium, Ca^{2+}	orange-red
copper, Cu^{2+}	green

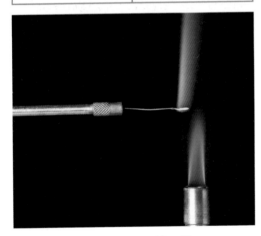

Figure 1 *A flame test can identify some metal ions in compounds*

Figure 2 *This distinctive blue precipitate that forms when you add sodium hydroxide solution tells you that Cu^{2+} ions are present*

Scientists working in environmental monitoring, industry, medicine, and forensic science need to analyse and identify substances. To identify unknown substances, there are a variety of different chemical tests.

Flame tests

Some metal ions (cations) produce flames with a characteristic colour. To carry out a flame test:

- A nichrome wire loop should be dipped in concentrated hydrochloric acid and then heated to clean it first. Then it should be dipped in the acid again before dipping it into the metal compound that is being tested.
- Then hold the loop in the roaring blue flame of a Bunsen burner.
- Use the colour of the Bunsen flame to identify the metal ion in the compound (see the table on the left).
- However, if the sample being tested contains a *mixture of metal ions*, then some flame colours can be masked. For example, the intense yellow colour of sodium ions can dominate other colours.

Metal cation tests with sodium hydroxide

Precipitation reactions : [handwritten annotation]

Reacting unknown compounds with sodium hydroxide solution can also help you identify some positive ions. Aluminium ions, calcium ions, and magnesium ions all form *white precipitates* with sodium hydroxide solution. So, if a white precipitate forms, you know that an unknown compound contains either Al^{3+}, Ca^{2+}, or Mg^{2+} ions.

For example, if aluminium sulfate solution is tested with sodium hydroxide solution, the precipitation reaction is:

$$Al_2(SO_4)_3(aq) + 6NaOH(aq) \rightarrow 3Na_2SO_4(aq) + 2Al(OH)_3(s)$$

The ionic equation for the reaction with aluminium ions is:

$$Al^{3+}(aq) + 3OH^-(aq) \rightarrow Al(OH)_3(s)$$

Higher

If you add more and more sodium hydroxide, then the precipitate formed with aluminium ions dissolves. However, the white precipitate formed with calcium or magnesium ions will not dissolve. Calcium and magnesium ions can be distinguished by a flame test. Calcium ions give a orange-red flame, but magnesium ions produce no colour at all.

Some metal ions form *coloured precipitates* with sodium hydroxide. If you add sodium hydroxide solution to a substance containing:

- copper(II) ions, a blue precipitate appears
- iron(II) ions, a green precipitate is produced
- iron(III) ions, a brown precipitate is formed.

can both ions be present? [handwritten annotation]

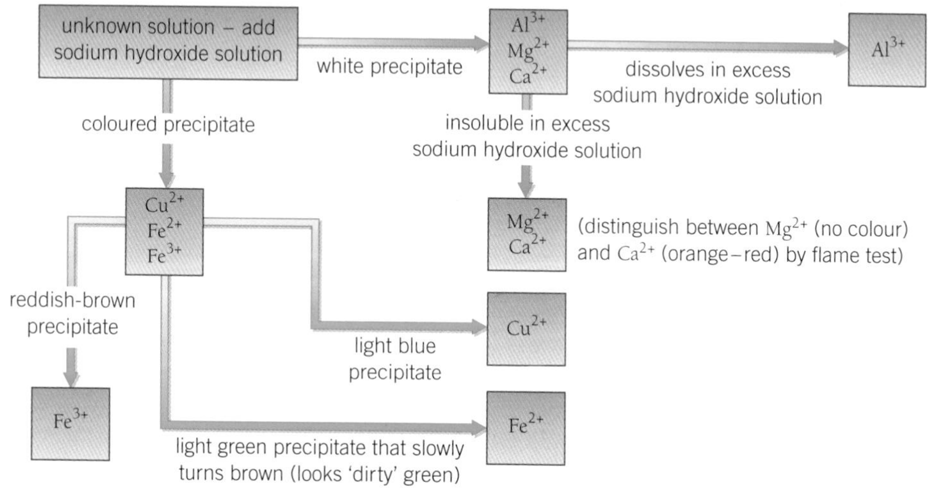

Figure 3 *Sodium hydroxide solution provides a very useful test for many positive ions*

For example, when iron(II) chloride solution is tested with sodium hydroxide solution:

$$FeCl_2(aq) + 2NaOH(aq) \rightarrow 2NaCl(aq) + Fe(OH)_2(s)$$

Higher

The ionic equation for the reaction with iron(II) ions is:

$$Fe^{2+}(aq) + 2OH^-(aq) \rightarrow Fe(OH)_2(s)$$

Identifying positive ions

Try to identify the metal ions in some unknown compounds provided for you.

Safety: Wear chemical splash-proof eye protection.

1 a What colour precipitate does dilute sodium hydroxide produce with aluminium, calcium, and magnesium ions? [1 mark]

clear
colourless ✓

   b How could you distinguish between these three metal ions? **◐** [6 marks]

2 a Draw a flow chart to describe how to carry out a flame test. [4 marks]

   b If a flame test was carried out on copper chloride, describe what you would see. [1 mark]

3 Complete the missing cells **a** to **j** in the table below:

Add sodium hydroxide solution	Flame test	Metal ion	
nothing observed	lilac	f	[1 mark]
white precipitate	orange-red	g	[1 mark]
a [1 mark]	d [1 mark]	Fe³⁺	
white precipitate, which dissolves in excess sodium hydroxide solution	nothing observed	h	[1 mark]
green precipitate, which slowly turns reddish-brown	nothing observed	i	[1 mark]
b [1 mark]	e [1 mark]	Na⁺	
c [1 mark]	crimson	j	[1 mark]

H 4 Write an ionic equation, including state symbols, for the precipitation of magnesium ions from solution by aqueous hydroxide ions. [2 marks]

Key points

- Some metal ions (including most Group 1 and 2 cations) can be identified in their compounds using flame tests.
- Sodium hydroxide solution can be used to identify metal ions that form insoluble hydroxides in precipitation reactions.

C12.5 Tests for negative ions

Learning objectives

After this topic, you should know:

- how to identify the following negative ions:
  - carbonates
  - halides
  - sulfates.

Synoptic link

For more information on carbonates, look back to Topic C5.6.

Study tip

In the test for a halide ion, add dilute **nitric** acid before the silver **nitrate**. Do not add any other acid, as they will produce precipitates with silver nitrate solution.

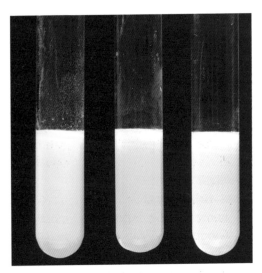

Figure 2 *Results of tests with dilute nitric acid and silver nitrate solution added to halide solutions. The precipitation reactions can tell you if an unknown substance contains iodide (first tube), bromide (middle tube), or chloride (end tube) ions*

You can also do chemical tests to identify some negative ions in ionic compounds.

Carbonates

If you add a dilute acid to a carbonate, it fizzes and produces carbon dioxide gas. This is a good test to see if an unknown substance is a carbonate.

Figure 1 *The test for a carbonate*

If the carbonate is magnesium carbonate, the reaction you get is:

$$\text{magnesium carbonate} + \text{hydrochloric acid} \rightarrow \text{magnesium chloride} + \text{water} + \text{carbon dioxide}$$

$$MgCO_3(s) + 2HCl(aq) \rightarrow MgCl_2(aq) + H_2O(l) + CO_2(g)$$

Note that most carbonates do not dissolve in water, but Group 1 carbonates, such as sodium or potassium carbonate, are soluble in water.

You can represent the reaction by just showing the ions that change in the reaction using an ionic equation:

$$\underset{\text{carbonate ion}}{CO_3^{2-}(s)} + \underset{\text{acid}}{2H^+(aq)} \rightarrow CO_2(g) + H_2O(l)$$

In limewater, the acidic carbon dioxide gas reacts with the alkaline calcium hydroxide. It forms a white precipitate of calcium carbonate, which turns the limewater milky (cloudy white).

Halides (chlorides, bromides, and iodides)

A very simple test shows whether chloride, bromide, or iodide ions are present in a compound. First, add dilute nitric acid and then silver nitrate solution. If a precipitate forms, there are halide ions present.

You add the nitric acid to dissolve the compound and to remove any carbonate ions, as they would also form a precipitate with the silver ions and so interfere with the test.

The colour of the precipitate tells you which halide ion is present (Figure 2):

- iodide ions, I^-, give a yellow precipitate.
- bromide ions, Br^-, give a cream precipitate
- chloride ions, Cl^-, give a white precipitate

Higher

If the unknown halide was sodium chloride, the precipitation reaction would be:

$$NaCl(aq) + AgNO_3(aq) \rightarrow NaNO_3(aq) + AgCl(s)$$

Here is the ionic equation, where X^- is the halide ion:

$$Ag^+(aq) + X^-(aq) \rightarrow AgX(s)$$

Sulfates

You can test an unknown compound for sulfate ions by adding dilute hydrochloric acid, followed by barium chloride solution. You add the dilute hydrochloric acid first to remove carbonate ions that would form a precipitate with the barium ions. A white precipitate tells you sulfate ions are present. The white precipitate is the insoluble salt barium sulfate, $BaSO_4$.

If the unknown compound was potassium sulfate, then the equation for the precipitation reaction would be:

$$K_2SO_4(aq) + BaCl_2(aq) \rightarrow 2KCl(aq) + BaSO_4(s)$$

Here is the ionic equation:

$$Ba^{2+}(aq) + SO_4^{2-}(aq) \rightarrow BaSO_4(s)$$

Figure 3 *The white precipitate of barium sulfate forming*

Identifying unknown ionic compounds

Now that you know the tests for some positive and negative ions, you can try to identify some unknown compounds.

Safety: Wear chemical splash-proof eye protection.

1 List the reagents you would need to collect in a lab to test a sample of an unknown substance for carbonate, halide, and sulfate ions. [5 marks]

2 An unknown compound is a white solid, which dissolves in water to produce a colourless solution. When this solution is acidified with nitric acid, and then silver nitrate is added, a yellow precipitate is produced. A flame test on the unknown compound produces a lilac flame. Deduce the name of the unknown compound and give your reasoning. [3 marks]

3 Write a word equation and a balanced symbol equation, including state symbols, for the following reactions:
   a magnesium chloride solution + silver nitrate solution [3 marks]
   b potassium carbonate powder + hydrochloric acid [3 marks]
   c aluminium sulfate solution + barium chloride solution. [3 marks]

Ⓗ 4 Write ionic equations, including state symbols, to summarise the reactions that help identify:
   a bromide ions [2 marks] b sulfate ions [2 marks]
   c carbonate ions. [2 marks]

5 In the test for halide ions, explain why nitric acid (rather than other acids) is added to a sample before you add silver nitrate solution. [3 marks]

Study tip

In the test for the sulfate ion, add *hydrochloric* acid before the barium *chloride*. Do not add sulfuric acid – it contains sulfate ions!

Key points

● You identify carbonates by adding dilute acid, which produces carbon dioxide gas. The gas turns limewater milky (cloudy).
● You identify halides by adding nitric acid, then silver nitrate solution. This produces a precipitate of silver halide (chloride = white, bromide = cream, iodide = yellow).
● You identify sulfates by adding hydrochloric acid, then barium chloride solution. This produces a white precipitate of barium sulfate.

C12.6 Instrumental analysis

Learning objectives

After this topic, you should know:

- some advantages of instrumental methods compared with traditional chemical tests
- how to interpret a result from flame emission spectroscopy, given appropriate data.

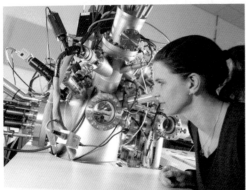

Figure 1 *Compared with the methods of 50 years ago (top photo), modern instrumental methods of analysis are quick, accurate and sensitive – three big advantages. They also need far fewer people to carry out the analysis than traditional laboratory analysis*

Synoptic link

Revisit Topic C1.5 to see how the existence of atomic energy levels was proposed by Niels Bohr, from his work looking at the energy released as electrons fall to lower energy levels.

Many industries need rapid and accurate methods for analysing their products, and to check on any emissions produced during the manufacturing process. They use modern instrumental analysis for this task, including flame emission spectroscopy.

Instrumental methods

Instrumental techniques are also important in the work of environmental agencies fighting pollution. Careful monitoring of the environment using sensitive instruments is now common. This type of analysis is also used frequently in healthcare. For example, it can be used to check if titanium alloy hip-replacement joints are still in good condition. The blood of patients that have received new hip joints can be tested for any traces of metal ions originating from the metal alloy.

Modern instrumental methods have a number of benefits over older methods:

- they are highly accurate and sensitive
- they are quicker
- they enable very small samples to be analysed.

Against this, the main disadvantages of using instrumental methods are that the equipment:

- is usually very expensive
- takes special training to use
- gives results that can often be interpreted only by comparison with data from known substances.

Flame emission spectroscopy

Chemists use flame emission spectroscopy to analyse samples for metal ions. The sample is heated in a flame. The energy provided excites electrons in the metal ions, making them jump into higher energy levels (or shells). When they fall back to lower energy levels, the energy is released as light energy. This explains the flame tests you did to identify metals in Topic C12.4.

In the spectrometer, the wavelengths of the light produced can be analysed by passing it through a spectroscope. Each type of metal ion absorbs and gives out its own characteristic pattern of radiation. This is called its line spectrum. The line spectrum can be used to identify the metal ions by comparing it with a database held on a computer.

The concentration of the metal ions present in a sample can also be determined. The spectrometer can measure the intensity (or absorbance) of light with a specific wavelength known to be characteristic of a particular metal ion. The machine can be calibrated using solutions of the metal ion of known concentrations and so give a value for the unknown concentration.

So flame emission spectroscopy, and similar methods, provide accurate ways to monitor water for metal ions. For example, the heavy metal mercury is a dangerous toxin that can build up in your body. Mercury can now be traced down to as little as 0.000 000 001 g (i.e., 1.0×10^{-9} g). Other pollutants, such as zinc, lead, and cadmium, can also be detected. Water companies can measure the concentration of calcium, magnesium, iron, and aluminium ions in the water they supply to us.

Go further

There are other important methods of instrumental analysis that you can find out about in A Level Chemistry. For example, chemists use gas chromatography to separate the substances in an unknown mixture, linked to a mass spectrometer that will identify each substance by 'fingerprinting' (matching its characteristic mass spectrum to those in a database).

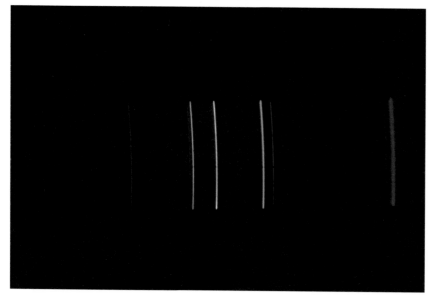

Figure 2 *This is the characteristic line spectrum of the heavy metal cadmium. It is used in the manufacture of batteries. If not disposed of responsibly at recycling depots, cadmium ions can cause pollution in water run-off from landfill sites*

Flame emission spectrometers are also used in the steel industry. Steel is a mixture of elements with iron. The steelmakers need to control carefully the amounts of trace metals present, as the quality of steel depends on this.

Study tip

Although simpler to use than traditional practical methods, instrumental methods still need trained technicians to operate them.

1 Describe the main advantages and disadvantages of using instrumental analysis compared with traditional practical methods. ✏ [5 marks]

2 Explain why metal ions heated in a flame emit light energy, and how this can be used to identify metal ions. ✏ [6 marks]

3 When analysing water samples, as little as 1.0×10^{-9} g of mercury, Hg, can be detected by flame emission spectroscopy.

Ⓗ a Look up the relative atomic mass of mercury, then calculate how many moles of mercury can be detected by flame emission spectroscopy. Give your answer in standard form to 2 significant figures. [2 marks]

b Mercury forms a compound with chlorine with the formula $HgCl_2$. Write the formula of a mercury ion. [1 mark]

Ⓗ c If 1.0×10^{-9} g of mercury was detected in 20 cm³ of solution, what would be the concentration of the solution given in mol/dm³? [3 marks]

Key points

- Modern instrumental techniques provide fast, accurate, and sensitive ways of analysing chemical substances.
- Flame emission spectroscopy is an example of an instrumental method.
- This method will tell us which metal ions are present from their characteristic line spectra, and also the concentration of the metal ions in a solution.

C12 Chemical analysis

Summary questions

1 As well as pigments (25% by mass), binders (30%), and solvents (40%), the formulation used to make a gloss paint has additives included in the mixture.

   a Calculate the percentage of additives in the formulation to make gloss paint. [1 mark]

   b Explain what a formulation is and why they are manufactured in the chemical industry. [2 marks]

   H c Titanium(IV) oxide, TiO_2, is used to give a brilliant white colour to paint. Assuming all the pigment is titanium(IV) oxide, how many moles of the oxide would be in a tin containing 6.00 kg of paint. Give your answer to 3 significant figures. [4 marks]

2 What would be entered in **a** to **f** to complete the table?

Add dilute acid	Add sodium hydroxide solution	Add dilute nitric acid followed by silver nitrate solution	Flame test	Substance
nothing observed	nothing observed	yellow precipitate	yellow	e [1 mark]
fizzing – gas turns limewater cloudy	pale blue precipitate formed	not needed	green flame	f [1 mark]
a [1 mark]	b [1 mark]	c [1 mark]	d [1 mark]	calcium chloride

3 An unknown compound gave the following positive tests when analysed:

   ● The Bunsen flame turned crimson in a flame test.

   ● When dissolved in dilute hydrochloric acid and a barium chloride solution was added, a white precipitate was formed.

   Name the unknown compound and write its chemical formula. [2 marks]

4 A sample of solid sodium nitrate, $NaNO_3$, was heated in a test tube. A gas was given off and collected over water in an inverted measuring cylinder. At the end of the experiment, 96 cm^3 of gas had been collected.

   a Complete and balance the equation:

   $$... NaNO_3 \xrightarrow{heat} ... NaNO_2 + ...$$
   [2 marks]

   b Draw the apparatus used to carry out the experiment. [3 marks]

   c Describe a positive test for the gas collected. [1 mark]

   d Explain what precaution should be taken before the heating is stopped. [3 marks]

H e Calculate the mass of $NaNO_3$ that decomposed in the experiment. [4 marks]

5 The label on a jar of white crystals has been damaged by water and cannot be read. A science technician thinks it is probably a jar of potassium bromide.

   a Describe how the technician could positively identify the crystals as potassium bromide. [2 marks]

   b Give the chemical formula of potassium bromide. [1 mark]

   c What type of bonding will be found in potassium bromide? [1 mark]

   d i Describe briefly the process that could be used to break down potassium bromide into its elements. [2 marks]

   H ii If the process in part **d i** is repeated with potassium bromide solution, explain the difference that would be observed, including any relevant half equations. [6 marks]

6 The chromatogram below is for two unknown substances, X and Y. Ethanol was used as the solvent.

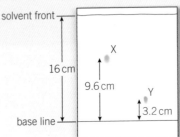

   a Using the chromatogram above, determine the R_f values of the unknown substances X and Y. [2 marks]

   b Explain the difference in the R_f values of X and Y. [2 marks]

   c Describe how you could use the R_f values to positively identify X and Y. [1 mark]

   d Before you could be certain of the identification made, what other condition, besides the solvent used, should be controlled when obtaining the R_f values for X and Y? [1 mark]

7 a Give two advantages and two disadvantages of using modern instrumental techniques of analysis. [2 marks]

   b i Name one modern instrumental technique used to detect metal ions. [1 mark]

   ii Describe how this technique could be used to test water from a river that runs near a landfill site for household waste. [2 marks]

Practice questions

01 A group of students analysed two solutions of ionic compounds. Their results are shown in **Table 1**.

Table 1

	Addition of dilute sodium hydroxide solution	Addition of dilute hydrochloric acid followed by barium chloride solution
Solution A	blue precipitate	white precipitate
Solution B	green precipitate	white precipitate

Identify Solutions **A** and **B**. [2 marks]

02 A teacher gave a group of students four colourless solutions in bottles that had been labelled solution **W**, solution **X**, solution **Y**, and solution **Z**.

The teacher told the students that the four colourless solutions were sodium carbonate, potassium carbonate, sodium chloride, and potassium chloride.

02.1 How could the students identify which solutions contained the carbonate ion, CO_3^{2-}. [2 marks]

02.2 What would the students add to identify the two solutions that contained the chloride ion Cl^-?
Give the result. [3 marks]

02.3 The teacher suggested using a flame test to identify which solutions contained sodium ions, Na^+, and potassium ions, K^+.
Outline a method for the flame test and state which colours the students would see. [4 marks]

02.4 What is the formula for potassium carbonate? [1 mark]

03 A class of students had been doing an experiment with colourless solutions of sodium compounds. At the end of the lesson the labels from the bottles had been removed and some of the solutions had been mixed together.

The bottles contained solutions of:

- sodium iodide, NaI
- sodium carbonate, Na_2CO_3
- sodium bromide, NaBr
- a mixture of sodium chloride, NaCl, and sodium carbonate, Na_2CO_3.

The teacher set a different class the task of identifying what the four bottles contained.

The teacher gave the students equipment for the flame test and solutions of nitric acid, calcium hydroxide (limewater), and silver nitrate.

The students started by doing a flame test.

03.1 Explain why the students were unable to identify any of the solutions using the flame test. [2 marks]

03.2 Plan a method to identify the four solutions. The teacher suggested adding dilute nitric acid to all four solutions before doing any other tests. Include the results of the tests. [5 marks]

04 A sample of drinking water was sent to a laboratory for analysis to see which metal ions it contained. The laboratory used flame emission spectroscopy to analyse the drinking water.

Figure 1 shows the flame emission spectra for several different metals.

Figure 1

lithium
sodium
potassium
rubidium
calcium
arsenic
wavelength 400 nm 450 nm 500 nm 550 nm 600 nm 650 nm 700 nm

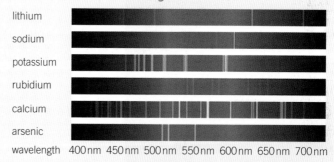

The flame emission spectrum that came back from the lab is shown in **Figure 2**.

Figure 2

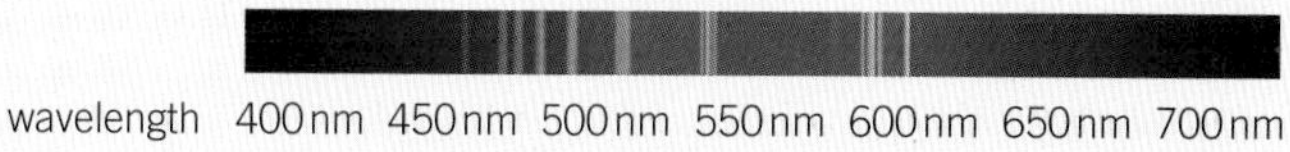

wavelength 400 nm 450 nm 500 nm 550 nm 600 nm 650 nm 700 nm

04.1 Which two metal ions does **Figure 2** show that the sample of water contains? [2 marks]

04.2 How does **Figure 2** show that the sample of drinking water does not contain the toxic metal arsenic? [1 mark]

04.3 Give **two** advantages of instrumental analysis compared to chemical methods of analysis. [2 marks]

04.4 Give **one** disadvantage of instrumental analysis. [1 mark]

04.5 It was suggested that both of the metal compounds in the water were sulfates. Give the chemical test for sulfate ions SO_4^{2-} and the result of the test. [3 marks]

Learning objectives

After this topic, you should know:

- a theory about how our atmosphere developed
- how to interpret evidence and evaluate different theories about the Earth's early atmosphere, given appropriate information.

Figure 1 *Volcanoes moved chemicals from inside the Earth to the surface and the newly forming atmosphere*

Figure 2 *The surface of one of Jupiter's moons, Io, with its active volcanoes releasing gases into its sparse atmosphere. This is likely to be what our own Earth was like billions of years ago*

Scientists think that the Earth was formed about 4.6 billion years ago. They think that to begin with, it was a molten ball of rock and minerals. For its first billion years it was a very hot, turbulent place. The Earth's surface was covered with volcanoes belching fire and gases into the atmosphere.

The Earth's early atmosphere

There are several theories about the Earth's early **atmosphere**, although there is little direct evidence to draw on from billions of years ago. However, scientists have reconstructed what they think the atmosphere must have been like, based on evidence from gas bubbles trapped in ancient rocks. They also use data gathered from the atmospheres of other planets and their moons in the solar system.

One theory suggests that volcanoes released carbon dioxide, CO_2, water vapour, H_2O, and nitrogen, N_2, and that these gases formed the early atmosphere. Water vapour in the atmosphere condensed as the Earth gradually cooled down, and fell as rain. Water collected in hollows in the crust as the rock solidified and the first oceans were formed. Another theory speculates that comets could also have brought water to the Earth. As icy comets rained down on the surface of the Earth, they melted, adding to its water supplies.

As the Earth began to stabilise, the atmosphere was probably mainly carbon dioxide. There could also have been some water vapour and nitrogen gas, and traces of methane, CH_4, and ammonia, NH_3. There would have been very little or no oxygen at that time. This resembles the atmospheres that are known to exist today on the planets Mars and Venus. Our nearest neighbours have atmospheres made up mainly of carbon dioxide with little or no oxygen.

After the initial violent years of the history of the Earth, the atmosphere remained quite stable. That is until life first appeared on Earth.

Oxygen in the atmosphere

There are many theories as to how life was formed on Earth billions of years ago. Scientists think that life began about 3.4 billion years ago, when the first simple organisms, similar to bacteria, appeared. These could use the breakdown of chemicals as a source of energy.

Then, about 2.7 billion years ago, bacteria and other simple organisms, such as algae, evolved. Algae could use the energy from the Sun to make their own food by photosynthesis. This produced oxygen gas as a waste product. Over the next billion years or so, the levels of oxygen rose steadily as the algae and bacteria thrived in the seas. More and more plants evolved – all of them were photosynthesising, removing carbon dioxide, and making oxygen.

$$\text{carbon dioxide} + \text{water} \xrightarrow{\text{(energy from sunlight)}} \text{glucose} + \text{oxygen}$$
$$6CO_2 + 6H_2O \longrightarrow C_6H_{12}O_6 + 6O_2$$

As plants evolved, they successfully colonised most of the surface of the Earth. So the atmosphere became richer in oxygen. This made it possible for the first animal forms to evolve. These animals could not make their own food like the algae and plants could. They relied on the algae and plants for their food and on oxygen to respire.

Figure 3 *Some of the first photosynthesising bacteria probably lived in colonies like these stromatolites. They grew in water and released oxygen into the early atmosphere*

On the other hand, many of the earliest living microorganisms could not tolerate a high oxygen concentration, because they had evolved without it. They largely died out, as there were fewer places where they could survive.

1 Name and give the chemical formula of five gases that scientists speculate were found in the Earth's early atmosphere. Display your answer in a table. [5 marks]

2 Describe how the Earth's early atmosphere was probably formed during its first billion years of existence. [2 marks]

3 a Suggest why scientists believe there was no life on Earth for its first billion years. [1 mark]
   b Suggest two possible sources of the water that collected and formed our early oceans. [2 marks]

4 Explain how the levels of oxygen in our atmosphere increased and why this was significant in the history of the Earth. Include any relevant chemical equations in your answer. ✓ [6 marks]

Ⓗ 5 Over a period of time, the algae in an ancient sea made 270 tonnes of glucose during photosynthesis.
   a Express 270 tonnes in grams using standard form. (1 tonne = 1000 kg) [1 mark]
   b Using the equation for photosynthesis at the top of this page, calculate the mass of oxygen gas produced by the algae over this period of time. Give your answer to 2 significant figures. [4 marks]

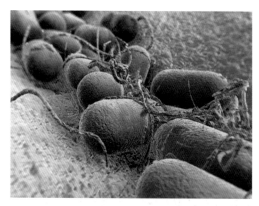

Figure 4 *Not only do bacteria such as these not need oxygen – they die if they are exposed to it. However, they can survive and breed in rotting tissue and other places where there is no oxygen*

Go further

Evaluating alternative theories

Scientists have found evidence from some of the oldest rocks on Earth that question assumptions that the early gases originated from volcanoes. Some scientists suggest that the mixture of gases could have been formed from solar debris, similar to comets, smashing into the Earth and vaporising around 500 million years after its formation.

Key points

- The Earth's early atmosphere was formed by volcanic activity.
- It probably consisted mainly of carbon dioxide. There may also have been nitrogen and water vapour, together with traces of methane and ammonia.
- As plants spread over the Earth, the levels of oxygen in the atmosphere increased.

C13.2 Our evolving atmosphere

Learning objectives

After this topic, you should know:

- the main changes in the atmosphere over time and some of the likely causes of these changes
- the relative proportions of gases in our atmosphere now.

Scientists think that the early atmosphere of the Earth contained a great deal of carbon dioxide. Yet the Earth's atmosphere today only has around 0.04% of this gas. So where has it all gone? The answer is mostly into living organisms and into materials formed from living organisms. As you saw in the Topic C13.1, algae and plants decreased the percentage of carbon dioxide in the early atmosphere by photosynthesis.

Carbon 'locked into' rock

Carbon dioxide, along with water, is taken in by plants and converted to glucose and oxygen during photosynthesis. The carbon in the glucose can then end up in new plant material. When animals eat the plants, some of this carbon can be transferred to the animal tissues, including their skeletons and shells.

Over millions of years, the skeletons and shells of huge numbers of these marine organisms built up at the bottom of vast oceans. There they became covered with layer upon layer of fine sediment. Under the pressure caused by being buried by all these layers of sediment, eventually the deposits formed sedimentary carbonate rocks such as limestone, a rock containing mainly calcium carbonate, $CaCO_3$.

Some of the remains of ancient living things (animals and plants) were crushed by large-scale movements of the Earth and were heated within the Earth's crust over very long periods of time. They formed the fossil fuels coal, crude oil, and natural gas.

- Coal is classed as a sedimentary rock, and was formed from thick deposits of plant material, such as ancient trees and ferns. When the plants died in swamps, they were buried, in the absence of oxygen, and compressed over millions of years.
- Crude oil and natural gas were formed from the remains of plankton deposited in muds on the seabed. These remains were covered by sediments that became layers of rock when compressed over millions of years. The crude oil and natural gas formed is found trapped beneath these layers of rock.

In this way, much of the carbon from the old carbon dioxide-rich atmosphere became locked up within the Earth's crust in rocks and fossil fuels.

Carbon dioxide gas was also removed from the early atmosphere by dissolving in the water of the oceans. It reacted, for example, with metal oxides, and made insoluble carbonate compounds. These fell to the seabed as sediments and helped to form more carbonate rocks.

Over the past 200 million years, the level of carbon dioxide in the atmosphere has not changed much. This is due to the natural cycle of carbon in which carbon moves between the oceans, rocks, and the atmosphere.

Figure 1 *There is clear fossil evidence in carbonate rocks of the organisms which lived millions of years ago*

Shelly carbonates

Carry out a test to see if crushed samples of shells contain carbonates. Think of the reaction that all carbonates undergo with dilute acid. How will you test any gas given off?

- Record your findings.

Ammonia and methane

Volcanoes also produced nitrogen gas, which gradually built up in the early atmosphere, and there may have also been small proportions of methane and ammonia gases.

Any methane and ammonia found in the Earth's early atmosphere reacted with the oxygen, formed by the evolving algae and plants:

$$CH_4 + 2O_2 \rightarrow CO_2 + 2H_2O$$
$$4NH_3 + 3O_2 \rightarrow 2N_2 + 6H_2O$$

This removed the methane and ammonia from the atmosphere. However, the levels of nitrogen gas, N_2, in the atmosphere could build up, as nitrogen is a very unreactive gas.

The atmosphere today

By 200 million years ago, the proportions of gases in the Earth's atmosphere had stabilised. These were much the same as they are today.

Look at the percentage of gases in the atmosphere today in the pie chart in Figure 2.

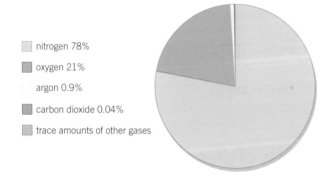

nitrogen 78%

oxygen 21%

argon 0.9%

carbon dioxide 0.04%

trace amounts of other gases

Figure 2 *The relative proportions of nitrogen, oxygen, and other gases in the Earth's atmosphere. The Earth's atmosphere also contains water vapour but the percentage in the atmosphere varies*

The noble gases are all found in air, with argon, Ar, the most abundant at about 0.9%. Neon, Ne, krypton, Kr, and xenon, Xe, together make up less than 0.1% of clean, dry air.

1 Complete the table to show the percentage proportions of gases in the Earth's atmosphere today. [3 marks]

Nitrogen	Oxygen	Argon	Carbon dioxide	Other gases

2 Explain the origins of nitrogen gas in the early atmosphere and suggest why its percentage of the composition of air remains so high. [4 marks]

3 Explain how most of the carbon dioxide in the Earth's early atmosphere was removed to arrive at a level of around 0.04% of today's atmosphere. ⬤ [4 marks]

Synoptic link

To see where the noble gases are situated in the periodic table, look back to Topic C2.2.

Key points

- Photosynthesis by algae and plants decreased the percentage of carbon dioxide in the early atmosphere. The formation of sedimentary rocks and fossil fuels that contain carbon also removed carbon dioxide from the atmosphere.
- Any ammonia and methane was removed by reactions with oxygen, once oxygen had been formed by photosynthesis.
- Approximately four-fifths (about 80%) of the atmosphere today is nitrogen, and about one-fifth (about 20%) is oxygen.
- There are also small proportions of various other gases, including carbon dioxide, water vapour, and noble gases.

C13.3 Greenhouse gases

Learning objectives

After this topic, you should know:

- how the greenhouse effect operates
- how to evaluate the quality of evidence in a report about global climate change, given appropriate information
- how to describe uncertainties in the evidence base
- the importance of peer review of results and of communicating results to a wide range of audiences.

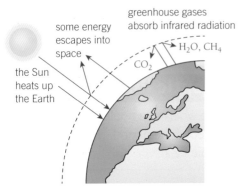

carbon dioxide, methane, and water vapour are the main greenhouse gases

Figure 1 *The molecules of a 'greenhouse gas' absorb the energy radiated by the Earth as it cools down at night. This increases the store of energy of the gases in the atmosphere and warms the Earth*

Go further

The levels of greenhouse gases in the atmosphere can be monitored using an instrumental technique called infrared spectroscopy. You can find out more about this in A Level Chemistry. Infrared radiation stimulates the bonds in molecules of carbon dioxide and methane to vibrate more vigorously, absorbing some of the radiation, which can be detected and displayed on an infrared spectrum.

You will have heard of the greenhouse effect and concerns, and some arguments, about global warming from reports in the media. However, these reports are sometimes biased or over-simplified, as they often seek out sensational headlines from scientific research without presenting the whole picture. Therefore, they can be misleading.

It is known that carbon dioxide along with methane and water vapour are the main 'greenhouse gases' in the Earth's atmosphere, that is, gases that absorb energy radiated from its surface. Without carbon dioxide in the Earth's atmosphere, the average temperature on Earth would be about −19 °C, and life as it is now could never have evolved in liquid water. So how do greenhouse gases warm up the Earth?

The Earth is heated by the Sun. Not all the energy reaching the Earth warms up our planet. Almost 30% is reflected back into space from the atmosphere and surface. The greenhouse gases let short-wavelength electromagnetic radiation (for example, ultraviolet light) pass through. The surface of the Earth cools down by emitting longer wavelength infrared (thermal) radiation. However, greenhouse gases absorb infrared radiation. The radiation stimulates the bonds in these molecules to vibrate, bend, and stretch more vigorously, raising their temperature. So some of the energy radiated from the surface of the Earth gets trapped in the atmosphere and the temperature rises. The higher the proportion of greenhouse gases in the air, the more energy is absorbed.

The increasing levels of greenhouse gases

Over the past century the amount of carbon dioxide released into the atmosphere has greatly increased. More fossil fuels than ever are used to make electricity, heat homes, and run cars. This has enormously increased the amount of carbon dioxide produced. Think about what happens when you burn fossil fuels. Carbon has been locked up for hundreds of millions of years in fossil fuels. It is released as carbon dioxide into the atmosphere when used as fuel. For example:

$$\text{propane} + \text{oxygen} \rightarrow \text{carbon dioxide} + \text{water}$$
$$C_3H_8 + 5O_2 \rightarrow 3CO_2 + 4H_2O$$

Methane gets into the atmosphere from swamps and rice fields. Another source of methane is emissions from the growing number of grazing cattle, and from their decomposing waste. The increasing human population produces more waste to dispose of in landfill sites, which are another source of methane gas.

There is no doubt amongst scientists that the levels of greenhouse gases, especially carbon dioxide, in the atmosphere are increasing. So you are experiencing an enhanced greenhouse effect, greater than the warming effect in pre-industrial times.

Figure 2 shows the data collected by scientists monitoring the proportion of carbon dioxide in the atmosphere at one location. The overall trend over the recent past has been ever upwards.

The balance between the carbon dioxide produced and the carbon dioxide absorbed by 'CO$_2$ sinks', such as tropical rainforests and the oceans, is affected by human activity. As more trees are cut down for timber and to clear land (deforestation), the carbon dioxide removed from the air as the trees photosynthesise is reduced. Also, as the temperature rises, carbon dioxide get less and less soluble in water. This makes the oceans less effective as 'CO$_2$ sinks'.

Weighing up the evidence

Most scientists agree that a trend in global warming has started. Their views are based on evidence presented in scientific journals. Such evidence has to be checked by other scientists working in the same area of expertise. However, a minority of scientists argue that rises observed are due to natural variations that have always happened throughout the long history of the Earth.

Scientists are searching for 'hard' evidence of the link between the levels of carbon dioxide and the climate. One source is ice cores drilled from Greenland's ice sheet, which have gases trapped inside.

Scientists can analyse the trapped air to find how the composition of the gases in the atmosphere has changed over time. Analysis suggests that the current levels of carbon dioxide are higher than at any time in the last 440 000 years. Figure 3 shows changes in temperature and the concentration of CO$_2$ in the atmosphere over the past 150 000 years.

Despite advances in science, you cannot predict with certainty the effects of increasing levels of greenhouse gases – even with the aid of the most powerful computers.

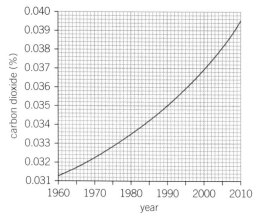

Figure 2 *The change in the levels of carbon dioxide in the atmosphere over time is shown by this graph*

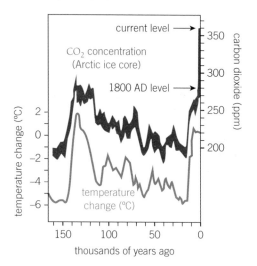

Figure 3 *Changes in temperature and the concentration of CO$_2$ in the atmosphere in the past 150 000 years, using data from gases trapped in ice core samples*

1 a Name three greenhouse gases. [3 marks]
   b Explain how increasing the levels of these gases in the atmosphere gases can result in a rise in the temperature of the Earth's atmosphere. [6 marks]
2 Explain how boiling an electric kettle may increase the amount of carbon dioxide in the Earth's atmosphere. [2 marks]
3 List three reasons why the amount of carbon dioxide in the Earth's atmosphere has increased so much in the recent past. [3 marks]
4 Look at the graph in Figure 2:
   a A closer look at the data would show annual variations to a peak and a trough in each year. Explain these variations. [5 marks]
   b Describe the overall trend shown by the data. [2 marks]
5 a Why should media reports about global warming be treated with caution? [1 mark]
   b Comment on the uncertainty of any conclusions drawn from the data in Figure 3. [4 marks]

Key points

- The amount of carbon dioxide in the Earth's atmosphere has risen in the recent past, largely due to the amount of fossil fuels now burnt.
- It is difficult to predict with complete certainty the effects on climate of rising levels of greenhouse gases on a global scale.
- However, the vast majority of peer-reviewed evidence agrees that increased proportions of greenhouse gases from human activities will increase average global temperatures.

C13.4 Global climate change

Learning objectives

After this topic, you should know:

- how emissions of carbon dioxide and methane can be reduced
- why actions to reduce greenhouse gas emissions may be limited
- how to discuss the scale, risk, and environmental implications of global climate change.

Figure 1 *Hurricanes could become more common in some areas as a consequence of global climate change*

Study tip

There have always been greenhouse gases in the Earth's atmosphere, making the planet warmer than it would otherwise have been. However, it is the rapid increase of the levels of greenhouse gases over the recent past that are enhancing this warming effect. This will have consequences for the global climate.

Synoptic link

To find out about assessing the life cycle of a product in terms of its environmental impact, see Topic C14.5.

Some scientists predict that global warming may mean that the Earth's average temperature could rise by as much as 5.8 °C by the year 2100. This would have a significant effect on weather patterns all over the world.

Consequences of rising levels of greenhouse gases

People are worried about changing global climates. For example, in Europe it has been estimated that winters are already almost two weeks shorter than they were 40 years ago. The changing weather patterns all over the world could have the following consequences:

- rising sea levels, as a result of melting ice caps and expansion of the warmer oceans. For example, the Arctic ice cap appears to be shrinking at a rate equivalent to the Netherlands melting away each year. This may cause the flooding of low-lying land and increased coastal erosion. Some islands could even disappear

- increasingly common extreme weather events, such as more frequent and severe storms

- changes in temperature and the amount, timing and distribution of rainfall. This could have impacts on the food-producing capacity of different regions. People have speculated that dry areas will get even drier and that monsoons in Asia will get heavier. It seems reasonable to assume that as some places get less suited to growing crops, others will become more suited. However, as there is no experience of such dramatic rises in temperature over such short timescales, nobody can yet be sure of the effects in different regions

- changes to the distribution of wildlife species, with some becoming extinct. Rapid changes in the global climate will put ecosystems around the world under stress.

Thinking of solutions

To tackle the problem of global climate change, it is widely agreed that levels of greenhouse gases must be controlled. You have probably heard that you should be thinking of ways to 'reduce your **carbon footprint**'.

The carbon footprint of a product, service, or event is the total amount of carbon dioxide and other greenhouse gases emitted over its full life cycle.

Most of the electricity used in the UK is made by burning fossil fuels. As you know, this releases carbon dioxide into the atmosphere. One solution would be to pump the carbon dioxide produced in fossil fuel power stations deep underground to be absorbed into porous rocks. This could be done in old, redundant oil fields. The technique is called **carbon capture and storage**. It is estimated that this would increase the cost of producing electricity by about 10%.

The methane produced from cattle could be decreased if there was less demand for beef. Plant-based diets offer a more efficient use of land, with farmers using their fields to grow crops and vegetables rather than feed animals.

World leaders meet regularly to negotiate limits on greenhouse emissions. Many governments are taking action, for example, by taxing fossil fuels and cars that burn a lot of petrol or diesel (the so-called 'carbon taxes'), and by funding research into alternative forms of energy. Governments can also support the use of biofuels. Biofuels are often made from plant material that absorbs carbon dioxide during photosynthesis, and effectively just return that to the atmosphere when they are burned, so they can be thought of as 'carbon neutral'. Incentives can also be given to improve home insulation to conserve energy. Companies that produce CO_2 in their processing of materials can offset carbon taxes on their emissions by planting trees. Other policies dictate that whenever trees are felled, new ones are planted to take their place.

Figure 2 *In some areas, people who car share (reducing their carbon footprint) can use special lanes that are less congested on busy roads*

Problems of reducing the carbon footprint

You have already read about the scientific disagreement over the causes and consequences of global climate change. This partly explains the incomplete international co-operation on setting targets for the reduction of greenhouse gas emissions. Reductions will have cost implications in all manufacturing and transport industries. Poorer countries were not originally included in negotiations, as they are not the main contributors to greenhouse gas emissions. However, restrictions could hinder their developing industries.

On a personal level, people need information about global climate change, so that every individual can make a positive contribution. If you can use less electricity, less fossil fuel will be used up and less CO_2 will be released. You can also use your cars less. Walking and cycling will not only make you healthier, but will reduce emissions. If you have to drive, it is more efficient to share lifts, or you could use public transport. You can also recycle your waste. However, to have any effect, more people must start to believe that their small contributions will help!

1 a List three possible consequences of global climate change.
    [3 marks]
  b Describe why these consequences are difficult to predict. [2 marks]
2 a Suggest two problems that representatives of countries face in reaching international agreements. [2 marks]
  b Suggest and explain the relationship between a nation's wealth and its emissions of carbon dioxide. Explain your reasoning. [3 marks]
3 a Explain the process of 'carbon capture and storage' (CCS) and why it could be important. [4 marks]
  b State an advantage and a disadvantage of using CCS. [2 marks]
  c State two other ways in which greenhouse gases can be reduced. [2 marks]
4 Suggest why hydrogen would be a better alternative to carbon-based fuels. Include a balanced symbol equation. [3 marks]

Key points

- Reducing greenhouse gases in the atmosphere relies on reducing the use of fossil fuels, mainly by using alternative sources of energy and conserving energy.
- The economies of developed countries are based on energy obtained from fossil fuels, so changes will cost money to implement.
- However, changes are needed because of the potential risks arising from global climate changes, such as rising sea levels, threats to ecosystems, and different patterns of food production around the world.

C13.5 Atmospheric pollutants

Learning objectives

After this topic, you should know:

- the products of combustion of a fuel, given the composition of the fuel and the conditions in which it is used
- the problems caused by increased amounts of pollutants in the air.

Figure 1 *Pollutants from fossil fuel power stations, especially old coal-fired power stations, can be deposited on land and waterways hundreds of miles away by acid rain and as tiny acidic particles*

Figure 2 *Motor vehicles cause air pollution. Modern engines are improving to meet governmental limits set for levels of different pollutants. However, there is evidence that the information on emissions published by some car manufacturers is unrealistically low. This is because their scientific tests are carried out in laboratories on roller tracks that do not resemble the conditions on real roads where higher levels of pollutants, such as nitrogen oxides, are measured*

Pollution from fuels

All fossil fuels – oil, coal, and natural gas – produce carbon dioxide and water when they burn in plenty of air. But as well as hydrocarbons, these fuels also contain other substances. Impurities containing sulfur found in fuels cause major problems.

All fossil fuels contain at least some sulfur. This reacts with oxygen when a fossil fuel is burned and forms a gas called sulfur dioxide. This acidic gas is toxic. This is bad for the environment, as it is a cause of acid rain which damages trees, as well as killing animal and plant life in lakes. Acid rain also attacks buildings, especially those made of limestone, and metal structures.

The sulfur impurities can be removed from a fuel *before* the fuel is burnt. This happens in petrol and diesel for cars, and in gas-fired power stations. In coal-fired power stations, sulfur dioxide can also be removed from the waste or 'flue' gases by reacting it with basic calcium oxide or calcium hydroxide.

When fuel burns in a car engine, even more pollution can be produced.

- When any fuel containing carbon is burned, it makes carbon dioxide. As discussed in Topic C13.3, carbon dioxide is the main greenhouse gas in the air. It absorbs energy released as radiation from the surface of the Earth. Most scientists think that this is causing global climate change.

- When there is not enough oxygen inside an engine, **incomplete combustion** occurs. Instead of all the carbon in the fuel turning into carbon dioxide, carbon monoxide gas, CO, is also formed. Carbon monoxide is a toxic gas. It is colourless and odourless, so you cannot tell that you are breathing it in. Your red blood cells pick up carbon monoxide and carry it around in your blood instead of oxygen. The carbon monoxide takes up the sites on haemoglobin in the red blood cells that usually bond to oxygen. So a victim of carbon monoxide poisoning will become starved of oxygen, get drowsy, lose consciousness, and then die if not removed from the source of the gas.

- The high temperature inside an engine also allows the normally unreactive nitrogen gas in the air to react with oxygen. This reaction makes **nitrogen oxides**. These are toxic and can trigger some people's asthma. Like sulfur dioxide, nitrogen oxides also cause acid rain.

● Diesel engines burn hydrocarbons with bigger molecules than those in petrol engines. When these large molecules react with oxygen in an engine, they do not always burn completely. Tiny solid particles containing carbon and unburnt hydrocarbons are produced. These **particulates** get carried into the air. They travel into the upper atmosphere, reflecting sunlight back into space, causing **global dimming**. Scientists also think that they may damage the cells in our lungs and even cause cancer.

Figure 3 *A combination of many cars in a small area and the right weather conditions can cause smog to be formed. This is a mixture of SMoke and fOG. Some of the yellowish brown colouration is caused by the presence of nitrogen dioxide gas, NO_2*

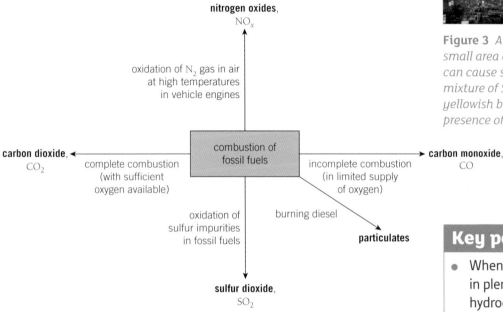

Figure 4 *A summary of the atmospheric pollutants produced when fossil fuels are burned under different conditions*

1 a What are the products of the complete combustion of a hydrocarbon and which environmental problems do they cause? [3 marks]

  b When fossil fuels burn, which element present in impurities can produce sulfur dioxide? [1 mark]

2 a Which pollution problem does sulfur dioxide gas contribute to? [1 mark]

  b Which other non-metal oxides released from cars also cause this pollution problem? [1 mark]

3 How are the following substances produced when fuels burn in vehicles:

  a sulfur dioxide [1 mark]

  b nitrogen oxides [2 marks]

  c particulates. [1 mark]

4 a Natural gas is mainly methane, CH_4. Write a balanced symbol equation for the complete combustion of methane, including state symbols, at the temperature in the flame. [3 marks]

  b When natural gas burns in a faulty gas heater, it can produce carbon monoxide (and water). Write a balanced symbol equation, including state symbols, to show this reaction. [3 marks]

  c Explain why carbon monoxide is so dangerous. [4 marks]

Key points

● When hydrocarbon fuels are burnt in plenty of air, the carbon and hydrogen in the fuel are completely oxidised. They produce carbon dioxide (the main greenhouse gas) and water.

● Sulfur impurities in fuels burn to form sulfur dioxide, which can cause acid rain. Sulfur can be removed from fuels before they are burned, or sulfur dioxide can be removed from flue gas.

● Changing the conditions in which hydrocarbon fuels are burnt can change the products made.

● In insufficient oxygen, poisonous carbon monoxide gas is formed. Particulates of carbon (soot) and unburnt hydrocarbons can also be produced, especially if the fuel is diesel. They can cause global dimming.

● At the high temperatures in engines, nitrogen from the air reacts with oxygen to form oxides of nitrogen. These cause breathing problems and can also cause acid rain.

C13 The Earth's atmosphere

Summary questions

1 The pie charts below show the atmosphere of a planet shortly after it was formed (A) and then millions of years later (B).

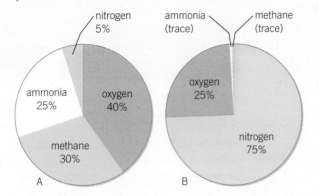

a Describe the changes in the planet's atmosphere over time. [2 marks]

b What might have caused the changes you described in part **a**? [2 marks]

c Copy and complete the word equations showing the chemical reactions that may have taken place in the atmosphere. [1 mark]

  i methane + _____ → carbon dioxide + _____ [1 mark]

  ii ammonia + _____ → nitrogen + _____ [1 mark]

  iii Write both the equations in part **c** as balanced symbol equations. [6 marks]

2 **a i** What gas is given off from fossil fuel power stations that can cause acid rain? [1 mark]

  ii Give *two* ways of stopping this acidic gas getting into the atmosphere. [2 marks]

  iii Name the other cause of acid rain, which comes from car engines, and how it arises. [2 marks]

b State the main reason why levels of carbon dioxide in the Earth's atmosphere have increased so sharply over the past 100 years. [1 mark]

3 Gases that cause global warming are called 'greenhouse gases'.

a Write the formulae of three of these gases and explain how they cause the temperature of the Earth to rise. [6 marks]

b Explain the effect of planting more trees on the levels of carbon dioxide in the Earth's atmosphere. [4 marks]

c Discuss how you can help reduce the levels of carbon dioxide in the air by changing your lifestyles. [4 marks]

d Explain why a minority of scientists are not convinced that the global warming observed in recent years is a result of human pollution of the atmosphere. [2 marks]

4 Core samples have been taken of the ice from Antarctica. The deeper the sample, the longer it has been there. It is possible to date the ice and to take air samples from it. The air was trapped when the ice was formed. It is therefore possible to test samples of air that has been trapped in the ice from many thousands of years ago.

This table shows some of these results. The more recent results are from actual air samples taken from a Pacific island.

Year	CO_2 concentration in ppm	Source
2005	379	Pacific island
1995	360	Pacific island
1985	345	Pacific island
1975	331	Pacific island
1965	320	Antarctica
1955	313	Antarctica
1945	310	Antarctica
1935	309	Antarctica
1925	305	Antarctica
1915	301	Antarctica
1905	297	Antarctica
1895	294	Antarctica
1890	294	Antarctica

ppm = parts per million

a If you have access to a spreadsheet program, enter the data and produce a line graph. [3 marks]

b Draw a line of best fit. [1 mark]

c What pattern can you detect? [3 marks]

d What conclusion can you make? [3 marks]

e Explain how the fact that the data came from two different sources might affect your conclusion. [3 marks]

Practice questions

01 Methane and carbon dioxide are both greenhouse gases. The percentage of carbon dioxide is increasing in the atmosphere.

01.1 Give one human activity that is causing carbon dioxide to increase in the atmosphere. [1 mark]

01.2 Suggest one way in which the amount of methane released into the atmosphere could be reduced. [1 mark]

A lot of carbon dioxide is released into the atmosphere by generating electricity in power stations.

01.3 Suggest one way a government could encourage energy companies to release less carbon dioxide into the atmosphere. [1 mark]

A diesel car produces 88 g of carbon dioxide (CO_2) per kilometre.

01.4 Calculate the number of moles of carbon dioxide formed in a 10 km journey. [3 marks]

Some cars use hydrogen as an alternative fuel to diesel.

01.5 Give the only product when hydrogen is burnt. [1 mark]

01.6 Suggest one problem with using hydrogen as a fuel in a car. [1 mark]

02 Most diesel is obtained by the fractional distillation of crude oil.

$C_{18}H_{38}$ is one of the compounds present in diesel. Biodiesel is made from plants. Plants are grown, harvested and their oil is made into biodiesel.

$C_{16}H_{33}COOCH_3$ is one of the compounds present in biodiesel.

02.1 Give the general formula of $C_{18}H_{38}$. [1 mark]

02.2 Explain why $C_{15}H_{31}COOCH_3$ is not a hydrocarbon. [2 marks]

02.3 Both $C_{18}H_{38}$ and $C_{15}H_{31}COOCH_3$ produce carbon dioxide when they are burnt. Carbon dioxide is a greenhouse gas.

Describe how greenhouse gases like carbon dioxide maintain temperatures on Earth that are high enough to support life.

You should refer to short and long wavelength radiation in your answer. [3 marks]

02.4 Complete the balanced equation for the complete combustion of $C_{18}H_{38}$.

$2C_{18}H_{38} + \ldots\ldots O_2 \rightarrow 36CO_2 + \ldots\ldots H_2O$ [2 marks]

02.5 Why does the equation show the complete combustion of $C_{18}H_{38}$? [2 marks]

02.6 Many scientists believe that increasing levels of greenhouse gases such as carbon dioxide are causing global temperatures to rise.

Name one other greenhouse gas. [1 mark]

02.7 Give **two** effects of increasing global temperatures. [2 marks]

02.8 Use the information provided and your own knowledge to explain why many scientists think that the use of biodiesel as a fuel causes less of an increase in carbon dioxide than the use of diesel as a fuel. [3 marks]

02.9 Carbon footprint is defined as the total amount of carbon dioxide and other greenhouse gases emitted into the atmosphere.

Other than using more biofuels, give **two** ways in which a family might reduce their carbon footprint. [2 marks]

03 The formulas of several products formed from the use of fuels are shown the box.

NO	CO	C	CO_2	SO_2

Choose a product from the box to answer questions **03.1** to **03.5**.

03.1 Which product would be formed from the complete combustion of a hydrocarbon such as C_8H_{18}? [1 mark]

03.2 Which product is a gas formed from the incomplete combustion of a hydrocarbon such as C_8H_{18}? [1 mark]

03.3 Which product is a solid that causes global dimming? [1 mark]

03.4 Which product is a toxic gas? [1 mark]

03.5 Which product is formed in car engines from the reaction of two gases that occur naturally in the air? [1 mark]

Learning objectives

After this topic, you should know:

- examples of natural products that are supplemented or replaced by agricultural and synthetic products
- how to distinguish between finite and renewable resources, given appropriate information
- how to extract and interpret information about resources from charts, graphs, and tables
- how to use orders of magnitude to evaluate the significance of data.

Natural resources	Use	Alterative synthetic product
wool	clothes, carpets	acrylic fibre (polyacrylonitrile), poly(propene)
cotton	clothes, textiles	polyester
silk	clothes	nylon
linseed oil	paint	acrylic resin
rubber	tyres, washers	various synthetic polymers, such as poly(butadiene)
wood	construction	PVC, composites (MDF)

Figure 1 *Brown bauxite is the finite resource from which we extract aluminium metal*

Synoptic link

For more information on orders of magnitude, see Maths skills M2d.

We all rely on the Earth's natural resources to live. They are used to make homes to live in, provide food, for the energy needed to cook, stay warm, and fuel transport. Humankind has found ever more ways to make use of the natural resources in the Earth's crust, oceans, rivers, lakes, and atmosphere.

People have always used natural products, gathered from their environment. The farming of plants and animals has increased the supply of these products. Not only that but chemists have developed synthetic alternatives to these natural products.

You can classify natural resources as finite or renewable.

Finite resources are those that are being used up at a faster rate than they can be replaced. So if we carry on using these resources at current rates, finite resources will eventually run out. Fossil fuels (coal, crude oil, and natural gas) are examples of finite resources.

Renewable resources are those that can be replaced at the same rate at which they are used up. The crops used to make biofuels are examples of renewable resources.

Examples of finite resources

The chemical industry uses natural resources as the raw materials to make new products. Consider the following examples:

- metal ores used to extract metals
- crude oil used to make polymers and petrochemicals
- limestone to make cement and concrete
- crude oil to make the petrol, diesel and kerosene that we use for transport.

Depending on the assumptions made, estimates of how long the finite resources will last differ by orders of magnitude. There are many uncertainties, for example, what will be the future rate of use? How accurately do we know the amounts of finite resources on the Earth? Will new sources be discovered?

Figure 2 shows estimates of how long resources will last based on two predictions. Neither prediction builds in the effect of recycling resources, which is likely to be significant in the future. So predictions like these may not be valid.

Examples of renewable resources

Wherever possible, industries are moving towards renewable resources to conserve finite resources and to improve sustainability. We can think of sustainability as developments that meet the needs of society now, without endangering the ability of future generations to meet their needs.

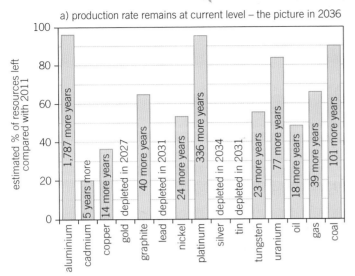
a) production rate remains at current level – the picture in 2036

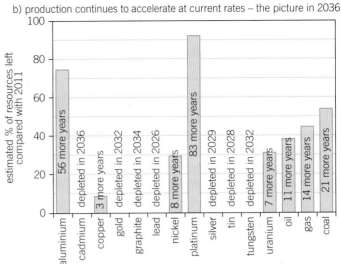
b) production continues to accelerate at current rates – the picture in 2036

Figure 2 *This data shows estimations of the percentage of finite reserves likely to be left in 2036 compared with estimates for 2011 if we assume that: a) we continue to use them at current rates, or b) our rate of use increases at current rates of acceleration*

For example, in the plastics industry, many of the polymers produced use ethene made from crude oil as a starting material. However, ethene can also be made from ethanol, and ethanol can be made by fermenting glucose from sugar cane or sugar beet. So using a renewable crop as the raw material for ethene makes plastics such as poly(ethene) more sustainable than ones using up finite supplies of crude oil.

Another example is the use of wood chips instead of fossil fuels to fuel power stations, linked to a programme of planting new trees.

Synoptic link

To remind yourself about poly(ethene) and how it is made, look back to Topic C11.1.

1 How do finite resources and renewable resources differ? [2 marks]

2 State two examples of:
  a finite resources [2 marks] b renewable resources. [2 marks]

3 a Explain how the raw materials for the manufacture of a polymer such as poly(ethene) can be described as finite or as renewable. [2 marks]
  b Describe why the renewable raw material in part **a** can also be described as the 'sustainable' option. [1 mark]

4 a As a rough estimate, there is 1.5×10^{16} metric tonnes of 'fossil carbon' on Earth. In 2013 it was also estimated that 9.2×10^9 metric tonnes of carbon were burned worldwide in that year. Assuming that the 2013 rate of carbon use was to continue, calculate an order of magnitude estimate of how long 'fossil carbon' will last. [2 marks]
  b However, there are only 5.5×10^{12} metric tonnes of fossil fuels that can be used as a useful resource existing on Earth. Assuming an estimated rate of fossil fuel use of 1×10^{10} metric tonnes per year, calculate an order of magnitude estimation of the time left before the fossil fuel reserves run out. [2 marks]
  c State two reasons why the estimations calculated should only be expressed in terms of order of magnitudes. [2 marks]

Key points

- We rely on the Earth's natural resources to make new products and provide us with energy.
- Some of these natural resources are finite – they will run out eventually if we continue to exploit them, e.g. fossil fuels.
- Others are renewable – they can be replaced as we use them up, e.g. crops used to make biofuels.
- Estimates of the time left before fossil fuels run out can only be rough estimates, because of the uncertainty involved in the calculations.

C14.2 Water safe to drink

Learning objectives

After this topic, you should know:

- the difference between potable water and pure water
- the differences in treatment of ground water and salty water
- how to carry out a simple distillation of salt solution and test the distillate to determine its purity.

Figure 1 *About 97% of the water on Earth is found its oceans, which is not potable water*

Potable water

Water is a vital and useful resource. We use it for agriculture and in industry. It is an important raw material, as a solvent and as a coolant. Other uses of water are for washing and cleaning – and of course, for drinking. Providing people with water that is fit to drink, called potable water, is a major issue all over the world.

You will be familiar with the natural circulation of water around our planet from previous work studying the water cycle. In countries such as the UK, rainwater falls to the ground, replenishing our supplies of fresh water in rivers and lakes. It also seeps down through soil and rocks to underground sources of water called aquifers. Fresh water can be obtained from these porous underground seams of rock by drilling a pipe down to form a water well.

The rainwater itself dissolves some gases from the air as it falls to the ground. Then once in contact with solid land, it will dissolve soluble substances as it passes over them. So water from natural sources will always contain dissolved minerals (salts), as well as microorganisms from soil and decaying matter. The levels of both these impurities must be reduced to meet strict safety standards for drinking water.

The best sources of fresh water contain low levels of minerals and microbes to start with. When water is taken from rivers or reservoirs, made to store

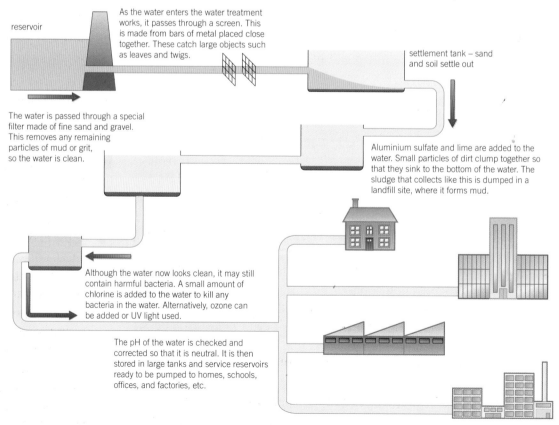

Figure 2 *From freshwater reservoir to end-user – the treatment of water from a reservoir to make potable water*

fresh water, it has to be treated to make it safe to drink. This treatment involves techniques, such as:

- passing the untreated water through filter beds made of sand and gravel to remove solid particles, and
- the addition of chlorine or ozone to sterilise the water by killing microorganisms or, without adding chemical sterilising agents, by passing ultra-violet light through the water.

Purifying salty water

In the UK, there is sufficient rain and natural supplies of fresh water to satisfy the needs of the population and industry. However, in countries with much drier climates and with few sources of natural fresh water, obtaining enough potable water can be difficult. Some of these places have to use sources of water that we would not consider. Water from any source, even seawater or salty water from marshes, can be made pure by distilling it. However, distillation is an expensive process. This is because of the energy costs involved in boiling large volumes of water, even though reduced pressure is used in a desalination plant. Under reduced pressure, water boils below 100 °C, saving on some of the energy costs. This process is called flash distillation.

Converting salty water to potable, useable water is called desalination. Desalination is used in the Middle East in some oil-rich nations, and on some islands with no natural sources of water apart from occasional rainwater. Besides distillation, a process called reverse osmosis can also be used to desalinate water. This uses membranes to separate the water and the salts dissolved in it. The membranes can remove 98% of dissolved salts from seawater. There is no heating involved, so it uses less energy than distillation. However, energy is still needed to pressurise the water passing through, and corrosion of pumps by salty water is also a problem.

1 Water that looks colourless and clear may contain microorganisms. Give three ways in which the water could be made potable. [3 marks]

2 a How can you convert water from a natural source into pure water? [1 mark]
 b How can you test that the water is pure? [1 mark]
 c Why aren't anhydrous copper sulfate or blue cobalt chloride used to test the purity of water? [1 mark]

3 Explain why bottled water sold in the supermarket should not be described as 'pure' water. [3 mark]

4 a Why is a shortage of water a problem for some hot countries, even though they have large coastlines? [1 mark]
 b Define the term 'desalination'. [1 mark]
 c i What is the main disadvantage of desalination using distillation? [1 mark]
  ii Name another process that can be used instead of distillation. [1 mark]

Analysis and purification of water samples

a Your teacher will give you a sample of salty water to test its pH, and another sample to desalinate by distillation (see Topic C1.3).

Using half the sample of distilled water collected, test for pure water by measuring its boiling point. Pure water boils at 100 °C.

Note that chemical tests for water (white anhydrous copper(II) sulfate turns blue or blue cobalt(II) chloride paper turns pink) only test for the presence of water. They do not tell you if the water is pure or not.

Using the other half of your distilled water sample, test its pH value. Record the results of your tests on salty water.

b Now, collect more water samples from different sources and find their pH and whether or not they contain any dissolved solids.

- Record your results in a table.
- Explain how you would ensure any samples collected are representative of that source.

Safety: Wear eye protection.

Key points

- Water is made fit to drink by passing it through filter beds to remove solids and adding chlorine, ozone, or by passing ultra-violet light through it (sterilising) to reduce microbes.
- Water can be purified by distillation, but this requires large amounts of energy, which makes it expensive.
- Reverse osmosis uses membranes to separate dissolved salts from salty water, but this method of desalination also uses energy to make the high pressures needed.

C14.3 Treating waste water

Learning objectives

After this topic, you should know:

- how waste water is made safe to release into the environment
- the relative ease of obtaining potable water from waste, ground and salt water.

Down the drain

Have you ever wondered what happens to all the waste that leaves our homes down the drains? Everything that drains from washing machines, dishwashers, sinks, baths and toilets flows down pipes and enters the larger sewer pipes. All this, along with waste water from businesses and industry is given the general name 'sewage'. This, together with waste water from farming activities, has to be treated at sewage treatment plants to make it safe before it can be returned to the environment, usually into rivers or piped out to sea.

Sewage treatment

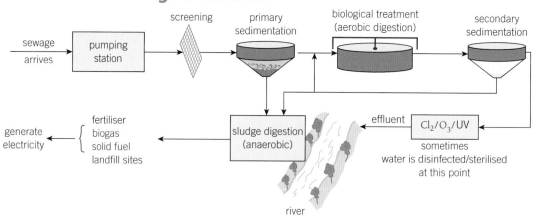

Figure 1 *The steps needed to make our waste water from urban and rural sources safe to return to the environment*

Sewage treatment involves a series of steps, which are described below and in Figure 1.

1   Screening

    Once the sewage arrives at the sewage treatment plant, the first step is to remove large solid objects and grit from the rest of the waste water. The sewage passes through a metal grid that traps the large objects.

2   Primary treatment

    In the first circular tank, the solid sediments are allowed to settle out from the mixture. Large paddles rotate, pushing the solids, called sludge, towards the centre of the tank. There the sludge is piped to a storage tank for further treatment.

    The watery liquid (effluent) above the sludge flows into the next tank. Although no solid matter is visible, this effluent still contains many potentially harmful microorganisms.

3   Secondary treatment

    In the second tank, useful bacteria feed on any remaining organic matter and harmful microorganisms still present, breaking them down aerobically (in the presence of oxygen). The tank is aerated

Figure 2 *A sewage treatment plant*

by bubbling air through the waste water. This can take from several hours to several days, depending on the quality of the waste water, size of the tank, rate of aeration and temperature.

4 Final treatment

In the last tank, the useful bacteria are allowed to settle out to the bottom of the tank as a sediment. The sediment is either recycled back into the secondary treatment tank or passed into the tank where the sludge is treated. At this point, the treated waste water is safe enough to be discharged back into rivers.

However, if the river is a particularly sensitive ecosystem, the water can be filtered one more time through a bed of sand. If necessary, the water can then be sterilised by ultraviolet light or by chlorine. However, the release of chlorine into rivers does cause concern, as toxic organic compounds of chlorine can be formed in the environment.

Treating the sewage sludge

The sludge separated off during the primary treatment of the sewage is not wasted. After further treatment, most can be dried and used as fertiliser on farmland to improve the soil or used as a source of renewable energy.

The sludge contains organic matter, including human waste, suspended solids, water, and dissolved compounds. It is digested anaerobically by microorganisms beneath the surface in the treatment tank.

This biological treatment can be carried out at a relatively high temperature of about 55 °C or a lower temperature of about 35 °C, which can take up to 30 days to complete. The higher temperature has the benefit of speeding up the breakdown of the organic matter, but energy has to be supplied to heat the sludge.

The breakdown products include biogas (a mixture of methane, carbon dioxide, and some hydrogen sulfide). Biogas can be burned and used to power the sewage treatment plant or provide electricity for the surrounding area. It can also be further cleaned to make methane, the main gas in natural gas, and piped into the gas supply.

Alternatively, the sludge can be dried out and turned into a crusty solid 'cake' that can be burnt to generate electricity (Figure 3).

1 Draw a basic flow diagram listing the main steps used in a sewage treatment plant to make waste water safe to discharge into the environment. [4 marks]
2 **a** Describe what takes place in a primary treatment tank. [2 marks]
   b State two uses of sewage sludge. [2 marks]
3 Describe how the processes involving microorganisms in a secondary treatment tank and a sewage sludge tank differ. [2 marks]
4 Using the information here and in the previous Topic C14.2, evaluate the use of waste water, salt water, and ground water from an aquifer as sources of potable water. [6 marks]

Figure 3 *Dried sludge can be used as a renewable energy source, along with biogas and biomethane. All of these are made from sewage*

Study tip

When sewage sludge is dried it takes up a lot less space, so it becomes easier to transport it away from the sewage treatment plant.

Key points

- Waste water requires treatment at a sewage works before being released into the environment.
- Sewage treatment involves the removal of organic matter and harmful microorganisms and chemicals.
- The stages include screening to remove large solids and grit, sedimentation to produce sewage sludge, and aerobic biological treatment of the safe effluent released into environment.
- The sewage sludge is separated, broken down by anaerobic digestion and dried. It can provide us with fertiliser and a source of renewable energy.

C14.4 Extracting metals from ores

Learning objectives

After this topic, you should know:

- how to evaluate alternative biological methods of metal extraction, given appropriate information.

Figure 1 *Mining copper ores can leave huge scars on the landscape. This quarrying of ores is called open-cast mining. About 90% of copper comes from open-cast mines. Our supplies of copper-rich ores are a limited, finite resource*

Extracting copper from copper-rich ores

Most copper is extracted from copper-rich ores. These are a finite resource and are in danger of running out.

There are two main methods used to obtain the copper metal from the ore.

- In one method sulfuric acid is used to produce copper sulfate solution, before extracting the copper metal.
- The other process is called smelting (roasting). Copper ore is heated to a high temperature in a furnace with air, to produce impure copper.

Then we use the impure copper as the positive electrode in electrolysis cells to make pure copper (Figure 3). About 80% of copper is still produced by smelting. Smelting and purifying copper ore uses huge amounts of energy and electricity. This costs a lot of money and will cause pollution of the environment.

Extracting copper from malachite

Malachite is a copper ore containing copper carbonate (harmful). To extract the copper, you first heat the copper carbonate in a boiling tube. **Thermal decomposition** takes place. Copper oxide is left in the tube. Which gas is given off?

You then add dilute sulfuric acid to the copper oxide (harmful). Stopper and shake the tube. This makes copper sulfate solution (harmful). Filter off any excess black copper oxide in the solution.

To extract the copper metal, either:

1 Put an iron nail into the copper sulfate solution. What happens to the iron nail?

or

2 Collect some extra copper sulfate solution and place it in a small beaker. Set up the circuit as shown in Figure 2. Turn the power on until you see copper metal collecting. Which electrode does the copper metal form on?

Safety: Wear eye protection. Chemicals used here are harmful.

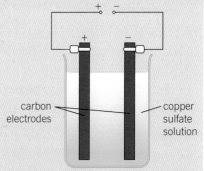

Figure 2 *Extracting copper metal from a solution containing $Cu^{2+}(aq)$ ions using electrolysis. In industry, copper electrodes are used to obtain very pure copper metal at the negative electrode*

Metal ions are always positively charged. Therefore, in electrolysis they are attracted to and deposited at the negative electrode. In industry, the electrolysis of copper is carried out in many cells running at once. This method gives the very pure copper needed to make electrical wiring. Electrolysis is also used to purify the impure copper extracted by smelting.

Look at Figure 3. We can show what happens with the half equations shown below. At the negative copper electrode (cathode) we get reduction:

$$Cu^{2+}(aq) + 2e^- \rightarrow Cu(s)$$

At the positive copper electrode (anode) we get oxidation:

$$Cu(s) \rightarrow Cu^{2+}(aq) + 2e^-$$

The copper can also be extracted from copper sulfate solution in industry by adding scrap iron. Iron can displace copper from its solutions:

iron + copper(II) sulfate → iron(II) sulfate + copper

The ionic equation is:

$$Fe(s) + Cu^{2+}(aq) \rightarrow Fe^{2+}(aq) + Cu(s)$$

Extracting copper from low-grade copper ores

Instead of extracting copper from our limited copper-rich ores, scientists are developing ways to get copper from low-grade ores. This would be uneconomical using traditional methods. New techniques use bacteria (**bioleaching**) or plants (phytomining) to help extract copper.

In phytomining, plants that can absorb copper ions are grown on soil containing low-grade copper ore. This could be on slag heaps of previously discarded waste from the processing of copper-rich ores. Then the plants are burnt and copper is extracted from copper compounds in the ash. The copper ions can be 'leached' (dissolved) from the ash by adding sulfuric acid. This makes a solution (leachate) of copper sulfate. Displacement by scrap iron and then electrolysis make pure copper metal.

In bioleaching, bacteria feed on low-grade metal ores. By a combination of biological and chemical processes, a solution of copper ions (leachate) can be obtained from waste copper ore. Once again, scrap iron and electrolysis is used to extract the copper from the leachate.

About 20% of copper comes from bioleaching. This is likely to increase as sources of copper-rich ores run out. Bioleaching is a slow process, so scientists are researching ways to speed it up. At present it can take years to extract 50% of the metal from a low-grade ore.

1   Why is copper important in our technological society?   [1 mark]
2   a   Give two traditional ways of extracting copper metal.   [2 marks]
    b   State one advantage of extracting copper using bacteria over traditional methods.   [1 mark]
    c   Why can copper occasionally be found native?   [1 mark]
    d   When copper is purified by electrolysis, which electrode does the pure copper collect at? Explain your answer.   [2 marks]
3   Write a balanced chemical equation, including state symbols, for the extraction of copper from copper sulfate solution by displacement.   [3 marks]
4   a   Write half equations for the reactions at each copper electrode in the electrolysis of copper sulfate solution.   [4 marks]
    b   Explain where reduction and oxidation occur in part **a**.   [4 marks]
5   Explain how copper is extracted by phytomining and why this method will become increasingly important.   🕐   [6 marks]

Synoptic links

For information on the half equations at electrodes in electrolysis, see Topic C6.2.

For information on the displacement (redox) reaction between copper ions in solution and iron metal, look back to Topic C5.2.

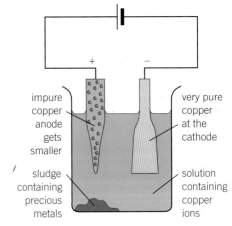

Figure 3 *Many of these cells operate at the same time in industry. The cathodes are removed about every two weeks*

impure copper anode gets smaller

very pure copper at the cathode

sludge containing precious metals

solution containing copper ions

Key points

- Most copper is extracted by smelting (roasting) copper-rich ores, although supplies of ores are becoming scarcer.
- Copper can be extracted from solutions of copper compounds by electrolysis or by displacement using scrap iron. Electrolysis is also used to purify impure copper, e.g. the copper metal obtained from smelting.
- Scientists are developing ways to extract copper that use low-grade copper ores. Bacteria are used in bioleaching and plants in phytomining.

C14.5 Life Cycle Assessments

Learning objectives

After this topic, you should know:

- how to carry out simple comparative Life Cycle Assessments for shopping bags made from plastic and paper
- how to interpret Life Cycle Assessments of materials or products, given appropriate information.

Figure 1 *Think about the raw materials and energy needed to manufacture these electricity cables containing the metals aluminium and copper*

You are probably familiar with the life cycles of animals and plants from studying Biology. However, the same principle of mapping the journey from 'cradle to grave' has now been adapted for manufactured products. The technique used by government agencies, business and industry is called **Life Cycle Assessment** (LCA). It is used to assess the impact on the environment caused by:

- getting and processing the raw materials
- making the product (and any packaging)
- using, reusing, and maintaining the product
- disposing of a product at the end of its useful 'life'.

The total energy needed to extract raw materials, make the product and distribute it, plus any other transport involved, are all taken into account.

An LCA is carried out by:

- listing all the energy and material inputs and all the outputs into the environment
- evaluating the potential environmental impacts from these inputs and outputs
- interpreting the results to help make decisions about using one material, process, product or service over another.

An LCA starts with the process of gathering raw materials needed to make the product. It ends when all the materials are returned to the environment. So an LCA provides an estimate of the total environmental impact resulting from all stages in the product's life cycle.

The outputs back into the environment include atmospheric emissions, waterborne wastes, solid wastes, energy dissipation to the surroundings, and any other products made in the process or product being assessed.

The stages in an LCA can be summarised as:

Raw material extraction → Manufacture → Use/Reuse/ Maintenance → Recycle/Waste management

However, the results of an LCA will always be open to debate. When considering the environmental impact (and health implications), it is common to convert data collected into a single 'impact score'. This requires subjective judgments to be made, usually by the people who paid for the study, an expert panel, or the analyst who devised that particular LCA.

The scale of the impacts is also important. But what weighting should be applied to global impacts (for example, global warming or the depletion of natural resources), compared with regional impacts (for example, acid rain or smog), and local impacts (for example, water loss from ground water or toxic emissions into a stream)? These are not objective scientific judgements.

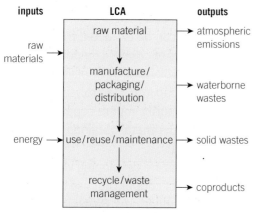

Figure 2 *This shows the inputs and outputs in a Life Cycle Assessment*

Such judgements are also made when trying to quantify all the inputs and outputs listed in the LCA. For example, when judging the impact of

5000 tonnes of sulfur dioxide gas against 2000 tonnes of nitrogen dioxide gas released into the atmosphere, questions to be asked include, what potential impact does the release of each gas have on smog, and on acid rain? What are the effects on asthma sufferers?

Sometimes there is no hard factual data available, so estimations are made, for example, from articles in journals or from a company's published figures. Calculations of data based on assumptions should include an indication of the uncertainty (answer ± uncertainty). Numerical values should only be used in LCAs where widely accepted data is available for energy, water, resources, and wastes.

LCAs will highlight environmental impact and health issues when comparing products or processes, but do not take into account differences in the cost or performance. The best practice should incorporate a peer review process into the LCA to check the data and validity of conclusions drawn. This is especially desirable if an LCA is conducted by the company that makes the product being assessed, and its results are then used to make claims in advertising.

Figure 3 *Using Life Cycle Assessments, should you use a plastic bag or a paper bag when shopping?*

Plastic or paper bags?

Your task is to carry out a simplified Life Cycle Assessment (LCA) for a supermarket that is deciding whether to use plastic, poly(ethene) bags or paper bags at its checkouts (Figure 3).

- List the inputs and outputs in terms of raw materials, energy, and environmental impacts that would have to be considered by the supermarket's management team.
- Try to give the environmental impacts a numerical rating (1 to 10, with 10 having the most serious consequences). What is the problem with this approach?
- Justify the choice you would make from a purely environmental point of view.
- What other considerations might the management team take into account before making the final decision?

1 List the stages in an LCA of a product. [4 marks]

2 Why should an LCA should be carried out on new products? [1 mark]

3 a Name the input shown on an LCA report that would be the raw material mined to produce the aluminium used in an alloy to make the wings of an aeroplane. [1 mark]

 b Name the output shown on an LCA that would be:
  i the greenhouse gas given off when a product is distributed from a factory to shops around the country on lorries [1 mark]
  ii the gas that causes acid rain given off as a result of using electricity generated in a coal-fired power station when making a product. [1 mark]

4 Explain why parts of some LCAs may not be totally objective. ✏ [4 marks]

Synoptic links

For information on crude oil, ethene and poly(ethene), see Topic C9.2, Topic C9.4 and Topic C11.1.

Key points

- Life Cycle Assessments (LCAs) are carried out to assess the environmental impact of products, processes or services.
- They analyse each of the stages of a life cycle, from extracting and processing raw material to disposal at the end of its useful life, including transport and distribution at each stage including all transport and distribution.
- Data is available for the use of energy, water, resources and production of some wastes.
- However, assigning numerical values to the relative effects of pollutants involves subjective judgements, so LCAs using this approach must make this uncertainty clear.

C14.6 Reduce, reuse, and recycle

Learning objectives

After this topic, you should know:

- how using less, reusing and recycling of materials decreases their environmental impact
- how to evaluate ways of reducing the use of limited supplies of metal ores, given appropriate information.

'Reduce, reuse, and recycle' is a message you might have seen in green campaigns asking us to take action to help the environment. The aim of these campaigns is to reduce:

- our use of limited resources
- our use of energy
- the waste we produce.

If we can manage this, then every individual's impact on the environment will be also be reduced.

For example, metals, glass, building materials, clay ceramics and most plastics are all produced from limited supplies of raw materials. To convert raw materials into the finished products requires a lot of energy. Much of the energy used in the processes also comes from limited resources, often fossil fuels.

However, some products, such as glass bottles or car parts, can be reused. Some other products cannot be reused, but can be recycled for a different use. For example, glass bottles can be crushed and melted to make different glass products, and the same applies to cars at the end of their life cycle.

Recycling aluminium

In the UK, each person uses on average around 8 kg of aluminium every year. This is why it is important to **recycle** aluminium to help conserve the Earth's reserves of aluminium ore. Aluminium is extracted from molten aluminium oxide at high temperatures using electrolysis. The process requires huge amounts of electrical energy. Recycling saves energy, and therefore money, since recycling aluminium does not involve electrolysis. When comparing recycled aluminium with aluminium extracted from its ore, there is a 95% energy saving.

Figure 1 *Recycling cans saves energy, as well as our limited supplies of metal ores. It also reduces pollution*

Synoptic link

To remind yourself about the production of aluminium, look back to Topic C6.3.

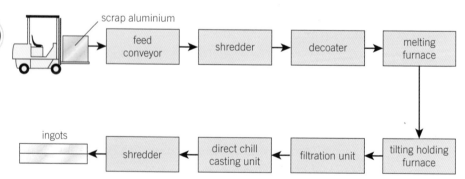

Figure 2 *The recycling of aluminium involves melting the scrap metal, but still uses a lot less energy than extracting aluminium from its ore, bauxite*

Recycling iron and steel

We also recycle iron and steel, for example from the bodywork and engines of scrap cars. 'Tin cans' are another source of scrap iron. These are usually steel cans with a very thin coating of tin to prevent rusting. The cans are easy for waste centres to separate from other domestic rubbish, as they are magnetic.

Using recycled steel saves about 50% of the energy used to extract iron and turn it into steel. Much of the energy in the production of steel from iron ore is supplied by burning fossil fuels, such as natural gas used to heat the air entering a **blast furnace**. Therefore, recycling helps save the dwindling supplies of the **non-renewable** fuels. There are also pollution problems that arise whenever fossil fuels are burned, so using less energy helps reduce these effects. The pollutants produced by fossil fuels include sulfur dioxide, carbon dioxide, carbon monoxide, unburnt hydrocarbons, and particulates.

Recycling copper

Copper is also recycled but the process is more difficult as copper is often alloyed with other metals, for example, in brass the copper is mixed with zinc. Impure copper from recycling has to be purified for use in electrical wiring, unless it has been reclaimed solely from old electricity wires. High-quality copper from wires can be recycled by melting and/or reusing it.

Environmental considerations

Recycling metals reduces the need to mine the metal ore and conserves the Earth's limited reserves of metal ores. It also prevents any pollution problems that arise from extracting the metal from its ore. For example, open-cast mining or quarrying is often used to get copper ore from the ground. The ores of iron and aluminium are also mainly mined like this. Huge pits that scar the landscape are made, creating noise and dust and destroying the habitats of plants and animals. The mines also leave large heaps of waste rock. The water in an area subjected to the mining of metal ores can also be affected. As rain drains through exposed ores and slag heaps of waste, the groundwater can become acidic.

Once ores are mined, they must be processed to extract the metals. For example, sulfide ores are heated to high temperatures in smelting. Any sulfur dioxide gas that escapes into the air will cause acid rain. In the extraction of iron, carbon dioxide is given off, which can contribute to the enhanced greenhouse effect and global warming issues. Reducing these factors is why recycling metals is so important.

1 Give two ways in which a mining company can help the environment when an open-cast mine is no longer economic. [2 marks]

2 Each person in the UK uses about 8 kg of aluminium each year. Recycling 1 kg of aluminium saves about enough energy to run a small electric fire for 14 hours. If you recycle 50% of the aluminium you use in one year, how long could you run a small electric fire on the energy you have saved? [1 mark]

3 Explain why the energy savings are so great when recycling aluminium compared with extracting aluminium from its ore. [3 marks]

4 Explain why it is difficult to recycle some metals, such as copper, and how this can be overcome. [4 marks]

5 Explain how pollution problems are reduced by recycling metals. [3 marks]

Synoptic links

For more information on iron and steels, see Topic C15.2.

For more information about the pollution caused by fossil fuels, look back to Topic C13.5.

Figure 3 *Scrap copper can be recycled as copper alloy, but for many uses it must be pure. Extraction by electrolysis works if the scrap has a high proportion of copper. If the proportion of copper is low, electrolysis only works if pure copper is added to increase the percentage of copper. This makes the process expensive*

Figure 4 *Recycling helps to reduce the pollution problems that result from the energy needed to process metal ores*

Key points

- There are social, economic and environmental issues associated with exploiting the Earth's limited supplies of raw materials, such as metal ores.
- Recycling metals saves energy and our limited, finite metal ores (and fossil fuels). The pollution caused by the mining and extraction of metals is also reduced by recycling.

C14 The Earth's resources

Summary questions

1 In some hot countries, getting sufficient fresh water is difficult. However, countries with large coastlines have plenty of seawater available. They can use desalination plants, such as the one in the photo below. These use a process called 'flash distillation' to turn the salty water into drinking water. Inside the desalination plant, seawater is boiled under reduced pressure, then the water vapour given off is cooled and condensed.

a Why is the pressure reduced before boiling the seawater? How does this keep costs down?   [2 marks]

b Rusting of the steel vessels and pipework in the desalination plant is a big problem. Give two reasons for the rapid rusting.   [2 marks]

c An alternative process uses 'reverse osmosis' to remove the salts from seawater. This passes seawater through a membrane. The latest membranes can remove 98% of the salts from seawater.
Why is reverse osmosis a better option than flash distillation for obtaining drinking water?   [2 marks]

Ⓗ 2 a What name is given to the method of extracting copper from an ore:
   i   using bacteria   [1 mark]
   ii using plants   [1 mark]
   iii by roasting   [1 mark]
   iv using electricity?   [1 mark]

b Which methods in part a are being developed to extract copper from low-grade copper ores?   [2 marks]

c Using the methods named in *both* part a iii *and* part a iv above, explain *in detail* whether copper in the ore is reduced or oxidised. Include half equations in your answer to both methods, assuming the method in part a iii is carried out on copper(I) sulfide.   [6 marks]

Ⓗ 3 Describe the advantages and disadvantages of using bioleaching to extract copper metal. ✔   [6 marks]

4 Scrap car dealers are required to recover 95% of all materials used to make a car. The table shows the metals in an average car:

Material	Average mass in kg	% mass
ferrous metal (steels)	780	68.3
light non-ferrous metal (mainly aluminium)	72	6.3
heavy non-ferrous metal (e.g. lead)	17	1.5

Other materials used include plastics, rubber, and glass.

a What is the average mass of metal in a car?   [1 mark]

b What percentage of a car's mass is made up of *non-metallic* materials?   [1 mark]

c i   What is the main metal found in cars?   [1 mark]
   ii Which of this metal's properties allows it to be separated from other scrap materials?   [1 mark]

5 Pure gold is said to be 24 carats. A carat is a twenty-fourth, so $24 \times \frac{1}{24} = 1$ or pure gold. So a 9-carat gold ring will have $\frac{9}{24}$ gold and $\frac{15}{24}$ of another metal, probably copper or sometimes silver.
How hard the 'gold' is will depend on the amount of gold and on the type of metal used to make the alloy.

Gold alloy in carats	Maximum hardness in BHN
9	170
14	180
18	230
22	90
24	70

BHN = Brinell Hardness Number

a i   In this investigation, which is the independent variable?   [1 mark]
   ii Which type of variable is 'the maximum hardness of the alloy' – continuous or categorical?   [1 mark]
   iii Plot a graph of the results.   [4 marks]
   iv What is the pattern in the results?   [2 marks]

b A Life Cycle Assessment (LCA) was commissioned by a jewellery manufacturer on the 18-carat wedding rings sold.
   i   If the copper used was obtained from ore containing copper(I) sulfide, Cu_2S, what would be an output of the LCA and why would it cause concern?   [2 marks]
   ii Besides the copper ore, state two other inputs into the process that would also cause depletion of natural resources.   [2 marks]

Practice questions

01 This question is about the treatment of water. Some of the stages in the treatment of reservoir water are shown in **Figure 1**.

Figure 1

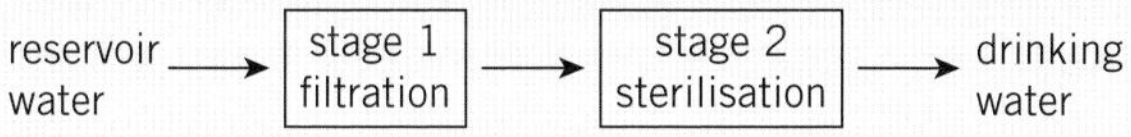

reservoir water → [stage 1 filtration] → [stage 2 sterilisation] → drinking water

01.1 What is removed from the reservoir water during Stage 1?
Tick (✓) one box

dissolved salts	
microbes	
solids	

[1 mark]

01.2 Name one chemical that can be added to sterilise the water in Stage 2. [1 mark]

01.3 Why is it not correct to describe the drinking water in **Figure 1** as pure? [1 mark]

01.4 Some countries have limited supplies of fresh water. Instead they have to treat salty water.
Figure 2 shows one method of obtaining drinking water from salty water.

Figure 2

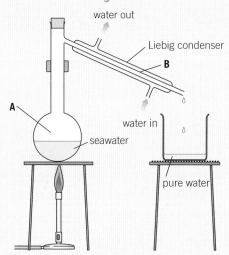

Name this technique. [1 mark]

01.5 Name the change of state that occurs at **A**. [1 mark]

01.6 Name the change of state that occurs at **B**. [1 mark]

01.7 This technique is not often used to produce drinking water from salty water. Suggest why. [1 mark]

01.8 The drinking water produced in **Figure 2** does not contain dissolved solids.

Describe how you could show that this water does not contain dissolved solids without using a chemical test. [2 marks]

02 Shopping carrier bags can be made from poly(ethene) that comes from crude oil, or from corn starch that comes from plants that are grown and harvested.

Table 1 contains information from a Life Cycle Assessment (LCA) comparing the two types of shopping bags.

Table 1

	Shopping bag made from corn starch	Shopping bag made from poly(ethene)
Raw materials	Corn on the cob plants.	Crude oil.
Manufacturing Process	Starch is extracted by reacting the corn with acid at 100 °C. Starch is polymerised at 50 °C at 1 atmosphere pressure. Manufacturing process takes six weeks.	Crude oil is heated to 400 °C and fractionally distilled. Ethene is produced by cracking which involves heating to 800 °C. Ethene is polymerised at 200 °C and 2000 atmospheres pressure. Manufacturing process takes less than one day.
Use during its lifetime	Can be reused until the bag splits or breaks. Will break down after 70 days.	Can be reused until the bag splits or breaks. Will not break down.
Disposal	Can be disposed of in compost or in landfill. Biodegradable. Can be recycled.	Landfill. Non-biodegradable. Difficult to recycle. Can be disposed of by burning.

Use the information in **Table 2** and your own knowledge to compare the advantages and disadvantages of the two types of carrier bag. [6 marks]

Learning objectives

After this topic, you should know:

● how experimental results can be used to show the conditions necessary for rusting

● how to protect iron from rusting.

Synoptic link

To revise the reactions of metals with substances found in the environment, such as oxygen, water, and acids, look back to Topic C5.1.

Figure 1 *Rusting stops machines working properly*

Study tip

Remember that steels are alloys in which the main metal is iron. Stainless steel is an alloy that does not rust (see Topic C15.2).

You will have seen evidence of the corrosion of metal objects when they are left outside without protection. Corrosion is caused by chemical reactions between the metal and substances in the environment. The products of corrosion can affect the strength of a metal, as well its appearance. In extreme cases the metal will be effectively destroyed. The corrosion of iron is called **rusting**. The rusting of iron costs society millions of pounds every year.

Rust forms on the surface of iron (and most steels). Unfortunately, this rust is a soft, crumbly substance. It soon flakes off, exposing fresh iron, so that more iron can rust. This is unlike the protective oxide layer formed on aluminium metal, which effectively protects the aluminium beneath it from further corrosion.

What causes iron to rust?

Set up the test tubes as shown below.

Tube A tests to see if air alone will make iron rust.

Tube B tests to see if water alone will make iron rust.

Tube C tests to see if air and water will make iron rust.

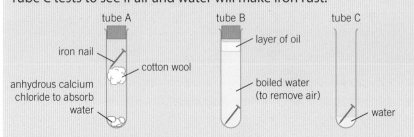

Figure 2 *Investigating the conditions needed for iron to rust*

Leave the tubes for a week.

● What do you observe in each test tube?

Safety: Calcium chloride is an irritant.

The experiment above shows that *both* air *and* water are needed for iron to rust.

Rust is a form of iron(III) oxide, Fe_2O_3. It has water loosely bonded in its structure. It is called hydrated iron(III) oxide.

The reaction can be summarised as:

iron + oxygen + water → hydrated iron(III) oxide

Preventing rust

Air and water are needed for iron to rust. Therefore, if these are kept away from iron, it cannot rust.

This can be done by coating the iron or steel with:

● paint (some types have rust inhibitors added)

● oil or grease

● plastic

- a less reactive metal
- a more reactive metal.

In the last two methods listed, the metal coating can be applied to the iron or steel by dipping the object in molten metal or by electroplating.

Most methods for preventing rust rely on keeping the iron away from air and water. A complete barrier is placed around the iron, usually as a protective coating on its surface. The iron will rust if there is even a tiny gap in the coating. Then the rust soon spreads under the coating.

However, this does not happen if you use a more reactive metal. Even if the coating is scratched, the iron does not rust. Zinc is often used to protect iron. The iron is **galvanised**. The zinc is *more reactive* than the iron. This is because zinc is a stronger reducing agent than iron, so it has a stronger tendency to form positive ions by giving away electrons. As the zinc atoms lose electrons they become oxidised. Therefore, any water or oxygen reacts with the zinc rather than the iron (protecting the iron atoms from oxidation). This is called **sacrificial protection**. The zinc is sacrificed to protect the iron.

Magnesium or aluminium can also be used instead of zinc to sacrificially protect iron. Sacrificial protection is used under harsh conditions, such as when the iron is in contact with seawater, which accelerates rusting. Examples are bars of magnesium connected to the legs of an iron pier or a ship's steel hull. The method is also used to protect underground pipes, which can be attached to the sacrificial metal by wires. The bars of the sacrificial metal eventually need replacing.

Sacrificial protection is also used where the coating is likely to be scratched, such as when emptying commercial wheelie bins. You can clearly see the coating of zinc metal on these bins (Figure 3) and on some lamp posts.

1 Explain in detail how you could show in an experiment that *both* air (oxygen) *and* water are needed for iron to rust. ⊘ [6 marks]

2 a What is the chemical name for rust? [1 mark]
  b Find out why it is difficult to write a chemical formula for rust. [1 mark]

3 a List five ways to prevent iron rusting. [5 marks]
  b Explain which method is best to protect:
    i a bicycle chain [2 marks]
    ii railings [2 marks]
    iii a gas pipeline along the seabed. [2 marks]

4 Plan an investigation you could do in the lab:
  a to show that iron rusts faster in seawater than in fresh water [3 marks]
  b to find out how the concentration of salt (sodium chloride) in water affects the rate of corrosion of iron. ⊘ [5 marks]

Ⓗ 5 A zinc coating is applied to a thin square sheet of steel, with sides of 2.0 m, by dipping it into molten zinc. This method produces a zinc coating of 300 g/m². Calculate the number of moles of zinc used to cover both sides of the steel sheet. [3 marks]

Figure 3 *The zinc coating will protect this bin, even when it gets scratched*

Key points

- Both air (oxygen) and water are needed for iron to rust.
- Providing a barrier between iron and any air (oxygen) and water protects the iron from rusting.
- Sacrificial protection provides protection against rusting, even when the iron is exposed to air and water. The iron needs to be attached to a more reactive metal (zinc, magnesium, or aluminium).

C15.2 Useful alloys

Learning objectives

After this topic, you should know:

- why metals are alloyed
- some examples of common alloys, including steels
- how to interpret and evaluate the composition and uses of alloys, given appropriate information.

Synoptic links

You first looked at the structure of metals in Topic C3.9.
To remind yourself of the structure of alloys, look back to Topic C3.10.

Figure 1 *The Statue of Liberty in New York contains over 80 tonnes of copper*

Figure 3 *An aluminium alloy of aluminium with magnesium and copper, called duralumin, is used to make aeroplanes, because of its strength and its low density for a metal. The copper makes the alloy hard but more susceptible to corrosion, so pure aluminium sheets are used as cladding over the alloy*

Pure metals, such as copper, gold, iron, and aluminium are relatively soft and easily shaped. Their regular layers of positive ions in their giant lattices can slide over each other when forces are applied to the metal.

Making mixtures of metals, called **alloys**, produces more useful materials. In the alloy, differently-sized metal ions (or other atoms, such as carbon) make it harder for the layers to slip – they are jammed in position. This means that alloys are much harder than the pure metals used to make them.

Copper alloys

Bronze was probably the first alloy made by humans, about 5500 years ago. It is usually made by mixing copper with tin. Bronze is used to make statues and decorative items. It is also used to make ship's propellers, because of its toughness and resistance to corrosion.

Brass is made by alloying copper with zinc. Brass is much harder than copper but it is workable. It can be hammered into sheets and pressed into intricate shapes. This property is used to make musical instruments, such as trumpets. It is also used for door fittings and taps.

Figure 2 *Brass (an alloy of copper and zinc) is used in door fittings and taps*

Aluminium alloys

Aluminium has a low density for a metal. It can be alloyed with a wide range of other elements. There are over 300 alloys of aluminium available. These alloys have very different properties. Lightweight but strong aluminium alloys are used to build aircraft, while others can be used as armour plating on tanks and other military vehicles.

Gold alloys

As with copper and aluminium, gold can be made harder by adding other elements. Gold is usually alloyed with copper when it is used to make jewellery. Pure gold wears away more easily than its alloy with copper. By varying the proportions of the two metals, it is also possible to get different shades of 'gold' objects.

The purity of gold is often expressed in 'carats', where 24-carat gold is almost pure gold (99.9%). If you divide the carat number by 24, you get the fraction of gold in your jewellery. So an 18-carat gold ring will contain $\frac{3}{4}$ (75%) gold.

Steels

Steels are alloys of iron with carbon and/or other elements. By carefully controlling the amounts of carbon and other elements, the properties of steels can be changed for their different uses.

Carbon steels

The simplest steels are the **carbon steels**. These are made by removing most of the carbon from the iron obtained from a blast furnace. In steel-making, the carbon content decreases from about 4% to as little as 0.03%. These are the cheapest steels to make. They are used in many products, such as the bodies of cars, machinery, ships, containers, and structural steel.

High carbon steel, with a relatively high carbon content, is very strong but brittle. On the other hand, low carbon steel is soft and easily shaped. It is not as strong, but is much less likely to shatter on impact with a hard object.

Alloy steels

Steels made with 1% to 5% of other metals are more expensive than carbon steels. Each of these metals produces a steel that is well-suited for a particular use.

Nickel-steel alloys are used to make long-span bridges, bicycle chains and military armour-plating. That is because they are very resistant to stretching forces. Tungsten steel operates well under very hot conditions, so it is used to make high-speed tools such as drill bits.

Steels that contain a much higher percentage of other metals are even more expensive. The chromium–nickel steels are known as **stainless steels**. They combine hardness and strength with great resistance to corrosion. Unlike most other steels, they do not rust! These properties make them ideal for use in cooking utensils and cutlery.

Stainless steels are also used in the chemical industry to make reaction vessels. This makes chemical plants expensive to set up, but the pipework and reaction vessels often have to withstand high temperatures and pressures, as well as corrosive chemicals.

1   Describe why copper alloys are more suitable for making coins than pure copper metal, which is easier to stamp patterns on.   [2 marks]

2   **a** Iron is extracted from its ore in a blast furnace. State why iron from a blast furnace is hard but very brittle.   [1 mark]
    b Describe how iron cast directly from a blast furnace differs from pure iron in its composition and properties.   [3 marks]

3   Make a table to summarise the main useful properties of low carbon steel, high carbon steel, and chromium–nickel alloy steel.   [3 marks]

4   **a** Describe why surgical instruments are made from steel containing chromium and nickel.   [2 marks]
    b Explain which type of steel is used to make: [2 marks]
       i car bodies   [2 marks]   **ii** railway tracks.   [2 marks]

5   **a** Explain why aluminium alloys are used in the aircraft industry.   [3 marks]
    b Explain, in terms of structure, why gold is alloyed with copper in many wedding rings.   *(i)*   [6 marks]

Figure 4 *Thin sheets of low carbon steel (containing about 1% carbon, also known as mild steel) are easily pressed into shapes*

Key points

- Alloys are harder than pure metals, because the regular layers in a pure metal are distorted by differently-sized atoms in an alloy.
- Copper, gold and aluminium are all alloyed with other metals to make them harder.
- Pure iron is too soft for it to be very useful.
- Carefully controlled quantities of carbon and other elements are added to iron to make steel alloys with different properties.
- Important examples of steels are:
    - high carbon steels, which are very hard but brittle
    - low carbon steels, which are softer and easily shaped
    - stainless steels, which are resistant to corrosion.

C15.3 The properties of polymers

Learning objectives

After this topic, you should know:

- that the properties of polymers depend on their monomers
- that changing reaction conditions can modify the polymers made
- the differences between thermosetting and thermosoftening polymers.

Figure 1 *The forces between the molecules in poly(ethene) are relatively weak, as there are no strong covalent bonds (cross-links) between the molecules. This means that this plastic softens fairly easily when heated*

Synoptic link

To remind yourself about the bonding in addition polymers such as poly(ethene), look back to Topic C11.1.

Go further

You will find out more about the structures of different polymers, including the proteins and DNA in your body, if you study A Level Chemistry.

As you know, we can make **polymers** from chemicals derived from crude oil. Small molecules called monomers join together to make much bigger molecules called polymers. As the monomers join together, they produce a tangled web of very long chain molecules. Poly(ethene) is an example.

The properties of a polymer depend on:

- the monomers used to make it
- the conditions chosen to carry out the reaction.

Different monomers

Polymer chains can be made from many different monomers. The monomers chosen make a big difference to the properties of the polymer made. Consider the properties of the polymers in the bag in Figure 1 and the electrical socket in Figure 4.

Different reaction conditions

There are two types of poly(ethene). One is called high density (HD) and the other low density (LD) poly(ethene). Both are made from ethene monomers, but they are formed under different reaction conditions.

- Using very high pressures and a trace of oxygen, ethene forms LD poly(ethene). The polymer chains are randomly branched and cannot pack closely together, hence its lower density.
- Using a catalyst at 50 °C and a slightly raised pressure, ethene makes HD poly(ethene). This is made up of straighter poly(ethene) chains. They can pack more closely together than branched chains, hence its higher density of HD poly(ethene). It also has a higher softening temperature and is stronger than LD poly(ethene).

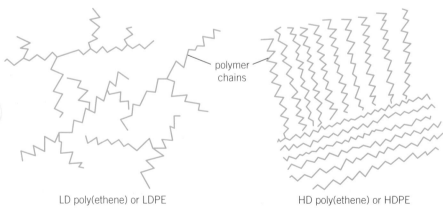

LD poly(ethene) or LDPE   HD poly(ethene) or HDPE

Figure 2 *The branched chains of LD poly(ethene) cannot pack as tightly together as the straighter chains in HD poly(ethene), giving the polymers different properties*

Thermosoftening and thermosetting polymers

You can classify polymers by looking at what happens to them when they are heated. Some will soften quite easily. They will then re-set when they cool down. These are called **thermosoftening polymers**. They are made up of individual polymer chains that are tangled together.

Other polymers do not melt when we heat them. These are called **thermosetting polymers**. These have strong covalent bonds forming 'cross-links' between their polymer chains (Figure 3).

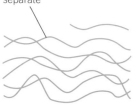
the tangled web of polymer chains are relatively easy to separate

thermosoftening polymer

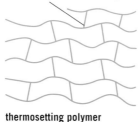
chains fixed together by strong covalent bonds – this is called cross-linking

thermosetting polymer

Figure 3 *Extensive cross-linking by covalent bonds between polymer chains makes a thermosetting plastic that is heat-resistant and rigid*

Forces between polymer chains

In thermosoftening polymers, the forces between the polymer chains are weak. When you heat the polymer, these weak intermolecular forces are broken. The polymer becomes soft. When the polymer cools down, the intermolecular forces bring the polymer molecules back together. Then the polymer hardens again. This type of polymer can be remoulded.

However, thermosetting polymers are different. Their monomers make covalent bonds between the polymer chains when they are first heated in order to shape them. These covalent bonds are strong, and they stop the polymer from softening. The covalent 'cross-links' between chains do not allow them to separate. Even if heated strongly, the polymer will still not soften. Eventually, the polymer will char at high enough temperatures.

1 a Describe how the polymer chains are arranged in a thermosoftening polymer. [1 mark]
  b i Describe the difference between the structures of a thermosoftening and a thermosetting polymer. [2 marks]
    ii Give two differences in the properties of the two types of polymer. [2 marks]
2 Describe why we use thermosetting rather than thermosoftening polymers to make the handles of pans. [2 marks]
3 Polymer A starts to soften at 100 °C, while polymer B softens at 50 °C. Polymer C resists heat, but eventually starts to char if heated to very high temperatures. Explain this, using ideas about intermolecular forces. [3 marks]
4 There are two types of poly(ethene), high density (HD) and low density (LD) poly(ethene).
  a State what is varied to produce the different types of poly(ethene). [1 mark]
  b Explain in detail the differing densities of HDPE and LDPE. ✏ [6 marks]

Modifying a polymer

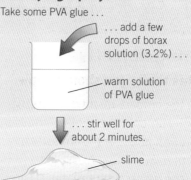
Take some PVA glue . . .
. . . add a few drops of borax solution (3.2%) . . .
warm solution of PVA glue
. . . stir well for about 2 minutes.
slime

The glue becomes slimy because the borax makes the long polymer chains in the glue link together to form a jelly-like substance.

● How could you investigate if the properties of the slime depend on how much borax you add?

Safety: Wear gloves and eye protection. Borax solution should not exceed 8%.

Figure 4 *Electrical sockets are made out of thermosetting polymers. If the plug or wires get hot, the socket will not soften*

Key points

● Monomers affect the properties of the polymers that they produce.
● Changing reaction conditions can also change the properties of the polymer that is produced.
● Thermosoftening polymers will soften or melt easily when heated, because their intermolecular forces are relatively weak. Thermosetting polymers will not soften, because of their 'cross-linking', but will eventually char if heated very strongly.

C15.4 Glass, ceramics, and composites

Learning objectives

After this topic, you should know:

- how to compare quantitatively the physical properties of glass and clay ceramics, polymers, composites, and metals
- how the properties of materials are related to their uses, and how to select appropriate materials.

Glass

Did you know that glass is made mainly from sand? As well as sand (mainly SiO_2), the other raw materials are limestone (mainly $CaCO_3$) and sodium carbonate – soda – (Na_2CO_3). This makes the most common form of glass, called soda-lime glass. Recycled glass is also becoming more important, making up to 30% of some glass-making mixtures.

These raw materials are heated to 1500 °C. At this high temperature, they melt and react to form molten glass. As it cools down, the glass turns into a solid. However, the particles do not form a regular pattern. It is as if the particles in the molten glass are frozen in place.

Look at Figure 1, showing the type of random arrangement in the structure of glass.

There are many different types of glass. These can be made by varying the glass-making mixture or by incorporating other material into the glass (Figure 4). For example, in borosilicate glass, the raw materials are sand and boron trioxide, B_2O_3. This type of glass is used for ovenware (and for test tubes), as it melts at higher temperatures than soda-lime glass.

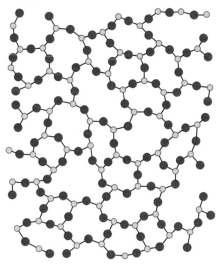

Figure 1 *This diagram represents the disorderly structure of a **glass***

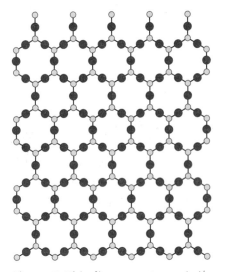

Figure 2 *This diagram represents the orderly giant structure of a **ceramic** material, which has crystalline regions*

Figure 3 *Clay ceramics are common materials used in homes*

Ceramics

Examples of ceramic objects made from clay are bricks, tiles, crockery, bathroom sinks, baths, and toilets. Clay ceramics are hard, but generally brittle, materials that are electrical insulators and resistant to chemical attack. They are made by moulding wet clay into the desired shapes, and then heating them in a furnace to around 1000 °C.

The clay contains compounds of metals (such as aluminium and potassium) and non-metals (silicon and oxygen), with ionic bonding between ions, but also has some covalent bonding between non-metal atoms. These ions and atoms are arranged in giant structures that form layers. When the clay is wet, the water molecules get between the layers of clay and make it slimy (as on a potter's wheel). However, when they are fired in a furnace, the water is driven out and strong bonds form between the layers in the giant structure, changing the properties dramatically. The higher the temperature inside the kiln or furnace, the harder the ceramic formed. Ceramics are brittle because a sharp blow can distort layers in their structure so that ions with like charges are adjacent and repel each other away, cracking the ceramic object.

Composites

Most composites are made of two materials, making a product with improved properties for a particular use. They are often a matrix (or binder) of one material surrounding and binding together fibres or fragments of the other material – a process called reinforcement. For example, you have seen above how glass and ceramics are both hard but brittle. However, a combination of the two, heated together, form a composite glass-ceramic which is hard but also very tough. It is no longer brittle as the glass melts between the crystals in the ceramic, so any cracks cannot spread through the whole structure.

Composites of ceramics with polymers as the binding material are tough and flexible, such as fibreglass. This is made of fine threads of glass embedded in a polymer resin that hardens once moulded into shape. It forms a tough, flexible, waterproof material with a low density, ideal for kayaks. Advanced composites are now being made that use carbon fibres or carbon nanotubes instead of glass fibres, with many new applications possible for tough, lightweight materials that can also be made to conduct electricity.

Other examples of composites include:

- Wood, for example, plywood, which is made of thin sheets of wood glued together, with the grain in successive layers running at right angles to each other. This means that plywood resists splitting along the grain. MDF (medium-density fibreboard) is another wood composite, made from woodchips, shavings, or sawdust compressed together and bound using a polymer resin. This can be cut into intricate shapes without splintering.

- Concrete, made from cement, sand, and gravel (small stones) mixed with water and left to set. Concrete is a very hard composite and is very strong in compression. It can be made more resistant to bending forces by setting it around a matrix of steel rods, forming 'reinforced concrete'.

Figure 4 *There are many different types of glass, including 'self-cleaning' glass, which has a very thin coating of titanium dioxide nanoparticles on its surface. This is an example of a composite material*

Carbon nanotubes can be thought of as rolled-up pieces of graphene. You first learned about graphene in Topic C3.8.

Key points

- Soda glass is made by heating a mixture of sand, limestone, and sodium carbonate. Borosilicate glass is made from sand and boron trioxide, and melts at a higher temperature than soda-lime glass.
- Clay ceramics include pottery and bricks. They are made by shaping wet clay then heating in a furnace.
- Composites are usually made of two materials, with one material acting as a binder for the other material, improving a desirable property that neither of the original materials could offer alone.

1 **a** Name the raw material that is common to both soda-glass and borosilicate glass. [1 mark]
   b i State two properties common to both types of glass. [2 marks]
   ii State one way in which they differ. [1 mark]

2 In some parts of the world homes are built from cob – a mixture of sandy sub-soil, clay, and straw. Water was added and the mixture was stamped on to mix it up before building the walls. Would you describe cob as a glass, a ceramic, a polymer, or a composite? [1 mark]

3 Explain how firing wet clay in a kiln can make a piece of pottery in terms of its structure. [3 marks]

4 Design a test to find out if a traditional fibreglass is more flexible than a more modern composite using carbon fibres instead of glass fibres embedded in a polymer resin. [5 marks]

C15.5 Making ammonia – the Haber process

Learning objectives

After this topic, you should know:

- why nitrogen-based fertilisers are needed to improve crop yields
- why ammonia is an important compound
- the raw materials and conditions used to manufacture ammonia.

Figure 1 *Plants are surrounded by nitrogen in the air. They cannot use this nitrogen, and rely on soluble nitrates in the soil instead. We supply these to crops by spreading fertiliser on the soil*

Nitrogen-based fertilisers

Each year, farmers add millions of tonnes of fertilisers to the soil to replenish nutrients needed by their crops. Most fertilisers, such as ammonium nitrate, NH_4NO_3, contain nitrogen.

Plants need nitrogen to grow, as it is one of the elements they need to make proteins. In nature there is a natural cycling of nitrogen, but this gets disturbed by farming crops. As the crop plants grow, they take in nitrogen in the form of soluble nitrate ions, $NO_3^-(aq)$, from the soil through their roots. Then the farmer harvests the crop, so most parts of the plant are not allowed to rot back into the soil. So the nitrogen absorbed from the soil during growth is not all replaced by the natural cycling of nitrogen. Therefore farmers need fertilisers to replace nitrogen and other nutrients in the soil before they sow their next crops.

Almost 80% of the air is nitrogen gas, so you might think that plants have plenty of nitrogen available. However, the gas is insoluble in water and most plants can only absorb a soluble form of nitrogen. Turning nitrogen gas from the air into nitrogen compounds that plants can absorb in solution through their roots is called 'fixing' nitrogen. The first step is to change nitrogen gas into ammonia, NH_3, in the Haber process.

The Haber process

The Haber process provides a way of turning nitrogen in the air into ammonia. The process is named after the German chemist, Fritz Haber. He first devised the reaction to make ammonia from its elements (nitrogen and hydrogen) about one hundred years ago.

Ammonia has many different uses. The most important of these is to make fertilisers.

The raw materials for the production of ammonia are:

- nitrogen from the air
- hydrogen, mainly from natural gas (which contains methane, CH_4).

The nitrogen and hydrogen are purified. Then they are passed over an iron catalyst at a high temperature (about 450 °C) and a high pressure (about 200 atmospheres). The product of this reversible reaction is ammonia.

- The reaction used in the Haber process is reversible. This means that the ammonia gas made breaks down again into nitrogen and hydrogen.
- The ammonia is removed by cooling the gases so that the ammonia liquefies. It can then be separated from the unreacted nitrogen gas and hydrogen gas.

● The unreacted nitrogen and hydrogen gases are recycled back into the reaction mixture. They are then re-compressed and heated before returning to the reaction vessel. There they have a chance to react again on the surface of the iron catalyst.

$$\text{nitrogen} \; + \; \text{hydrogen} \quad \underset{\text{iron catalyst}}{\rightleftharpoons} \quad \text{ammonia}$$

$$N_2(g) \; + \; 3H_2(g) \quad \rightleftharpoons \quad 2NH_3(g)$$

The Haber process is carried out under carefully chosen conditions to give a reasonable yield of ammonia as quickly as possible. You will look at how these conditions are chosen in Topic C15.6.

Synoptic link

To remind yourself about reversible reactions, look back to Topic C8.6 and Topic C8.7.

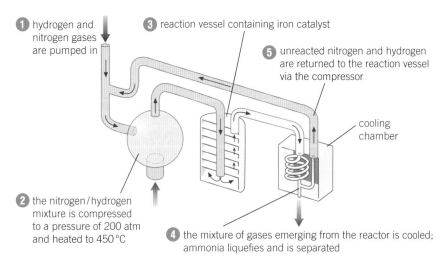

① hydrogen and nitrogen gases are pumped in

③ reaction vessel containing iron catalyst

⑤ unreacted nitrogen and hydrogen are returned to the reaction vessel via the compressor

cooling chamber

② the nitrogen/hydrogen mixture is compressed to a pressure of 200 atm and heated to 450 °C

④ the mixture of gases emerging from the reactor is cooled; ammonia liquefies and is separated

Figure 2 *The Haber process*

1   What are the main raw materials used to get the nitrogen gas and hydrogen gas needed for the production of ammonia in industry?
[1 mark]

2   a   Write a word equation and balanced symbol equation, including state symbols, for the production of ammonia in the Haber process. [3 marks]
    b   State the conditions chosen for the Haber process, including the catalyst used. [3 marks]

3   a   How is ammonia separated from the reaction mixture in the Haber process? [2 marks]
    b   What happens to unreacted nitrogen and hydrogen that leaves the reaction vessel? [1 mark]

4   Draw a flowchart to show how the Haber process is used to make ammonia. [4 marks]

5   a   What is the atom economy of the Haber process? [1 mark]
    b   Explain why your answer to part **a** makes the Haber process more sustainable than many other industrial processes. [3 marks]
    c   If you were to complete a Life Cycle Assessment (LCA) for the Haber process, explain in detail why the process would have some negative impact on the environment. Read the introduction to Topic C15.6 to help you answer this question in more detail. ⬤
[6 marks]

Key points

● Ammonia is an important chemical for making other products, including fertilisers.
● Ammonia is made from nitrogen and hydrogen in the Haber process.
● We carry out the Haber process under conditions of about 450 °C and 200 atmospheres pressure, using an iron catalyst.
● Any unreacted nitrogen and hydrogen are recycled back into the reaction vessel in the Haber process.

C15.6 The economics of the Haber process

Learning objectives

After this topic, you should know:

- how the commercially used conditions for the Haber process are related to the availability and cost of raw materials and energy supplies, control of equilibrium position, and rate.

Synoptic links

Look back at Topic C8.8 and Topic C8.9 to revise chemical equilibrium and the effect of reaction conditions on equilibrium.

Figure 1 *It is very expensive to build chemical plants that operate at very high pressures*

You have seen in Topic C15.5 how ammonia is made in the Haber process. The nitrogen gas needed for the process is extracted from air. This is a free resource but there are energy costs in separating the nitrogen from the other gases in the air. This is done by the fractional distillation of liquid air. The cooling down of the air to temperatures of about −200 °C to liquefy it requires energy. In the cooling process, air needs to be compressed by high-pressure pumps which are expensive to run.

The hydrogen for the process is made by reacting methane gas with steam at very high temperatures:

methane + steam → hydrogen + carbon monoxide

The water is cheap but there are costs involved in heating the reaction mixture to make super-heated steam However, the main cost in making ammonia is the price of methane gas. This has to be bought from the gas industry as natural gas (a diminishing fossil fuel).

The methane, or the hydrogen it produces, also provides another way to obtain nitrogen for the Haber process. The flammable gases are mixed with air and react with oxygen in a reaction vessel. This removes the oxygen from air, leaving mainly nitrogen gas.

The effect of pressure

Nitrogen and hydrogen react to make ammonia in a reversible reaction:

$$N_2(g) + 3H_2(g) \rightleftharpoons 2NH_3(g)$$

As the balanced equation above shows, there are four molecules of gas on the left-hand side of the equation (N_2 and $3H_2$), but on the right-hand side there are only two molecules of gas ($2NH_3$). This means that the volume of the reactants is greater than the volume of the products. So, an increase in pressure will tend to shift the position of equilibrium to the right, producing more ammonia in order to reduce the pressure. This is an application of Le Chatelier's Principle.

To get the maximum possible yield of ammonia, you would need to make the pressure as high as possible. But very high pressures need lots of energy to compress the gases. Very high pressures also need expensive reaction vessels and pipes. They have to be strong enough to withstand very high pressures, otherwise there is always the danger of an explosion. To avoid the higher costs of building a stronger chemical plant, the Haber process uses a pressure of 200 atmospheres. This is a good compromise. This pressure gives a lower yield than it would with even higher pressures, but it reduces cost and helps produce a reasonable rate of reaction between the gases.

The effect of temperature

The effect of temperature on the Haber process is more complicated than the effect of pressure. The forward reaction is exothermic:

$$N_2(g) + 3H_2(g) \rightleftharpoons 2NH_3(g)$$

Energy transferred to the surroundings = 93 kJ = 46.5 kJ/mol of ammonia produced.

Lowering the temperature would increase the amount of ammonia in the reaction mixture at equilibrium. This happens because the forward reaction to form ammonia transfers energy to the surroundings, thereby raising the temperature of the surroundings (opposing the change introduced).

However, at a low temperature, the rate of the reaction would be very slow, because the gas molecules would collide less frequently and less energetically. Running a chemical plant is expensive, so to make ammonia commercially, the reaction needs to go quickly.

This is achieved in the Haber process by another compromise. A reasonably high temperature is used to get the reaction going at a reasonable rate, even though this reduces the yield of ammonia. Look at the graph in Figure 2.

A lower temperature would also reduce the effectiveness of the iron catalyst.

The effect of a catalyst
An iron catalyst is also used in the Haber process to speed up the reaction. The catalyst speeds up the rate of both the forward and reverse reactions by the same amount. Therefore it does not affect the actual yield of ammonia, but it does cause ammonia to be produced more quickly, which is an important economic consideration in industry.

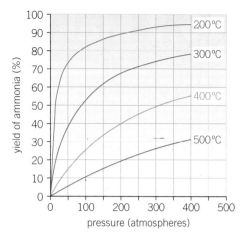

Figure 2 *The conditions for the Haber process are a compromise between getting a reasonable yield of ammonia and getting the reaction to take place at a fast enough rate*

1 Why is an iron catalyst used in the Haber process? [1 mark]

2 Explain the effect of increasing the pressure in the Haber process on:
   a the yield of ammonia [2 marks]
   b the rate of production of ammonia. [3 marks]

3 Look at Figure 2.
   a What is the approximate yield of ammonia at a temperature of 500 °C and 400 atmospheres pressure? [1 mark]
   b What is the approximate yield of ammonia at a temperature of 500 °C and 100 atmospheres pressure? [1 mark]
   c What is the approximate yield of ammonia at a temperature of 200 °C and 400 atmospheres pressure? [1 mark]
   d What is the approximate yield of ammonia at a temperature of 200 °C and 100 atmospheres pressure? [1 mark]
   e Given your answers to parts **a–d**, explain why the Haber process is carried out at around 200 atmospheres and 450 °C. [3 marks]

4 a Assuming 100% conversion of nitrogen gas and hydrogen gas into ammonia, calculate the maximum volume of ammonia that could be produced from 1200 dm³ of hydrogen, measured at room temperature and pressure. (1 mole of any gas occupies 24 dm³ at room temperature and pressure.) [2 marks]
   b Using your answer to part **a**, calculate the mass of ammonia that could be produced at room temperature and pressure. [3 marks]
   c If the actual yield of ammonia was 113 g, calculate the percentage yield. [2 marks]

Key points

- The Haber process uses a pressure of around 200 atmospheres to increase the amount of ammonia produced.
- Although higher pressures would produce higher yields of ammonia, they would make the chemical plant too expensive to build and run.
- A temperature of about 450 °C is used for the reaction. Although lower temperatures would increase the yield of ammonia, it would be produced too slowly.

C15.7 Making fertilisers in the lab

Learning objectives

After this topic, you should know:

- how ammonia can be neutralised by acids to make fertilisers
- how to prepare a fertiliser in the laboratory.

Ammonia and fertilisers

The ammonia made in the Haber process can be used as a fertiliser itself in liquid form. However, most is changed into compounds of ammonia. These compounds have advantages over ammonia as fertilisers. About 10% of the ammonia made in the Haber process is converted in another process into nitric acid. Then the nitric acid can react with ammonia to make ammonium nitrate fertiliser:

$$\text{ammonia} + \text{nitric acid} \rightarrow \text{ammonium nitrate}$$
$$NH_3(aq) + HNO_3(aq) \rightarrow NH_4NO_3(aq)$$

We can make other solid fertiliser salts by reacting ammonia (an alkali) with different acids. Look at the flow diagram in Figure 1 below for the manufacture of fertilisers by reacting ammonia with different acids:

Study tip

Remember that most plants cannot use nitrogen gas directly from the air as a nutrient source – they have to take it in from the soil in a soluble form.

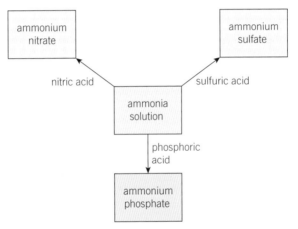

Figure 1 *Ammonia can be made into different ammonium salts that can act as fertilisers*

Here is the neutralisation reaction between ammonia solution and phosphoric acid:

$$\text{ammonia} + \text{phosphoric acid} \rightarrow \text{ammonium phosphate}$$
$$3NH_3(aq) + H_3PO_4(aq) \rightarrow (NH_4)_3PO_4(aq)$$

Remember that salts are formed when acids and alkalis (such as ammonia solution) react together in a **neutralisation** reaction. When sulfuric acid neutralises ammonia solution, the salt made is called ammonium sulfate:

$$\text{ammonia} + \text{sulfuric acid} \rightarrow \text{ammonium sulfate}$$
$$2NH_3(aq) + H_2SO_4(aq) \rightarrow (NH_4)_2SO_4(aq)$$

You can make this fertiliser in the lab using the titration of dilute sulfuric acid against ammonia solution:

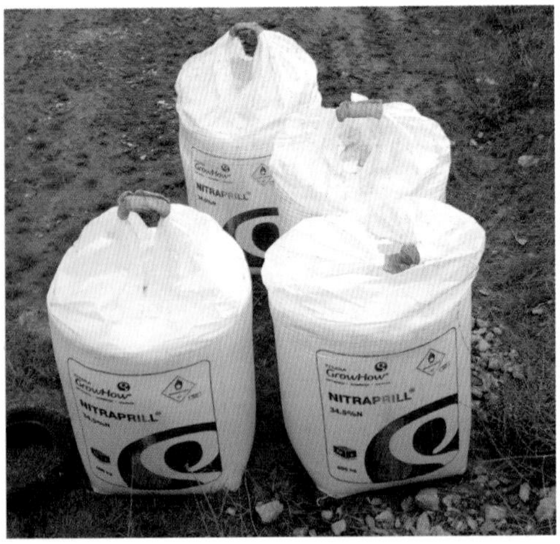

Figure 2 *This ammonium nitrate fertiliser puts nitrogen back into the a soil*

Making ammonium sulfate fertiliser in the lab

Collect 25 cm³ of ammonia solution in a small conical flask. Use a pipette and filler to measure this accurately.

Add dilute sulfuric acid, 1 cm³ at a time, from a burette. After adding each cm³ of acid, swirl your flask (Figure 3).

Dip a glass rod into the solution. Then test a drop of the solution on a small piece of blue litmus paper on a spotting tile.

Keep adding acid until the litmus just turns pink. Repeat the titration until you get two volumes of sulfuric acid within 0.1 cm³ of each other. These are called concordant results.

● How much sulfuric acid did you need to neutralise the ammonia solution?

Then pour the solution into an evaporating dish. Heat it on a water bath until about half of the water from the solution has evaporated off. (Do not let it boil dry.)

Leave the rest of the solution to evaporate off slowly to leave crystals of ammonium sulfate. Alternatively, before all the water has evaporated off, you can filter off any crystals and dab them dry with another piece of filter paper. Then leave them to dry under a clean piece of filter paper.

● Plan an investigation to see if your ammonium sulfate affects the growth of a seedling.

Safety: Wear eye protection

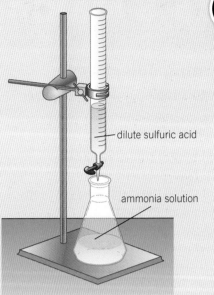

dilute sulfuric acid

ammonia solution

Figure 3 *Titrating ammonia solution against dilute sulfuric acid*

Figure 4 *Evaporating ammonium sulfate solution to the point of crystallisation (when the solution becomes saturated)*

---

1 Name and give the formula of the fertiliser made from:
   a ammonia and nitric acid [2 marks]
   b ammonia and sulfuric acid [2 marks]
   c ammonia and phosphoric acid. [2 marks]
2 Draw a 2D scientific diagram to show the titration in Figure 3. [2 marks]
3 Write the word equation and balanced symbol equation for the reaction of ammonia solution with dilute nitric acid. Include state symbols in your answer [2 marks]
4 a Write a balanced symbol equation, including state symbols for the reaction between ammonia solution and dilute sulfuric acid. [2 marks]
   b Name the type of reaction in part a. [1 mark]
   c Describe how to collect pure, dry crystals of ammonium sulfate from its solution. [3 marks]
H 5 In a titration, a student found that it takes 15.15 cm³ of 0.20 mol/dm³ dilute sulfuric acid to neutralise 25.0 cm³ of ammonia solution. Calculate the concentration of the ammonia solution. [3 marks]

Synoptic link

To remind yourself about the practical technique of titration, look back to Topic C4.7.

Key points

● Ammonia is used to make nitric acid.
● The nitric acid made can then be reacted with more ammonia to make ammonium nitrate fertiliser.
● Ammonia can also be neutralised by sulfuric acid to make ammonium sulfate fertiliser, and with phosphoric acid to make ammonium phosphate fertiliser.

C15.8 Making fertilisers in industry

Learning objectives

After this topic, you should know:

- compounds of nitrogen, phosphorus and potassium are used as fertilisers to improve crop production
- the processes involved in obtaining the compounds used in manufactured fertilisers
- how to compare the industrial production of fertilisers with laboratory preparations of the same compounds, given appropriate information.

Figure 1 *This fertiliser provides all three of the essential plant 'macro-nutrients'. The numbers 21 : 8 : 11 indicate the proportions of the elements in the order N, then P, and finally K – so this fertiliser provides lots of nitrogen as well as some phosphorus and potassium. Plants also need other 'micro-nutrients' in smaller quantities to remain healthy, such as magnesium and zinc*

Figure 2 *This potash mine in Belarus supplies potassium compounds to make the formulations used in NPK fertilisers*

Making NPK fertilisers in industry

As well as nitrogen, N, crops also need significant amounts of the nutrients phosphorus, P, and potassium, K, for healthy growth. Farmers can buy fertilisers that provide compounds of nitrogen, phosphorus and potassium. No compound containing all three elements has yet been made for use as a fertiliser. Therefore, bags of NPK fertiliser contain formulations of compounds to provide all three of the 'macro-nutrients'.

The sources of phosphorus are deposits of phosphate-containing rock, which is dug or mined from the ground. It cannot be used directly on the soil, as it is insoluble in water, so the rock is treated with acids to make fertiliser salts. Phosphate rock is treated:

- with nitric acid to produce phosphoric acid, H_3PO_4, and calcium nitrate $Ca(NO_3)_2$. Then the phosphoric acid is neutralised with ammonia to produce ammonium phosphate, $(NH_4)_3PO_4$
- with sulfuric acid to produce single superphosphate, a mixture of calcium phosphate, $Ca_3(PO_4)_2$, and calcium sulfate, $CaSO_4$
- with phosphoric acid to produce triple superphosphate, which is calcium phosphate, $Ca_3(PO_4)_2$.

The potassium salts potassium chloride, KCl, and potassium sulfate, K_2SO_4, are mined from the ground (all potassium compounds are soluble in water, so they can be separated from impurities and used directly)

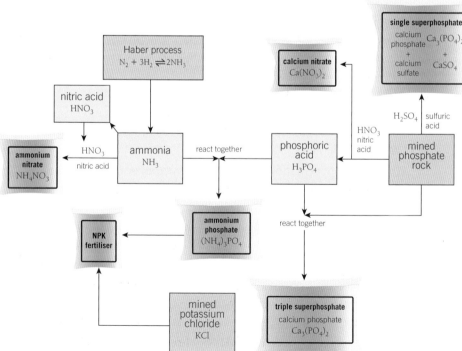

Figure 3 *A summary of the integrated processes used to manufacture fertilisers*

Comparing the production of fertiliser

In your experiment to make ammonium sulfate in Topic C15.7:

- the ammonia solution and sulfuric acid solutions used were provided for you and added together in relatively small quantities, to make a small 'batch' of ammonium sulfate crystals
- the concentration of the solutions used were relatively dilute for safety reasons
- the apparatus used was made of glass
- the titration was carried out slowly and carefully until the reaction was just completed
- the crystals of ammonium sulfate were slowly crystallised and collected (to test their effectiveness as a fertiliser).

Contrast that with the manufacture of ammonium sulfate fertiliser on an industrial scale:

- the process starts with the raw materials to make ammonia and sulfuric acid, which are needed on a large scale (tonnes) – Figure 2 shows the many interrelated reactions and products made at a fertiliser plant
- having made the ammonia and sulfuric acid, they can undergo neutralisation to make ammonium sulfate, in an adjoining part of the fertiliser plant.

In a continuous process, the reactants are piped into large reaction towers, where concentrated sulfuric acid is sprayed into anhydrous ammonia gas at 60 °C. This is a potentially hazardous operation (there have been serious explosions at fertiliser factories, killing many people)

- the pipes and reaction vessels need to be made of strong stainless steel to withstand high pressures and the corrosive nature of many of the raw materials, reactants and product
- the energy transferred in the exothermic reaction is used to heat a neighbouring tower. Here the slurry (paste) of ammonium sulfate is heated, with air blown up through it from the base, to make granules (pellets) of the fertiliser. These granules have to be of equal size to aid their spreading on fields. (The needle-like crystals of ammonium sulfate tend to clump together and can block nozzles of spraying equipment.)

Draw a table to compare the industrial manufacture of ammonium sulfate with its laboratory preparation. In your table comment on scale, safety, equipment needed, reaction conditions and method used to obtain solid ammonium sulfate.

---

1. **a** Name and give the chemical formula of the fertiliser produced when phosphate rock is treated:
   - **i** directly with nitric acid [2 marks]
   - **ii** with phosphoric acid [2 marks]
   b Name and write the formula of:
   - **i** two potassium compounds that are used as fertilisers [2 marks]
   - **ii** the two compounds found in 'single superphosphate' fertiliser. [2 marks]

2. **a** Using the information in Figure 3, name two elements, besides phosphorus, that must be present in phosphate rock. [2 marks]
   b Write the formula of a phosphate ion. [1 mark]
   H c Write a balanced symbol equation, including state symbols, to show the complete ionisation of dilute phosphoric acid, H_3PO_4. [2 marks]

3. State one advantage that using the fertiliser ammonium phosphate has over using ammonium nitrate. [1 mark]

H 4 Calculate the percentage of nitrogen in each of the fertilisers urea, $CO(NH_2)_2$, ammonium sulfate and ammonium nitrate, to decide which provides most nitrogen to the soil for a given mass of fertiliser. Give your percentages to three significant figures. [6 marks]

Key points

- Fertilisers are used to supply nitrogen, phosphorus and potassium to plants. These can all be added to the land at the same time in mixtures of compounds called NPK fertilisers.
- The nitrogen comes from ammonia, made in the Haber process, which is reacted with acids to make fertilisers, such as ammonium nitrate and ammonium sulfate.
- The source of phosphorus is phosphate rock, which is mined and then treated with acids to form fertilisers, such as ammonium phosphate and calcium phosphate.
- The potassium comes from potassium salts mined from the ground for use as fertilisers, such as potassium chloride and potassium sulfate.

Summary questions

1 The experiment below shows an investigation into the factors that are needed for iron to rust.

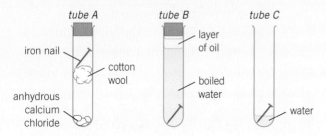

tube A   tube B   tube C

iron nail, cotton wool, anhydrous calcium chloride, layer of oil, boiled water, water

a Which test tube shows an iron nail in contact with:
  i only air [1 mark]
  ii water and air [1 mark]
  iii only water? [1 mark]

b Why is the water in tube B boiled? [1 mark]

c Explain why the nail in tube A is suspended on cotton wool above the anhydrous calcium chloride in the bottom of the tube. [1 mark]

d What will be the conclusion drawn from this experiment? [1 mark]

e Car owners who live near the coast often complain that their cars have more problems with rust than those in other places.
Which chemical compound causes their cars to rust more quickly? [1 mark]

f Describe a simple experiment that you could do to show that the substance in part **e** makes an iron nail rust more quickly. [4 marks]

g Explain the method used to prevent rusting in:
  i cogs in a machine [3 marks]
  ii food cans [2 marks]
  iii underground pipes. [4 marks]

2 Most alkenes are used to make polymers.

a State two features of polymer molecules that alkenes do not have. [2 marks]

b Poly(ethene) is made in two forms, depending on the conditions used for polymerisation. These are described as LD and HD forms.
  i State the meanings of the terms LD and HD. [1 mark]
  ii Both forms are described as *thermosoftening*. What is the meaning of this term? [1 mark]

c Another type of polymer is described as thermosetting.
  i How does a thermosetting polymer behave when heated? [1 mark]

  ii Describe how the bonding in a thermosetting polymer differs from the bonding in a thermosoftening polymer. [1 mark]

H 3 The table below gives data on the yield of ammonia at equilibrium in the reversible reaction with nitrogen and hydrogen:

Percentage of ammonia present at equilibrium					
Pressure in atmospheres	Temperature in °C				
	100	200	300	400	500
10	88	51	15	4	1
25	92	64	27	9	3
50	94	74	39	15	6
100	97	82	52	25	11
200	98	89	67	39	18
400	99	95	80	55	32
1000	99.9	98	93	80	57

a i What is the effect of increasing the pressure on the yield of ammonia? [1 mark]
  ii What is the effect of increasing the temperature on the yield of ammonia? [1 mark]
  iii Explain how you used the table to judge the effect of changing the temperature and the pressure on the yield of ammonia. [2 marks]

b Explain why the conditions in the Haber process (200 atmospheres pressure and a temperature of 450 °C) are described as a compromise. [6 marks]

c How does the use of an iron catalyst affect the yield of ammonia? [1 mark]

4 Here is the equation for the reaction used in the Haber process:

$$N_2(g) + 3H_2(g) \xrightleftharpoons{\text{Fe(s) catalyst}} 2NH_3(g)$$

Energy transferred to the surroundings is 92 kJ in the forward reaction.

a Draw a reaction profile diagram for this reaction, showing the reaction pathway with and without the use of the iron catalyst. [2 marks]

H b In 1909 Fritz Haber demonstrated his method to make ammonia on a laboratory scale. He could make 125.0 cm³ of ammonia per hour. Using this rate, calculate the mass of ammonia he could make in one whole day, assuming volume measurements were taken at room temperature and pressure.
(1 mole of gas occupies 24 dm³ at r.t.p.) [3 marks]

Practice questions

01 This question is about iron.

Iron can be produced by reacting iron(III) oxide with carbon:

$$2Fe_2O_3(s) + 3C(s) \rightarrow 3CO_2(g) + 4Fe(s)$$

01.1 Explain why iron can be produced by reacting iron(III) oxide with carbon. [2 marks]

01.2 The equation for percentage atom economy is:

$$\frac{\text{relative formula mass of desired product from the equation}}{\text{sum of the relative formula masses of all reactants from the equation}} \times 100$$

Calculate the percentage atom economy for the production of iron.

Give your answer to 1 decimal place. [3 marks]

01.3 Why is it important that the carbon dioxide is not released into the atmosphere? [1 mark]

01.4 Pure iron is too soft for most uses. Most iron is mixed with carbon to make an alloy called steel. The structures of iron and steel are shown in **Figure 1**.

Figure 1

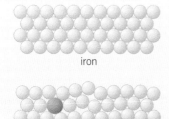

iron

alloy

Use **Figure 1** to explain why steel is much harder than pure iron. [3 marks]

01.5 Iron can rust. Rusting is a type of corrosion. There are several methods of rust prevention. Explain how painting iron prevents rusting. [2 marks]

01.6 Explain how attaching magnesium helps prevent iron rusting. [2 marks]

02 This question is about different types of material. Poly(ethene) is an addition polymer that is made from ethene.

02.1 Complete the equation to show the formation of poly(ethene) from ethene. [4 marks]

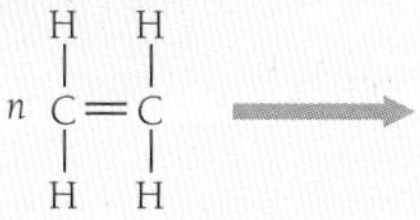

$$n \; \underset{\overset{|}{H} \; \overset{|}{H}}{\overset{\overset{H} \; \overset{H}}{C = C}} \longrightarrow$$

02.2 The properties of polymers depend on their structure.

Figure 2 shows a thermosoftening polymer and a thermosetting polymer.

Figure 2

thermosoftening polymer   thermosetting polymer

Use **Figure 2** and your knowledge of structure and bonding to explain why the thermosoftening polymer melts at a low temperature when heated. [3 marks]

02.3 Explain why concrete reinforced with metal is a composite. [2 marks]

Ⓗ 03 Ammonia is manufactured industrially from nitrogen and hydrogen in the Haber process:

$$N_2(g) + 3H_2(g) \rightleftharpoons 2NH_3(g)$$

The forward reaction is exothermic.

03.1 What is meant by the symbol $\rightleftharpoons$? [1 mark]

03.2 Explain what is meant by equilibrium. [2 marks]

03.3 In industry it is important that both the yield of a chemical reaction and the rate of a chemical reaction are high.

In industry the conditions used for the Haber process are:

- 450 °C
- 200 atmospheres pressure
- iron catalyst.

Use the equation and your knowledge to explain why these conditions are used in the Haber process. You need to consider both rate and yield in your answer. [6 marks]

03.4 Most ammonia from the Haber process is used to manufacture fertilisers.

Ammonia is first converted into ammonium nitrate. Which acid does ammonia need to be reacted with to make ammonium nitrate? [1 mark]

03.5 The ammonium nitrate is then used to make NPK fertilisers which contain the three main nutrients needed for plant growth. NPK fertilisers are formulations.

Explain why NPK fertilisers are formulations. [2 marks]

Paper 1 questions

01 A molecule of methane is shown in **Figure 1**.

Figure 1

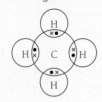

01.1 What is the chemical formula of methane.
Tick (✓) the correct answer.

C₄H	
CH₄	
C4H	

[1 mark]

01.2 In **Figure 1** what does •× represent?
Tick (✓) the correct answer.

a pair of electrons	
a pair of neutrons	
a pair of protons	

[1 mark]

01.3 What particles are present in the nucleus of a carbon atom?
Tick (✓) the correct answer.

neutrons and electrons	
protons and electrons	
protons and neutrons	

[1 mark]

Figure 2 shows another way of representing a molecule of methane.

Figure 2

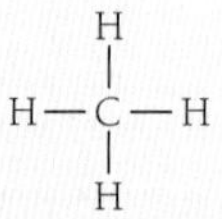

$$\begin{array}{c} \text{H} \\ | \\ \text{H} - \text{C} - \text{H} \\ | \\ \text{H} \end{array}$$

01.4 What type of bond is represented by each line?
Tick (✓) the correct answer.

covalent bond	
ionic bond	
metallic bond	

[1 mark]

01.5 The bonds in methane are very strong but methane has a low boiling point.

Why does methane have a low boiling point?
Tick (✓) the correct answer.

Methane has a giant structure	
Methane has only four bonds per molecule that need to be broken	
Methane has weak intermolecular forces between its molecules	

[1 mark]

02 Lithium is represented as:

$$^{7}_{3}\text{Li}$$

02.1 What is the atomic number of lithium? [1 mark]

02.2 Complete the diagram to show the electronic structure of a lithium atom.

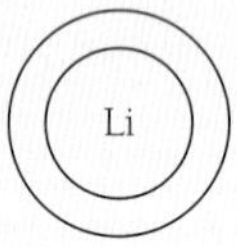

[1 mark]

02.3 Describe how a lithium atom changes into a lithium ion, Li⁺. [2 marks]

02.4 *In this question you will be assessed on using good English, organising information clearly, and using specialist terms where appropriate.*
A teacher adds lithium to water.
The equation for the reaction is:
$$2\text{Li(s)} + 2\text{H}_2\text{O(l)} \rightarrow 2\text{LiOH(aq)} + \text{H}_2\text{(g)}$$
Describe what the equation means.
Your description should include what the symbols and numbers represent. [6 marks]

02.5 Explain, in terms of structure and bonding, why lithium chloride is a solid with a high melting point. [4 marks]

AQA, 2014

03 **Figure 3** shows one way of producing iron.

Figure 3

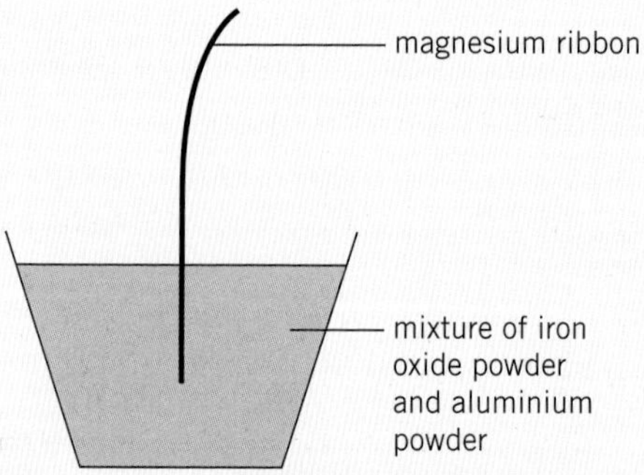

magnesium ribbon

mixture of iron oxide powder and aluminium powder

Iron oxide reacts with aluminium to produce iron.
The symbol equation for the reaction is:
$$\text{Fe}_2\text{O}_3 + 2\text{Al} \rightarrow 2\text{Fe} + \text{Al}_2\text{O}_3$$

03.1 Complete the word equation for this reaction.
iron oxide + aluminium → iron + ... [1 mark]

03.2 The magnesium ribbon is lit to start the reaction. Why does the burning magnesium ribbon start the reaction? [1 mark]

03.3 In industry, iron is produced in the blast furnace when iron oxide is heated with carbon.
The iron from the blast furnace is called cast iron.
Cast iron contains carbon.
Figure 4 shows the structure of pure iron and cast iron.

Figure 4

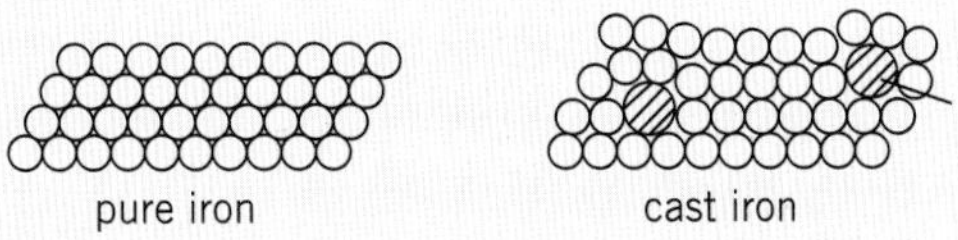

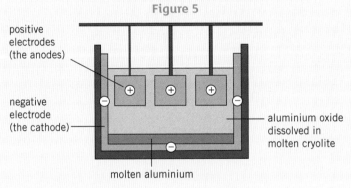

pure iron   cast iron

Draw a ring around the correct answer to complete the sentence.

Pure iron is an element because pure iron

| contains only one sort of atom. |
| is magnetic. |
| is a metal. |

[1 mark]

03.4 Suggest why cast iron is harder than pure iron. [2 marks]

03.5 Aluminium is extracted by electrolysis using the ionic compound aluminium oxide.

Figure 5

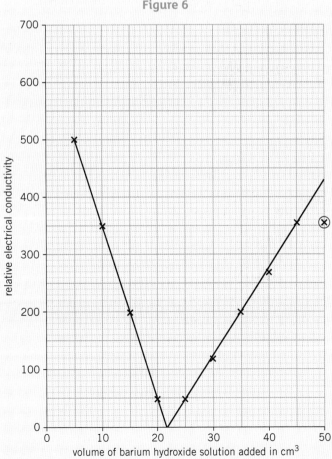

positive electrodes (the anodes)

negative electrode (the cathode)

aluminium oxide dissolved in molten cryolite

molten aluminium

Aluminium cannot be extracted by heating aluminium oxide with carbon.
Suggest why. [1 mark]

03.6 Why is aluminium oxide dissolved in molten cryolite? [1 mark]

H **03.7** Aluminium metal is produced at the negative electrode (cathode).
Complete the half equation for the process:
$Al^{3+} + \ldots\ldots e^- \rightarrow Al$ [1 mark]

H **03.8** Use the half equation to state why Al^{3+} ions are reduced. [1 mark]

03.9 Explain why the positive electrodes (anodes) burn away.

Use your knowledge of the products of electrolysis to help you. [4 marks]
AQA, 2013

04 Barium sulfate is an insoluble salt.
Barium sulfate can be made by adding barium hydroxide solution to dilute sulfuric acid.
The balanced chemical equation for the reaction is:
$H_2SO_4(aq) + Ba(OH)_2(aq) \rightarrow BaSO_4(s) + 2H_2O(l)$
A student investigated how the electrical conductivity of dilute sulfuric acid changed as barium hydroxide solution was added.
This is the method she used.

Step 1: Place 25.0 cm³ of dilute sulfuric acid in a conical flask.

Step 2: Add 5.0 cm³ of barium hydroxide solution.

Step 3: Stir the mixture.

Step 4: Use a conductivity meter to measure the electrical conductivity of the mixture.

Step 5: Repeat Step 2, Step 3, and Step 4 until 50 cm³ of barium hydroxide solution have been added.

The student's results are shown on the graph in **Figure 6**.

Figure 6

relative electrical conductivity

volume of barium hydroxide solution added in cm³

04.1 The ringed point on the graph is anomalous. What could have happened to cause the anomalous point?
Tick (✓) one box.

No more barium hydroxide solution was added	
Too much barium hydroxide solution was added	
Too much dilute sulfuric acid was used	

[1 mark]

04.2 Use the graph in Figure 6 to estimate the relative electrical conductivity of the dilute sulfuric acid before any barium hydroxide solution was added. Show your working on the graph. [2 marks]

04.3 Explain why dilute sulfuric acid conducts electricity. [2 marks]

04.4 What was the volume of barium hydroxide solution added when the relative electrical conductivity of the mixture was zero? [1 mark]

04.5 Suggest why the relative electrical conductivity became zero. [1 mark]

04.6 The student did another experiment using the same solutions as she used before.
She used the same volume (25.0 cm³) of dilute sulfuric acid in the conical flask.
She then added an unknown volume of barium hydroxide solution.
She found that the relative electrical conductivity of the mixture was 260.
This is the student's conclusion:
13 cm³ of barium hydroxide solution must have been added.
Why may the student's conclusion **not** be correct?
[1 mark]

04.7 The student said that she could check whether she was correct by adding something to the mixture. What could she add to the mixture? How would this tell her whether she was correct? [3 marks]

AQA, 2015

05 Some students were investigating the rate at which carbon dioxide gas is produced when metal carbonates react with an acid.
One student reacted 1.00 g of calcium carbonate with 50 cm³, an excess, of dilute hydrochloric acid.
The apparatus used is shown in **Figure 7**.

Figure 7

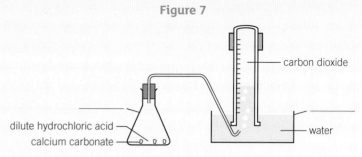

carbon dioxide

dilute hydrochloric acid

calcium carbonate

water

05.1 Complete the **two** labels for the apparatus on the diagram. [1 mark]

05.2 The student measured the volume of gas collected every 30 seconds.
Table 1 shows the student's results.

Table 1

Time in seconds	Volume of carbon dioxide collected in cm³
30	104
60	
90	198
120	221
150	232
180	238
210	240
240	240

Figure 8 Shows what the student saw at 60 seconds.

Figure 8

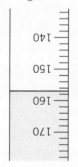

What is the volume of gas collected? [1 mark]

05.3 Why did the volume of gas stop changing after 210 seconds? [1 mark]

H **05.4** Another student placed a conical flask containing 1.00 g of a Group 1 carbonate (M_2CO_3) on a balance.
He then added 50 cm³, an excess, of dilute hydrochloric acid to the flask and measured the mass of carbon dioxide given off.
The equation for the reaction is:
$$M_2CO_3 + 2HCl \rightarrow 2MCl + H_2O + CO_2$$

The final mass of carbon dioxide given off was 0.32 g.

Calculate the amount, in moles, of carbon dioxide in 0.32 g carbon dioxide.

Relative atomic masses A_r: C = 12; O = 16

[2 marks]

(H) 05.5 How many moles of the metal carbonate are needed to make this number of moles of carbon dioxide? [1 mark]

(H) 05.6 The mass of metal carbonate used was 1.00 g. Use this information, and your answer to **05.5**, to calculate the relative formula mass M_r of the metal carbonate.

If you could not answer **05.5**, use 0.009 43 as the number of moles of metal carbonate. This is not the answer to **05.5**. [1 mark]

(H) 05.7 Use your answer to **05.6** to calculate the relative atomic mass A_r of the metal in the metal carbonate (M_2CO_3) and so identify the Group 1 metal in the metal carbonate.

If you could not answer **05.6**, use 230 as the relative formula mass of the metal carbonate. This is not the answer to **05.6**.

Remember, you must show your working.

[3 marks]

(H) 05.8 Two other students repeated the experiment in **05.4**.

When the first student did the experiment some acid sprayed out of the flask as the metal carbonate reacted.

Explain the effect this mistake would have on the calculated relative atomic mass of the metal. [3 marks]

(H) 05.9 The second student used 100 cm³ of dilute hydrochloric acid instead of 50 cm³.

Explain the effect, if any, this mistake would have on the calculated relative atomic mass of the metal. [3 marks]

AQA, 2014

(H) 06 Ethane reacts with bromine at room temperature in sunlight.

The displayed structural formulae of the products and reactants are shown in **Figure 9**.

Figure 9

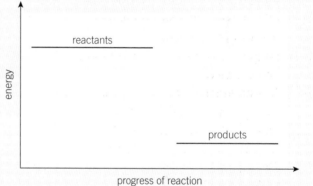

06.1 Suggest why this reaction will not take place at room temperature in the dark. [1 mark]

06.2 The bond energies are shown in **Table 2**.

Table 2

Bond	Bond energy in kJ/mol
C—C	347
C—H	413
Br—Br	193
C—Br	290
H—Br	366

Show that the energy transferred to the surroundings by the reaction is 50 kJ/mol. [3 marks]

06.3 Complete the reaction profile. Show arrows to label the activation energy E_a and the energy transferred to the surroundings.

[3 marks]

06.4 The HBr produced was dissolved in water to give a solution of hydrobromic acid.

A student wanted to check the concentration of the solution of hydrobromic acid.

She had a solution of sodium hydroxide of known concentration. She pipetted 25.00 cm³ of sodium hydroxide into a conical flask.

Describe how the student should use the method of titration to obtain accurate results. [5 marks]

06.5 Sodium hydroxide neutralises hydrobromic acid as shown in the equation:

$NaOH(aq) + HBr(aq) \rightarrow NaBr(aq) + H_2O(l)$

The results of the titration are shown in **Table 3**.

Table 3

Concentration of NaOH in mol/dm³	Volume of NaOH in cm³	Concentration of HBr in mol/dm³	Volume of HBr in cm³
0.150	25.00		10.45

Calculate the concentration of the hydrobromic acid in mol/dm³.

Give your answer to three significant figures. [3 marks]

Paper 2 questions

01 This question is about gases in the Earth's atmosphere.

01.1 Complete **Table 1** to show the current composition of the Earth's atmosphere.

Table 1

Percentage of the gas in the Earth's atmosphere	Name of gas
79	
20	
1	carbon dioxide, water vapour, and noble gases

[2 marks]

Three billion years ago the temperature on Earth was much higher, as were the percentages of carbon dioxide and water vapour in the atmosphere.

01.2 Explain why the percentage of water vapour decreased. [2 marks]

01.3 Give **two** processes that caused the percentage of carbon dioxide to decrease. [2 marks]

02 Iron will rust in damp air.
Iron reacts with water and oxygen to produce rust.

02.1 As iron rusts there is a colour change.
Draw a ring around the correct answer to complete the sentence.
During the reaction iron changes from grey to:

blue   brown   green

[1 mark]

02.2 A student set up the apparatus shown in **Figure 1**.

Figure 1

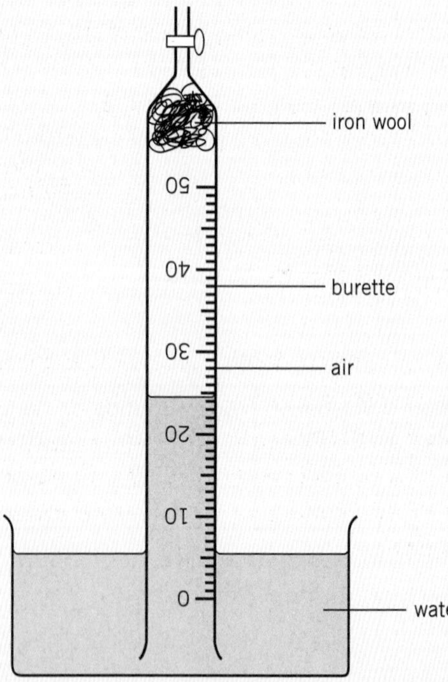

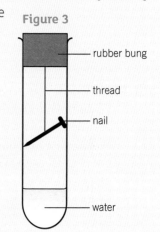

The student left the apparatus for a few days. The water level in the burette slowly went up and then stopped rising.

Figure 2 shows the water level in the burette at the start of the experiment and after a few days.

Figure 2

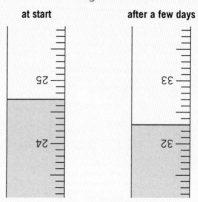

Complete **Table 2** to show the reading on the burette after a few days.

Table 2

Burette reading at start	24.7 cm³
Burette reading after a few days	____ cm³

[1 mark]

02.3 Calculate the volume of oxygen used up in the reaction. [1 mark]

02.4 The percentage of air that is oxygen can be calculated using the equation:

$$\text{percentage of air that is oxygen} = \frac{\text{volume of oxygen used up}}{\text{volume of air at start}} \times 100$$

The student cannot use his results to calculate the correct percentage of air that is oxygen.
Explain why? [2 marks]

02.5 A student investigated the rusting of an iron nail at different temperatures.
This is the method the student used:

- measure the mass of a nail
- set up apparatus as shown in **Figure 3**
- leave for three days
- measure the mass of the rusted nail.

Figure 3

The student repeated the experiment at different temperatures using a new, identical, nail each time.

The student's results are shown on the graph in **Figure 4**.

Figure 4

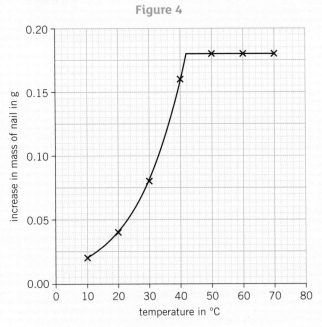

Why does the mass of the nail increase when it rusts? [1 mark]

02.6 Use the graph to describe the relationship between the temperature and the increase in mass of the nail. [3 marks]

02.7 The increase in mass of the nail after three days is a measure of the rate of rusting.
The student's graph does not correctly show how increasing the temperature above 42 °C changes the rate of rusting.
How could the experiment be changed to show the effect of temperatures above 42 °C on the rate of rusting?
Give a reason for your answer. [2 marks]

AQA, 2014

03 Chromatography can be used to separate components of a mixture.

03.1 A student used paper chromatography to analyse a black food colouring.
The student placed spots of known food colours, **A**, **B**, **C**, **D**, and **E**, and the black food colouring on a sheet of chromatography paper.
The student set up the apparatus as shown in **Figure 5**.

Figure 5

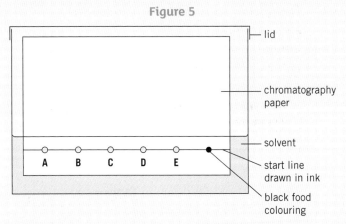

The student made **two** errors in setting up the apparatus.
Identify the **two** errors and describe the problem each error would cause. [4 marks]

03.2 A different student set up the apparatus without making any errors.
The chromatogram in **Figure 6** shows the student's results.

Figure 6

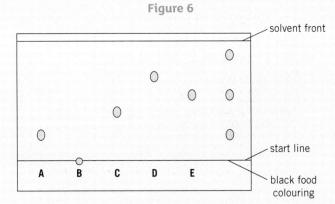

What do the results tell you about the composition of the black food colouring? [2 marks]

03.3 Use **Figure 6** to complete **Table 3**.

Table 3

	Distance in mm
Distance from start line to solvent front	_____
Distance moved by food colour **C**	_____

[2 marks]

03.4 Use your answers in **03.3** to calculate the R_f value for food colour **C**. [1 mark]

03.5 **Table 4** gives the results of chromatography experiments that were carried out on some known food colours, using the same solvent as the students.

Table 4

Name of food colour	Distance from start line to solvent front in mm	Distance moved by food colour in mm	R_f value
Ponceau 4R	62	59	0.95
Carmoisine	74	45	0.61
Fast red	67	21	0.40
Erythrosine	58	17	0.29

Which of the food colours in **Table 4** could be food colour **C** from the chromatogram?
Give the reason for your answer. [2 marks]
AQA, 2014

04 Hydrogen gas is produced by the reaction of methane and steam.

04.1 **Figure 7** represents a molecule of hydrogen.

Figure 7

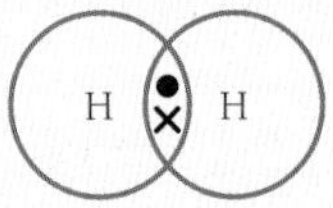

What type of bond joins the atoms of hydrogen?
Tick (✓) the correct answer.

covalent	
metallic	
ionic	

[1 mark]

04.2 A catalyst is used in the reaction.
Draw a ring around the correct answer to complete the sentence.

A catalyst
- increases the rate of reaction.
- increases the temperature.
- increases the yield of a reaction.

[1 mark]

04.3 The equation for the reaction of methane and steam is:
$$CH_4(g) + H_2O(g) \rightleftharpoons CO(g) + 3H_2(g)$$
What is meant by the symbol $\rightleftharpoons$? [1 mark]

ⓗ 04.4 Lowering the pressure reduces the rate of reaction.
Explain why, in terms of particles. [2 marks]

ⓗ 04.5 The graph shows the yield of hydrogen at different temperatures.
The forward reaction is endothermic.
How does **Figure 8** show that the forward reaction is endothermic? [1 mark]

Figure 8

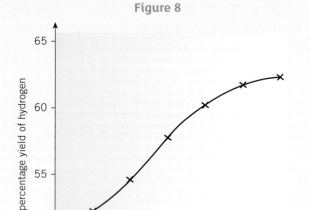

ⓗ 04.6 Why is a higher yield produced if the reaction is repeated at a lower pressure? [1 mark]

04.7 *In this question you will be assessed on using good English, organising information clearly, and using specialist terms where appropriate.*
Car engines are being developed that use hydrogen gas as a fuel instead of petrol.
Table 5 compares the two fuels.

Table 5

	Hydrogen	Petrol
Energy	5700 kJ per litre	34 000 kJ per litre
State	gas	liquid
Equation for combustion	$2H_2 + O_2 \rightarrow 2H_2O$	$2C_8H_{18} + 25O_2 \rightarrow 16CO_2 + 18H_2O$
How fuel is obtained	Most hydrogen is produced from coal, oil or natural gas. Hydrogen can be produced by the electrolysis of water or the solar decomposition of water.	Fractional distillation of crude oil.

Use the information in **Table 5** and your knowledge of fuels to evaluate the use of hydrogen instead of petrol as a fuel.
You should describe the advantages and disadvantages of using hydrogen instead of petrol. [6 marks]
AQA, 2014

05 This question is about identification of chemicals.

05.1 A student had samples of two gases – hydrogen, H_2, and oxygen, O_2. The samples are unlabelled. Give a test the student could use to identify **one** of the gases. [2 marks]

05.2 The student also had white solids that could be aluminium sulfate, $Al_2(SO_4)_3$, or magnesium sulfate, $MgSO_4$. Give a test the student could use to identify **one** of the compounds. [2 marks]

05.3 Another group of students had four different colourless solutions in four unlabelled beakers. The students knew that the beakers contained:
- calcium chloride
- potassium chloride
- calcium carbonate
- potassium carbonate.

They do not know which solution was in each beaker.

The students were provided with:
- splints soaked in each of the solutions for the flame test
- dilute nitric acid
- test tubes.

Plan a method to identify the four solutions. [6 marks]

06 **Figure 9** shows how crude oil can be separated into fractions **A, B, C, D, E,** and **F**.

Figure 9

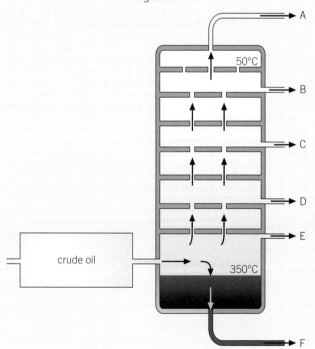

06.1 Which fraction contains hydrocarbons with the lowest boiling point? [1 mark]

06.2 Which fraction contains hydrocarbons with the longest carbon chains? [1 mark]

06.3 Fractions **B, C, D,** and **E** are liquids at room temperature. Which fraction contains the hydrocarbons that are the most flammable? [1 mark]

06.4 Which fraction **B, C, D,** or **E** would be the most viscous? [1 mark]

06.5 Hydrocarbon **V** was cracked to form four different hydrocarbons: **W, X, Y,** and **Z**. The equation for the cracking process is shown below.

$$\begin{array}{ccccccccc} \mathbf{V} & & \mathbf{W} & & \mathbf{X} & & \mathbf{Y} & & \mathbf{Z} \\ C_{28}H_{58} & \rightarrow & C_{14}H_{30} & + & C_{9}H_{18} & + & C_{2}H_{4} & + & C_{3}H_{6} \end{array}$$

Describe the conditions for cracking. [2 marks]

06.6 Give the general formula for hydrocarbon **V**. [1 mark]

06.7 Hydrocarbon **W** and hydrocarbon **X** are both colourless liquids. Describe a chemical test and the result of the test to identify hydrocarbon **X**. [2 marks]

06.8 Hydrocarbon **Y** is ethene. Complete the equation for the formation of ethanol from ethene.

$C_2H_4 + \dots \rightarrow C_2H_5OH$ [1 mark]

06.9 Copy and complete the displayed structural formula of ethanol.

$$\begin{array}{ccc} & H & H \\ & | & | \\ H - & C & C \\ & | & | \\ & H & H \end{array}$$

[3 marks]

Maths skills for Chemistry
MS1 Arithmetic and Numerical Computation

Learning objectives

After this topic, you should know how to:

- recognise and use expressions in decimal form
- recognise and use expressions in standard form
- use ratios, fractions, and percentages
- make estimates of the results of simple calculations.

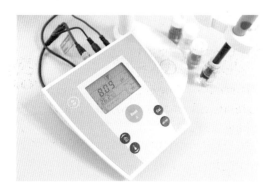

Figure 3 *If you use a pH meter to measure the pH of a solution, your reading could be a decimal number*

How big is an atom? How many atoms are in 12 g of carbon? What is the size of a nanoparticle?

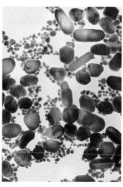

Figure 1 *How big is a nanoparticle?*

Figure 2 *There are 6.02×10^{23} atoms in this 12 g of carbon*

Scientists use maths all the time – when collecting data, looking for patterns, and making conclusions. This chapter includes the maths you need for your GCSE chemistry course. The rest of the book gives you many opportunities to practise using maths when it is needed as you learn about chemistry.

1a Decimal form

There will always be a whole number of atoms in a molecule, and a whole number of protons, neutrons, or electrons in an atom.

However, when you make measurements in science the numbers may *not* be whole numbers but numbers *in between* whole numbers. These are numbers in decimal form, for example, the volume of acid used in a titration could be 22.35 cm^3, or the mass of a powder could be 8.7 g.

The value of each digit in a number is called its place value. For example, in the number 4512.345:

thousands	hundreds	tens	units	.	tenths	hundredths	thousandths
4	5	1	2	.	3	4	5

1b Standard form

Place values can help you to understand the size of a number, however some numbers in science are too large or too small to understand when they are written as ordinary numbers. For example, the number of atoms, ions or molecules in a mole of substance, 602 000 000 000 000 000 000 000, or the diameter of the nucleus of a hydrogen atom, 0.000 000 000 000 001 75 m.

Standard form is used to show very large or very small numbers more easily.

In standard form, a number is written as $A \times 10^n$.

- A is a decimal number between 1 and 10 (but not including 10), for example, 6.02 or 1.75.
- n is a whole number. The power of ten can be positive or negative, for example, 10^{23} or 10^{-15}.

This gives you a number in standard form, for example, 6.02×10^{23}/mol or 1.75×10^{-15}.

Figure 4 *What do 18 g of water, 108 g of gold, and 4 g of helium have in common? They all have 6.02×10^{23} number of particles*

Table 1 explains how you convert numbers to standard form.

Table 1 *How to convert numbers into standard form*

The number	The number in standard form	What you did to get to the decimal number	...so the power of ten is...	What the *sign* of the power of ten tells you
1000 m	1.0×10^3 m	You moved the decimal point 3 places to the *left* to get the decimal number	+3	The positive power shows the number is *greater* than one.
0.01 s	1.0×10^{-2} s	You moved the decimal point 2 places to the *right* to get the decimal number	−2	The negative power shows the number is *less* than one.

When carrying out multiplications or divisions using standard form, you should add or subtract the powers of ten to work out roughly what you expect the answer to be. This will help you to avoid mistakes.

Multiplying numbers in standard form

You can use a scientific calculator to calculate with numbers written in standard form. You should work out which button you need to use on your own calculator (it could be **EE**, **EXP**, **10ˣ**, or **×10ˣ**).

Study tip

Always remember to add any relevant units, for example, metres (written as 'm').

Synoptic link

To see how chemists use numbers in standard form, see Topic C1.7 and Topic C2.4.

Study tip

Check that you understand the power of ten, and the sign of the power.

Figure 5 *You can use a scientific calculator to do calculations involving standard form*

Figure 6 *Airships were common before the 1940s when a series of high-profile accidents and their use was eclipsed by aeroplanes*

Worked example: Standard form

The Avogadro constant N_a tells you how many particles are in a mole of any particles. Its value is 6.02×10^{23} per mol.

Calculate how many atoms of helium there are in an airship containing about 2.58×10^4 moles of helium gas.

Solution

Step 1: Write down the formula for the number of atoms.

number of atoms = N (/mol) × number of moles (mol)

Step 2: Substitute numbers into the formula for number of atoms.

number of atoms of helium = $(6.02 \times 10^{23}/\text{mol}) \times (2.58 \times 10^4 \,\text{mol})$

$= 1.55 \times 10^{28}$ atoms in standard form

1c Ratios, fractions, and percentages
Ratios

A **ratio** compares two quantities. A ratio of 2:4 of carbon atoms to oxygen atoms means that for every two carbon atoms, there are four oxygen atoms.

You can describe the number of carbon atoms in relation to the number of oxygen using many different ratios, for example, 2:4, 1:2, and 0.5:1. All of the ratios are equivalent – they mean the same thing.

You can simplify a ratio so that both numbers are the lowest whole numbers possible.

Figure 7 *You should always add acid to water. Mixing acid and water is an exothermic process. If you add water to acid so much energy can be transferred that boiling acid splashes out of the mixing vessel*

Worked example: Simplifying ratios

A student mixed $15\,\text{cm}^3$ of acid with $90\,\text{cm}^3$ of water. Calculate the simplest ratio of the volume of acid to the volume of water.

Solution

Step 1: Write down the ratio of *acid : water*

Step 2: Both 15 and 90 have a common factor, 5.
Divide both numbers by 5.

$$÷5 \left(\begin{array}{c} 15:90 \\ \\ 3:18 \end{array} \right) ÷5$$

Step 3: Both 3 and 18 have a common factor, 3.
Divide both numbers by 3.

$$÷3 \left(\begin{array}{c} \\ 1:6 \end{array} \right) ÷3$$

To get the simplest form of the ratio, you have divided by 15 (i.e., 3×5), which is the highest common factor of 15 and 90.

Synoptic link

To see examples of how chemists use ratios, look at Topic C3.11 and Topic C8.2.

Fractions

A fraction is a part of a whole.

$\frac{1}{3}$ The numerator tells you how many parts of the whole you have.

The denominator tells you how many equal parts the whole has been divided into.

To convert a fraction into a decimal, divide the numerator by the denominator.

$\frac{1}{3} = 1 \div 3 = 0.33333\ldots = 0.\dot{3}$ (the dot shows that the number 3 recurs, or repeats over and over again).

To convert a decimal to a fraction, use the place value of the digits, then simplify. For example, the smallest place value in 0.045 is a thousandth, so $0.045 = \frac{45}{1000}$. This can be simplified to $\frac{9}{200}$.

Worked example: Calculating the fraction of a quantity

A student has a 25 g sample of sodium chloride. Calculate the mass of $\frac{2}{5}$ of this sample.

Solution

$\frac{2}{5}$ of 25 is the same as $\frac{2}{5} \times 25$ so:

Step 1: Divide the total mass of the sample by the denominator.

$25\,g \div 5 = 5\,g$

Step 2: Multiply by the numerator.

$5\,g \times 2 = 10\,g$

Figure 8 *One square of this chocolate bar represents* $\frac{1}{24}$*. A column of four squares represents* $\frac{4}{24}$*, which can be simplified to* $\frac{1}{6}$

Study tip

Place values were introduced in 1a Decimal form.

Percentages

A **percentage** is a number expressed as a fraction of 100, for example:

$77\% = \frac{77}{100} = 0.77$.

Worked example: Calculating a percentage

A student found that a 7.5 g sample of limestone contained 7.2 g of calcium carbonate.

Calculate the percentage by mass of calcium carbonate in the sample.

Solution

Step 1: Calculate the fraction of calcium carbonate in the sample.

$\frac{\text{mass of calcium carbonate}}{\text{mass of limestone}} = \frac{7.2}{7.5}$

Step 2: Convert the fraction to a decimal.

$\frac{7.2}{7.5} = 7.2 \div 7.5 = 0.96$

Step 3: Multiply the decimal by 100%.

$0.96 \times 100\% = 96\%$

Figure 9 *One of the biggest areas of limestone in the UK is in the Yorkshire Dales National Park*

Figure 10 *Old lime kilns are a common sight in the Yorkshire Dales*

Figure 11 *Copper(II) carbonate*

Synoptic link

You can find examples of how chemists use percentages in Topic C4.4, Topic C4.5, and Topic C13.2.

You may also need to calculate a percentage of a quantity.

Worked example: Using a percentage to calculate a quantity

3.2 tonnes of limestone are put into a lime kiln. After an hour, 25% of this was converted into lime. How many tonnes of limestone were converted into lime?

Solution

Step 1: Convert the percentage of limestone to a decimal.

$$25\% = \frac{25}{100} = 0.25$$

Step 2: Multiply by the total mass of limestone.

$$0.25 \times 3.2 \text{ tonnes} = 0.8 \text{ tonnes}$$

Finally, you may need to calculate a percentage increase or decrease in a quantity from its original value.

Worked example: Calculating a percentage change

A student heats a 4.75 g sample of copper carbonate. The mass decreases to 3.04 g. Calculate the percentage change in mass.

Solution

Step 1: Calculate the decrease in mass.

$$4.75 \text{ g} - 3.04 \text{ g} = 1.71 \text{ g}$$

Step 2: Divide the decrease in mass by the original mass.

$$\frac{1.71}{4.75} = 0.36$$

Step 3: Convert the decimal to a percentage.

$$0.36 \times 100\% = 36\%$$

Remember that in this case the answer is a percentage *decrease*.

1d Estimating the result of a calculation

When you use your calculator to work out the answer to a calculation you can sometimes press the wrong button and get the wrong answer. The best way to make sure that your answer is correct is to estimate the answer in your head first.

Worked example: Estimating an answer

A reaction produces gas at a constant rate of 34 cm³/min. Find the volume of gas produced after eight minutes. Estimate the answer and then calculate it.

Solution

Step 1: Round each number up or down to get a whole number multiple of 10.

34 cm³/min is about 30 cm³/min

8 min is about 10 min

Step 2: Multiply the numbers in your head.

30 cm³/min × 10 min = 300 cm³

Step 3: Do the calculation and check it is close to your estimate.

Distance = 34 cm³/min × 8 min = 272 cm³

This is quite close to 300 so it is probably correct.

Notice that you could do other things with the numbers:

$$34 + 8 = 42 \qquad \frac{34}{8} = 4.3 \qquad 34 - 8 = 26$$

Not one of these numbers is close to 300. If you got any of these numbers you would know that you needed to repeat the calculation.

1   If the mass of 1 mole of chlorine atoms is 35.5 g, calculate the mass of 6.0 moles of chlorine atoms.   [1 mark]

2   **a** The concentration of a dilute acid is 0.000 038 g/dm³. Express this concentration in standard form.   [1 mark]

   b A metal ore can be processed at a plant at a rate of about 5×10^4 tonnes per year. If this rate is maintained, how long will it be before 1.6×10^6 tonnes will be used up?   [1 mark]

3   **a** What is the simplest ratio of masses (sulfur : oxygen) in a mixture that contains 48 g of carbon and 144 g of sulfur?   [1 mark]

   b A 300 tonne batch of bauxite was found to contain 60% aluminium oxide. What mass of aluminium oxide is in the batch?   [1 mark]

   c The approximate percentage of nitrogen gas in the air is 80%. Express this percentage as a fraction.   [1 mark]

MS2 Handling Data

Learning objectives

After this topic, you should know how to:

- use an appropriate number of significant figures
- find arithmetic means
- construct and interpret frequency tables and bar charts
- make order of magnitude calculations.

Figure 1 *In 2009, the UK consumed 1 611 000 barrels (four significant figures) or 1.6×10^6 (two significant figures) of oil per day*

Synoptic link

To find out how to use this type of calculation, you can find examples in Topic C8.1 and Topic C8.2. There are also many examples of applying significant figures in the calculations in Chapter 4.

2a Significant figures

Numbers are rounded when it is not appropriate to give an answer that is too precise.

When rounding to **significant figures (s.f.)**, count from the first non-zero digit.

These masses each have three significant figures. The significant figures are underlined in each case.

$$15\underline{3}\,g \qquad 0.\underline{153}\,g \qquad 0.00\underline{153}\,g$$

Table 1 below shows some more examples of measurements given to different numbers of significant figures.

Table 1 *The number of significant figures – the significant figures in each case are underlined. Notice that zeros at the end of decimal numbers are significant - see the last example in the table.*

Number	0.0$\underline{5}$ s	$\underline{5.1}$ nm	0.$\underline{775}$ g/s	$\underline{23.50}$ cm^3
Number of significant figures	1	2	3	4

In general, you should give your answer to the same number of significant figures as the data in the question that has the lowest number of significant figures.

Remember that rounding to significant figures is *not* the same as decimal places. When rounding to decimal places, count the number of digits that follow the decimal point. For example, 0.00 153 has three significant figures but five decimal places.

> **Worked example: Significant figures**
>
> Calculate the mean rate of a reaction that gives off 25 cm^3 of gas in 7.85 seconds.
>
> **Solution**
>
> **Step 1:** Write down what you know.
>
> volume of gas given off = 25 cm^3 (2 s.f.)
>
> time = 7.85 s (3 s.f.)
>
> You should give your answer to 2 s.f.
>
> **Step 2:** Write down the equation that links the quantities you know and the quantity you want to find.
>
> $$\text{mean rate of reaction (cm}^3/\text{s)} = \frac{\text{amount of product formed (cm}^3)}{\text{time } t \text{ (s)}}$$
>
> **Step 3:** Substitute values into the equation.
>
> $$\text{speed} = \frac{25\,\text{cm}^3}{7.85\,\text{s}}$$
>
> $$= 3.184\,713\,375\,\text{cm}^3/\text{s}$$
>
> $$= 3.2\,\text{cm}^3/\text{s to 2 s.f.}$$

2b Arithmetic means

How to calculate the mean

To calculate the **mean** (or average) of a series of values:

● add together all the values in the series to get a total

● divide the total by the number of values in the data series.

You will often need to do this with your sets of repeat readings when conducting investigations. This helps you to obtain more accurate data from sets of repeat readings where you have some random measurement errors.

Figure 2 *When carrying out a titration, you should repeat the experiment then take a mean of your results*

> **Worked example: Calculating a mean**
>
> A student measured the time to collect 20 cm³ of hydrogen gas from the reaction of magnesium ribbon in excess dilute sulfuric acid.
>
> Their results were as follows:
>
> 18.7 s   19.5 s   18.5 s   19.2 s
>
> Calculate the mean time to collect 20 cm³ of the gas.
>
> **Solution**
>
> **Step 1:** Add together the recorded values.
>
> 18.7 s + 19.5 s + 18.5 s + 19.2 s = 75.9 s
>
> **Step 2:** Then divide by the number of recorded values (in this case, 4 times were taken).
>
> $\dfrac{75.9\,s}{4} = 18.9\,s$ (3 s.f.)
>
> The mean time to collect 20 cm³ of hydrogen gas was 18.9 s. (3 s.f.)

2c Frequency tables and bar charts

Frequency tables and bar charts

The word data describes observations and measurements that are made during experiments or research.

Qualitative data is non-numerical data, such as colours or elements. For example, the chemical elements can be divided into the categories metals, metalloids and non-metals.

The frequency table (Table 2) shows the number of metal, non-metals, or metalloids in the first 20 elements.

Table 2 *A frequency table for the number of metals, non-metals, and metalloids in the first 20 elements*

Type of element	Frequency
metal	7
non-metal	11
metalloid	2

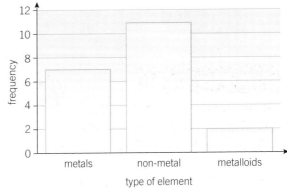

Figure 3 *Number of metals, non-metals, and metalloids in the first twenty elements*

The height of the bars in the bar chart represent the frequency of each category.

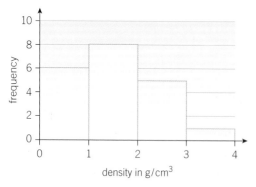

frequency

density in g/cm³

Figure 4 *Density of the first twenty elements*

Quantitative data is numerical measurements.

Discrete data can only take exact whole number (integer) values (usually collected by counting).

Continuous data can take any value (usually collected by measuring), such as mass, volume, or density.

The frequency table (Table 3) shows the density of the first 20 elements. Density cannot be measured exactly, so the measurements are grouped into intervals.

Table 3 *A frequency table for the density of the first 20 elements*

Density d in g/cm³	Frequency
$0 < d < 1$	6
$1 \leq d < 2$	8
$2 \leq d < 3$	5
$3 \leq d < 4$	1

There are no gaps between the bars when the data is continuous.

2h Estimates and order of magnitude

Being able to make a rough estimate is helpful. It can help you to check that a calculation is correct by knowing roughly what you expect the answer to be. A simple estimate is an **order of magnitude** estimate, which is an estimate to the nearest power of 10.

For example, to the nearest power of 10, you are probably 1 m tall and can run 10 m/s.

Figure 5 *Orders of magnitude can be useful but take care with them. Your height to the nearest power of 10 is probably 1 m but the average height of a 15 year old boy is 1.7 m and a 15 year old girl is 1.6 m*
Similarly, to the nearest power of 10 you can run 10 m/s, but the mean speed of the eight finalists of the men's 100-metre sprint in the 2008 Beijing Olympics was 9.92 m/s – a lot faster than most of us can actually run

You, your desk, and your chair are all of the order of 1 m tall. The diameter of a molecule is of the order of 1×10^{-9} m, or 1 nanometre.

Synoptic link

You can practise making orders of magnitude calculations in Topic C3.11, Topic C5.8, and Topic C14.1.

Worked example: Comparing orders of magnitude

If the size of a molecule is of the order 1×10^{-9} m, and the size of a nanoparticle is of the order 1×10^{-8} m, estimate how the nanoparticle's size compares with the size of the molecule to the nearest order of magnitude.

Solution

Step 1: Divide the order of magnitude of the nanoparticles by the order of magnitude of the molecule.

$$\frac{1 \times 10^{-8}}{1 \times 10^{-0}} = 10$$

Step 2: Interpret the answer to the calculation.

The nanoparticle is one order of magnitude (10^1) larger than the molecule.

---

1 How many significant numbers are the following numbers quoted to?

  a 33.0 [1 mark] **b** 0.02 [1 mark]
  c 250 [1 mark] **d** 13.35 [1 mark]
  e 0.225 [1 mark] **f** 3×10^5 [1 mark]
  g 1.673×10^{-6} [1 mark] **h** 6.02×10^{23} [1 mark]

2 **a** A student timed how long it took for a pencil mark on a piece of paper under a conical flask took to disappear as a precipitate formed in the flask. They repeated the experiment three times and obtained the following results:

   1st test 184 s   2nd test 203 s   3rd test 196 s

   Calculate the mean time taken for the pencil mark to disappear, giving your answer to the appropriate number of significant figures. [2 marks]

  b In a titration experiment a student got the following results:

   1st titre 23.15 cm³

   2nd titre 20.40 cm³

   3rd titre 20.30 cm³

   Explain how the student would calculate the mean titre and express the mean to 4 significant figures (2 decimal places). [3 marks]

3 A student tested how long it took to collect 10 cm³ of oxygen gas when four different metal oxides (labelled as A to D) were added to catalyse the breakdown of identical solutions of hydrogen peroxide. The student wrote the results as follows:

   A took 18 seconds; B took 27 seconds; C took 12 seconds; D took 35 seconds

  a What type of variable is 'type of metal oxide' – qualitative or quantitative? [1 mark]

  b Draw a table and record the student's results in it. [2 marks]

  c Display the results graphically. [4 marks]

4 **a** Some estimates state that our current supplies of coal will run out in about 300 years. How many orders of magnitude is this estimate made to? [1 mark]

  b Describe how you would estimate the answer to a calculation involving numbers expressed in standard form. Use the case of an annual usage of a metal ore of 1.6×10^8 tonnes/year, the estimated reserves of the metal ore of 7.5×10^{10} tonnes, and the calculation of how long the metal ore will last at the current rate of usage. [3 marks]

MS3 Algebra

Learning objectives

After this topic, you should know how to:

- understand and use the symbols: =, <, <<, >>, >, ∝, ~
- change the subject of an equation
- substitute numerical values into algebraic equations using appropriate units for quantities.

Figure 1 *The diameter of a round bottom flask is very much bigger than the diameter of even the biggest atom*

3a Mathematical symbols

You have used lots of different symbols in maths, such as +, −, ×, ÷. There are other symbols that you might meet in chemistry. These are shown in Table 1.

Table 1 *The symbols you will meet whilst studying chemistry*

Symbol	Meaning	Example
=	equal to	$2\,g/s \times 2\,s = 4\,g$
<	is less than	The mean diameter of an atom < the mean diameter of a nanoparticle
<<	is very much less than	The diameter of an atom << the diameter of a round-bottom flask
>>	is very much bigger than	The number of atoms in a polymer molecule >> the number of atoms in a hydrogen molecule
>	is greater than	The pH of an alkali > 7
∝	is proportional to	pressure of a gas ∝ its concentration
~	is approximately equal to	$272\,cm^3 \sim 300\,cm^3$

3b Changing the subject of an equation

An equation shows the relationship between two or more variables. You can change an equation to make *any* of the variables become the subject of the equation.

To change the subject of an equation, you can do an opposite (inverse) operation to both sides of the equation to get the variable that you want on its own.

This means that:

- subtracting is the opposite of adding (and adding is the opposite of subtracting)
- dividing is the opposite of multiplying (and multiplying is the opposite of dividing)

Worked example: Changing the subject

Look at the equation for the number of moles of a substance.

$$\text{number of moles } n = \frac{\text{mass } m}{\text{relative atomic mass } A_r}$$

Change the equation to make mass m the subject.

Solution

Step 1: Multiply both sides by the relative atomic mass A_r.

$$n \times A_r = \frac{m}{A_r} \times A_r$$

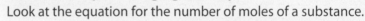

Step 2: Cancel out any repeating variables.

The two A_r variables on the right hand side of the equation can cancel out.

$$n \times A_r = \frac{m}{\cancel{A_r}} \times \cancel{A_r}$$

This give the final equation:

$$m = n \times A_r$$

Synoptic link

You can see examples of changing the subject of equations in the chemical calculations in Chapter 4.

3c Quantities and units
SI Units

When you take a measurement in science you need to include a number *and* a unit. When you do a calculation your answer should also include both a number *and* a unit. There are some special cases where the units cancel but usually they do not.

Everyone doing science, including you, needs to use the **SI system of units**.

Table 2 shows some of the quantities that you will use, along with their units.

Table 2 *Some quantities, and their units, you will meet during your chemistry GCSE*

Quantity	Base Unit
time	second, s
temperature	kelvin, K
amount of substance	mole, mol
energy	Joule, J
pressure	Pascal, Pa
electric potential difference	Volt, V

Figure 2 *Measurements should have units. 3 is not a measurement, but $3\,cm^3$ is*

For example, 1.5 seconds is a *measurement*. The number 1.5 is not a measurement because it does not have a unit.

Some quantities that you *calculate* do not have a unit because they are relative values, for example, relative atomic mass.

Metric prefixes

You can use metric prefixes to show large or small multiples of a particular unit. Adding a prefix to a unit means putting a letter in front of the unit. It shows you that you should multiply your value by a particular power of 10 for it to be shown in an SI unit.

For example, 3 kilometres $= 3\,km = 3 \times 10^3\,m$. To convert the unit from metre to kilometres, a 'k' is put in front of the 'm'.

Most of the prefixes that you will use in science involve multiples of 10^3. However, when dealing with volumes in chemistry you will often deal with decimetres cubed (dm^3), where a decimetre (dm) is one tenth of a metre (or $1 \times 10^{-1}\,m$).

Synoptic link

You can see examples of changing between metres and nanometres in Topic C1.7, and conversions between cm³ and dm³ in Topic C4.9.

Table 3 *Common prefixes you will use in your units*

Prefix	Symbol	Multiplying factor
giga	G	10^9
mega	M	10^6
kilo	k	10^3
deci	d	10^{-1}
centi	c	10^{-2}
milli	μ	10^{-3}
micro	μ	10^{-6}
nano	ν	10^{-9}

Converting between units

It is helpful to use standard form when you are converting between units. To do this, it is best to consider how many of the 'smaller' units are contained within one of the 'bigger' units. For example:

- there are 1 000 000 000 nm in 1 m. So, $1\,\text{nm} = \dfrac{1}{1\,000\,000\,000\,\text{m}}$

$$= 1 \times 10^{-9}\,\text{m}$$

- there are $1000\,\text{cm}^3$ in $1\,\text{dm}^3$. So, $1\,\text{dm}^3 = 1000\,\text{cm}^3$

$$= 1 \times 10^3\,\text{cm}^3.$$

1 How would you read the following expressions as a sentence?
  a The pH of an acid < 7 [1 mark]
  b rate of reaction $\propto$ the concentration of reactant A [1 mark]
  c $22\,\text{dm}^3 \sim 24\,\text{dm}^3$ [1 mark]

2 Here is an equation chemists use to calculate the 'atom economy' of a reaction:

percentage atom economy

$= \dfrac{\text{relative formula mass of the desired product in an equation} \times 100}{\text{sum of relative formula masses of all the reactants in the equation}}$

  a Rearrange the equation above to make its subject the 'relative formula mass of the desired product' in an equation. [1 mark]

  b Rearrange the equation above to make its subject the 'sum of the relative formula masses of all the reactants' in the equation. [1 mark]

3 **a** Express 58 nm in metres (m), using standard form. [1 mark]
  b Express $25.6\,\text{dm}^3$ (decimetres cubed) in cm³, using standard form. [1 mark]

MS4 Graphs

During your GCSE course you will collect data in different types of experiment or investigation. In investigations, the data is collected from a practical where you have changed *one* independent variable and measured its effect on a dependent variable.

4a Collecting data by changing a variable

In many investigations you change one variable (the independent variable) and measure the effect on another variable (the dependent variable). In a fair test, the other variables are kept constant.

For example, you can vary the concentration of sodium chloride in water (independent variable) and measure the effect on the temperature of the boiling point (dependent variable).

A scatter diagram lets you show the relationship between two numerical values.

- The independent variable is plotted on the *x-axis* (horizontal axis).

- The dependent variable is plotted on the *y-axis* (vertical axis).

The line of best fit is a line that goes roughly through the middle of all the points on the scatter graph. The **line of best fit** is drawn so that the points are evenly distributed on either side of the line.

If the gradient of the line of best fit is:

- **positive** it means as the independent variable *increases* the dependent variable *increases*

- **negative** it means as the independent variable *increases* the dependent variable *decreases*

- **zero** it means changing the independent variable has no effect on the dependent variable.

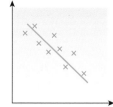 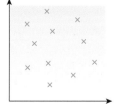

You say that the relationship between the variables is positive or negative, or that there is no relationship.

For example:

- As you increase the concentration of a reactant, the rate of reaction increases.

- As you increase temperature of water, the time it takes sugar to dissolve decreases.

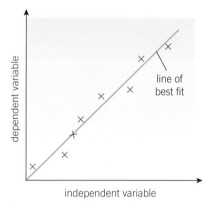

Figure 1 *A scatter graph*

Study tip

Use a transparent ruler to help you draw a straight line of best fit so you make sure that there are the same number of points on either side of the line.

The presence of a relationship does not always mean that changing the independent variable *causes* the change in the dependent variable. In order to claim a causal relationship, you must use science to predict or explain *why* changing one variable affects the other.

4b Graphs and equations

If you are changing one variable and measuring another you are trying to find out about the relationship between them. A straight line graph tells you about the mathematical relationship between variables but there are other things that you can calculate from a graph.

Straight line graphs

The equation of a straight line is $y = mx + c$, where m is the **gradient** and c is the point on the y-axis where the graph intercepts, called the y-intercept.

Straight line graphs that go through the origin (0,0) are special. For these graphs, y is directly proportional to x, and $y = mx$. If two quantities are directly proportional, as one quantity increases, the other quantity increases by the same proportion.

In science, plotting a graph usually means plotting the points then drawing a line of best fit.

When you describe the relationship between two *physical* quantities, you should think about the reason why the graph might (or might not) go through (0,0).

For example, if you are measuring the volume of a gas produced in a reaction over time, when the time = 0 s (at the start of the reaction), the volume of gas produced at that point will be obviously be 0 cm³.

However, if you are measuring the mass of reactants over time, in a reaction that gives off a gas, at time = 0 s, the y-intercept will not be zero but the starting mass of the reactants.

4c Plotting data

When you draw a graph you choose a scale for each axis.

- The scale on the x-axis should be the *same* all the way along the x-axis but it can be *different* to the scale on the y-axis.

- Similarly, the scale on the y-axis should be the *same* all the way along the y-axis but it can be *different* to the scale on the x-axis.

- Each axis should have a label and a unit, such as time in s.

4d Determining the gradient of a straight line

When you are studying rates of reaction you might need to calculate a gradient from a graph of either:

- the amount of reactant as it decreases with time, or

- the amount of product as it increases with time.

For all graphs where the quantity on the *x-axis* is time, the gradient will tell you the *rate of change* of the quantity on the *y-axis* with time.

The gradient is calculated using the equation:

$$\text{gradient} = \frac{\text{change in } y}{\text{change in } x}$$

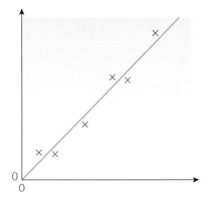

Figure 2 *A line of best fit that passes through the origin*

4e Using tangents

The graphs that you investigate may be curved lines.

To find the rate of reaction at a point **T** on the curve:

● Draw a tangent to the curve. The line should pass through point **T** and have the same slope as the curve at that point.

● Make a right-angled triangle with your line as the hypotenuse. Make sure that the triangle is large enough for you to calculate sensible changes in values.

● Use the triangle to read off the *change in y* and *the change in x*.

● Calculate gradient $= \dfrac{\text{change in } y}{\text{change in } x}$

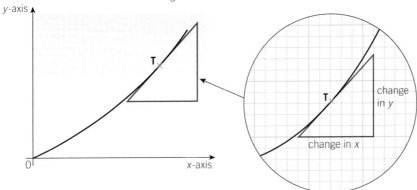

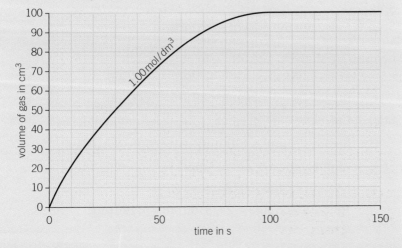

Synoptic link

For a worked example of how to calculate the initial rate of a reaction from a line graph, see in Topic C8.4.

Figure 3 *You find the gradient by drawing a tangent*

1 Sketch a line graph, labelling the *x*- and *y*-axes, that shows:
   a a positive, constant gradient that passes through the origin (0,0). [1 mark]
   b a negative gradient – the gradient decreases as *x* increases. [1 mark]

2 a Write the general equation that describes a straight line on a graph, using the letters *y*, *x*, *m* and *c*. [1 mark]
   b State what letters *m* and *c* represent on the straight line graph. [2 marks]
   c Write the general equation that describes a straight line on a graph that passes through the origin (0,0). [1 mark]

3 Calculate the gradient of the line at **30** seconds.

MS5 Geometry and Trigonometry

Learning objectives

After this topic, you should know how to:

- visualise and represent 2D and 3D forms including two dimensional representations of 3D objects
- calculate areas of rectangles, and surface areas, and volumes of cubes

Synoptic link

For some examples of the 2D and 3D representations of structures and molecules, see Chapter 3 and Topic C10.2, Topic C11.3, Topic C11.4, Topic C15.2, Topic C15.3, and Topic C15.4.

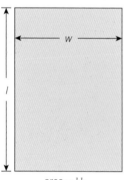

area = hb

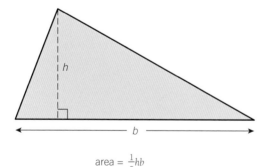

area = $\frac{1}{2}hb$

Figure 2 *Calculating the area of a rectangle and a triangle*

5b Shapes and structures

An important part of chemistry is visualising and representing the shapes and structures of elements and compounds. Throughout this book you will see 2D representations and models of the 3D shapes that make up all substances. Although you are not be expected to reproduce the more complex structural diagrams, you should be able to interpret what given structures represent.

600 ml

Figure 1 *When drawing experimental setup, a cross-section diagram is used instead of a diagram that shows the perspective*

5c Area, surface area, and volumes
Surface area

You should remember the formulae for the area of rectangles and triangles.

- area of a rectangle = base b × height h
- area of a triangle = $\frac{1}{2}$ × base b × height h

The surface area of a 3D object is equal to the total surface area of all its faces. In a cuboid, the areas of any two opposite faces are equal. This allows you to calculate the surface area of the cuboid without having to draw a net.

Worked example: Calculating the surface area of a grain of sodium chloride

A tiny grain of salt is a cuboid, measuring 15 μm × 20 μm × 80 μm. Calculate its surface area.

Solution

Step 1: Calculate the area of each face.

area of face 1 = 15 μm × 20 μm = 300 μm²

area of face 2 = 15 μm × 80 μm = 1200 μm²

area of face 3 = 20 μm × 80 μm = 1600 μm²

Step 2: Calculate the total area of the three different faces.

area = 300 μm² + 1200 μm² + 1600 μm² = 3100 μm²

Step 3: Multiply the answer to **Step 2** by 2 because the opposite sides of a cuboid have equal areas.

total surface area = 2 × 3100 μm² = **6200 μm²**

Volumes

Use this expression to calculate the volume of a cuboid:

volume of cuboid = length l × width w × height h

You can calculate the volume in different units depending on the units of length, width, and height.

Worked example: Volume of a cuboid

Calculate the volume of a ceramic block of length 15 cm, width of 6 cm and depth of 1.5 cm, expressing your answer in cm^3 and m^3.

Solution

Step 1: Calculate the volume using the equation.

volume = length × width × height

$\quad\quad$ = 15 cm × 6 cm × 1.5 cm

$\quad\quad$ = 135 cm^3

$\quad\quad$ = 140 cm^3 (2 s.f.)

Step 2: Convert the measurements to metres.

length = 0.15 m

width = 0.06 m

height = 0.015 m

Step 3: Use the equation to calculate the volume.

volume = length × width × height

$\quad\quad$ = 0.15 m × 0.06 m × 0.015 m

$\quad\quad$ = 0.000135 m^3

$\quad\quad$ = **0.00014 m^3** (2 s.f.)

Figure 3 *What is the surface area of a grain of salt?*

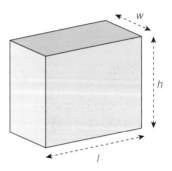

volume = l × h × w

Figure 4 *Calculating the volume of a cuboid*

Synoptic link

For chemists, the concept of surface area to volume ratio (SA : V) is important when explaining the effect of particle size on the rate of a reaction, see in Topic C3.11 and Topic C8.2.

---

1 Look at the 3D model of methane, CH_4, in Figure 5.
   a Use the 3D model to draw a 2D ball and stick model of methane.
   [1 mark]
   b Draw a 3D model of methane. Find out and use the actual
   H – C – H bond angles in CH_4 in your answer. [2 marks]

2 A nanoparticle is made that has a cubic shape of side 20 nm.
   a Calculate the surface area of the nanoparticle cube, in nm^2. [1 mark]
   b Calculate the volume of the nanoparticle cube, in nm^3. [1 mark]
   c Calculate the surface area to volume ratio of the nanoparticle cube. The unit will be 'per nm' (/nm). [1 mark]

Figure 5 *A 3D model of methane*

Working Scientifically

Figure 1 *All around you, everyday, there are many observations you can make. Studying science can give you the understanding to explain and make predictions about some of what you observe*

WS1 Development of scientific thinking

Science works for us all day, every day. Working as a scientist you will have knowledge of the world around you, particularly about the subject you are working with. You will observe the world around you. An enquiring mind will then lead you to start asking questions about what you have observed.

Science usually moves forward by slow steady steps. Each small step is important in its own way. It builds on the body of knowledge that we already have. In this book you can find out about:

- how scientific methods and theories change over time (Topics C1.5, C12.6, C13.1)
- the models that help us to understand theories (Chapter 3)
- the limitations of science, and the personal, social, economic, ethical and environmental issues that arise (Topics C3.12, C9.3, C13.3, C14.1, C14.5, C14.6, C15.6)
- the importance of peer review in publishing scientific results (Topic C14.5)
- evaluating risks in practical work and in technological applications (Topics WS2, C3.12, C14.6, C15.6).

The rest of this section will help you to work scientifically when planning, carrying out, analysing and evaluating your own investigations.

WS2 Experimental skills and strategies

Deciding on what to measure
Variables are quantities that change or can be changed. It helps to know about the following two types of variable when investigating many scientific questions:

A **categoric variable** is one that is best described by a label, usually a word. For example, the type of metal used in an experiment is a categoric variable.

A **continuous variable** is one that you measure, so its value could be any number. For example, temperature, as measured by a thermometer or temperature sensor, is a continuous variable. Continuous variables have values (called quantities). These are found by taking measurements and S.I. units such as grams (g), metres (m), and joules(J) should be used.

Making your data repeatable and reproducible
When you are designing an investigation you must make sure that you, and others, can trust the data you plan to collect. You should ensure that each measurement is **repeatable**. You can do this by getting consistent sets of repeat measurements and taking their mean. You can also have more confidence in your data if similar results are obtained by different investigators using different equipment, making your measurements **reproducible**.

You must also make sure you are measuring the actual thing you want to measure. If you don't, your data can't be used to answer your original question. This seems very obvious, but it is not always easy to set up. You need to make sure that you have controlled as many other variables as you can. Then no-one can say that your investigation, and hence the data you collect and any conclusions drawn from the data, is not **valid**.

How might an independent variable be linked to a dependent variable?

- The **independent variable** is the one you choose to vary in your investigation.

- The **dependent variable** is used to judge the effect of varying the independent variable.

These variables may be linked together. If there is a pattern to be seen (e.g., as one thing gets bigger the other also gets bigger), it may be that:

- changing one has caused the other to change

- the two are related (there is a correlation between them), but one is not necessarily the cause of the other.

Starting an investigation

Scientists use observations to ask questions. You can only ask useful questions if you know something about the observed event. You will not have all of the answers, but you will know enough to start asking the correct questions.

When you are designing an investigation you have to observe carefully which variables are likely to have an effect.

An investigation starts with a question and is followed by a **prediction**, and backed up by scientific reasoning. This forms a **hypothesis** that can be tested against the results of your investigation. You, as the scientist, predict that there is a **relationship** between two variables.

You should think about carrying out a preliminary investigation to find the most suitable range and interval for the independent variable.

Making your investigation safe

Remember that when you design your investigation, you must:

- look for any potential **hazards**

- decide how you will reduce any **risk**.

You will need to write these down in your plan:

- write down your plan

- make a risk assessment

- make a prediction and hypothesis

- draw a blank table ready for the results.

Study tip

Observations, measurements, and predictions backed up by creative thinking and good scientific knowledge can lead to a hypothesis.

Figure 2 *Safety precautions should be appropriate for the risk. Chlorine gas is toxic but you do not need to wear a gas mask when only a small amount of chlorine gas is produced in an investigation carried out in a well-ventilated laboratory or fume cupboard*

Figure 3 *Imagine you wanted to investigate the effect pollution from a chemical factory has on nearby plants. You should choose a control group that is far away enough from the chemical plant to not be affected by the pollution, but close enough to be still experiencing similar environmental conditions*

Study tip

Trial runs will tell you a lot about how your investigation might work out. They should get you to ask yourself:

● do I have the correct conditions?

● have I chosen a sensible range?

● have I got sufficient readings that are close enough together? The minimum number of points to draw a line graph is generally taken as five.

● will I need to repeat my readings?

Study tip

Just because your results show precision it does not mean your results are accurate.

Imagine you carry out an investigation into the energy value of a type of fuel. You get readings of the amount of energy transferred from the burning fuel to the surroundings that are all about the same. This means that your data will have precision, but it doesn't mean that they are necessarily accurate.

Different types of investigation

A **fair test** is one in which only the independent variable affects the dependent variable. All other variables are controlled and kept constant.

This is easy to set up in the laboratory, but almost impossible in fieldwork. Investigations in the environment are not that simple and easy to control. There are complex variables that are changing constantly.

So how can we set up the fieldwork investigations? The best you can do is to make sure that all of the many variables change in much the same way, except for the one you are investigating. For example, if you are monitoring the effects of pollution on plants, they should all be experiencing the same weather, together – even if it is constantly changing.

If you are investigating two variables in a large population then you will need to do a survey. Again, it is impossible to control all of the variables. For example, imagine scientists investigating the effect of a new drug on diabetes. They would have to choose people of the same age and same family history to test. Remember that the larger the sample size tested, the more valid the results will be.

Control groups are used in these investigations to try to make sure that you are measuring the variable that you intend to measure. When investigating the effects of a new drug, the control group will be given a placebo. The control group think they are taking a drug but the placebo does not contain the drug. This way you can control the variable of 'thinking that the drug is working', and separate out the effect of the actual drug.

Designing an investigation
Accuracy

Your investigation must provide **accurate** data. Accurate data is essential if your results are going to have any meaning.

How do you know if you have accurate data?

It is very difficult to be certain. **Accurate results are very close to the true value.** However, it is not always possible to know what the true value is.

Sometimes you can calculate a theoretical value and check it against the experimental evidence. Close agreement between these two values could indicate accurate data.

You can draw a graph of your results and see how close each result is to the line of best fit.

Try repeating your measurements and check the spread or range within sets of repeat data. Large differences in a repeated measurement suggest inaccuracy. Or try again with a different measuring instrument and see if you get the same readings.

Precision

Your investigation must provide data with sufficient **precision** (i.e., **close agreement within sets of repeat measurements**). If it doesn't then you will not be able to make a valid conclusion.

Precision versus accuracy

Imagine measuring the temperature after a set time when a fuel is used to heat a fixed volume of water. Two students repeated this experiment, four times each. Their results are marked on the thermometer scales in Figure 4:

- A **precise** set of results is grouped closely together.
- An accurate set of results will have a mean (average) close to the true value.

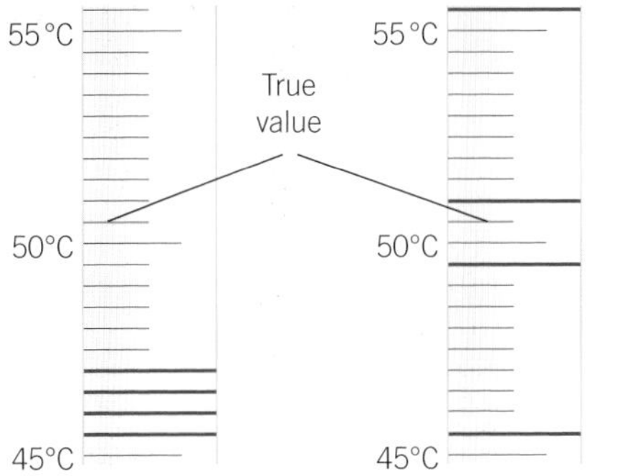

How do you get precise, repeatable data?

You have to repeat your tests as often as necessary to improve repeatability.

You have to repeat your tests in exactly the same way each time.

You should use measuring instruments that have the appropriate scale divisions needed for a particular investigation. Smaller scale divisions have better resolution.

Making measurements
Using measuring instruments

There will always be some degree of uncertainty in any measurements made (WS3). You cannot expect perfect results. When you choose an instrument you need to know that it will give you the accuracy that you want (i.e., it will give you a true reading). You also need to know how to use an instrument properly.

Some instruments have smaller scale divisions than others. Instruments that measure the same thing, such as mass, can have different resolutions. The resolution of an instrument refers to the smallest change in a value that can be detected (e.g., a ruler with centimetre increments compared to a ruler with millimetre increments). Choosing an instrument with an inappropriate resolution can cause you to miss important data or make silly conclusions.

But selecting measuring instruments with high resolution might not be appropriate in some cases where the degree of uncertainty in a measurement is high, for example, judging when an 'X' under a conical flask disappears (see Topic C8.1). In this case, a stopwatch measuring to one hundredths of a second is not going to improve the accuracy of the data collected.

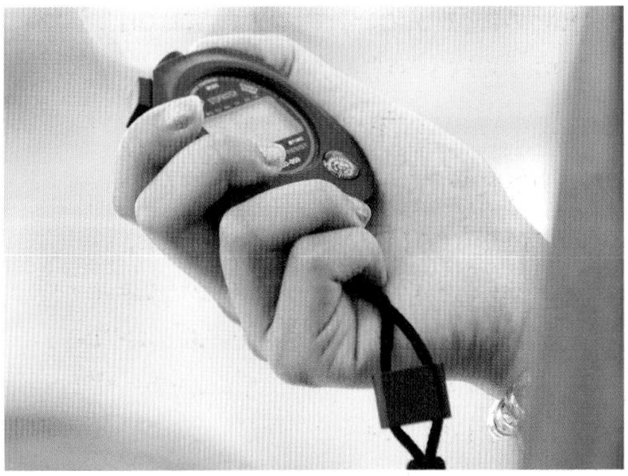

Figure 4 *The green line shows the true value and the pink lines show the readings two different groups of students measured. Precise results are not necessarily accurate results*

Figure 5 *Despite the fact that a stopwatch has a high resolution, it is not always the most appropriate instrument to use for measuring time*

WS3 Analysis and evaluation

Errors

Even when an instrument is used correctly, the results can still show differences. Results will differ because of a **random error**. This can be a result of poor measurements being made. It could also be due to not carrying out the method consistently in each test. Random errors are minimised by taking the mean of precise repeat readings, looking out for any outliers (measurements that differ significantly from the others within a set of repeats) to check again, or omit from calculations of the mean.

The error may be a systematic error. This means that the method or measurement was carried out consistently incorrectly so that an error was being repeated. An example could be a balance that is not set at zero correctly. Systematic errors will be consistently above, or below, the accurate value.

Presenting data
Tables
Tables are really good for recording your results quickly and clearly as you are carrying out an investigation. You should design your table before you start your investigation.

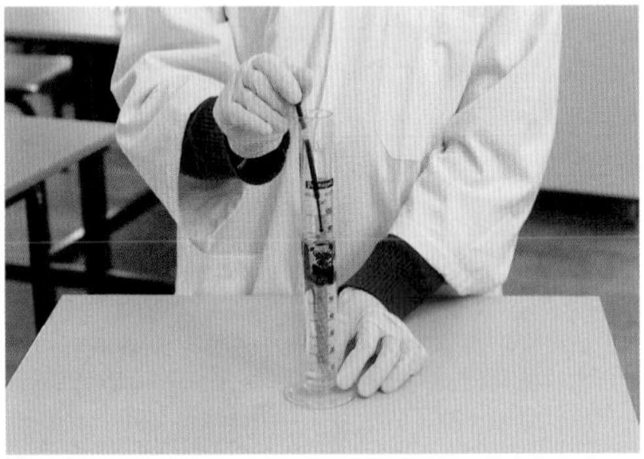

The range of the data
Pick out the maximum and the minimum values and you have the **range**. You should always quote these two numbers when asked for a range. For example, the range is between the lowest value in a data set, and the highest value. *Don't forget to include the units.*

The mean of the data
Add up all of the measurements and divide by how many there are. As seen in Topic WS2, you can ignore outliers in a set of repeat readings when calculating the mean, if found to be the result of poor measurement.

Bar charts
If you have a categoric independent variable and a continuous dependent variable then you should use a **bar chart**.

Line graphs
If you have a continuous independent and a continuous dependent variable then use a **line graph**.

Scatter graphs
These are used in much the same way as a line graph, but you might not expect to be able to draw such a clear line of best fit. For example, to find out if the melting point of an element is related to its density you might draw a scatter graph of your results.

Using data to draw conclusions
Identifying patterns and relationships
Now you have a bar chart or a line graph of your results you can begin looking for patterns. You must have an open mind at this point.

Firstly, there could still be some anomalous results. You might not have picked these out earlier. How do you spot an anomaly? It must be a significant distance away from the pattern, not just within normal variation.

A line of best fit will help to identify any anomalies at this stage. Ask yourself – 'do the anomalies represent something important or were they just a mistake?'

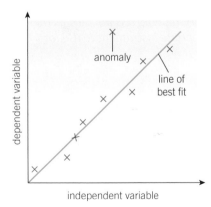

Figure 6 *How you record your results will depend upon the type of measurements you are taking*

Figure 7 *A line of best fit can help to identify anomalies*

Secondly, remember a line of best fit can be a straight line or it can be a curve – you have to decide from your results.

The line of best fit will also lead you into thinking what the relationship is between your two variables. You need to consider whether the points you have plotted show a linear relationship. If so, you can draw a straight line of best fit on your graph (with as many points above the line as below it, producing a 'mean' line). Then consider if this line has a positive or negative gradient.

A **directly proportional** relationship is shown by a positive straight line that goes through the origin (0, 0).

Your results might also show a curved line of best fit. These can be predictable, complex or very complex. Carrying out more tests with a smaller interval near the area where a line changes its gradient will help reduce the error in drawing the line (in this case a curve) of best fit.

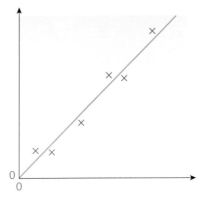

Figure 8 *When a straight line of best fit goes through the origin (0, 0) the relationship between the variables is directly proportional*

Drawing conclusions

Your graphs are designed to show the relationship between your two chosen variables. You need to consider what that relationship means for your conclusion. You must also take into account the repeatability and the reproducibility of the data you are considering.

You will continue to have an open mind about your conclusion.

You will have made a prediction. This could be supported by your results, it might not be supported, or it could be partly supported. It might suggest some other hypothesis to you.

You must be willing to think carefully about your results. Remember it is quite rare for a set of results to completely support a prediction or be completely repeatable.

Look for possible links between variables, remembering that a positive relationship does not always mean a causal link between the two variables.

Your conclusion must go no further than the evidence that you have. Any patterns you spot are only strictly valid in the range of values you tested. Further tests are needed to check whether the pattern continues beyond this range.

The purpose of the prediction was to test a hypothesis. The hypothesis can:

- be supported,
- be refuted
- lead to another hypothesis.

You have to decide which it is on the evidence available.

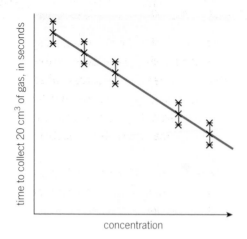

Figure 9 *Indicating levels of uncertainty. These are all the results of a group that chose five different concentrations to test and repeated each test three times.*

Making estimates of uncertainty

You can use the range of a set of repeat measurements about their mean to estimate the degree of uncertainty in the data collected.

For example, in a test that looked at the effect of concentration on the rate of reaction between calcium carbonate and acid, a student got these results:

40 cm³ of acid and 10 cm³ of water gave off 20 cm³ of carbon dioxide in 45 s (1ˢᵗ attempt), 49 s (2ⁿᵈ attempt), 44 s (3ʳᵈ attempt) and 48 s (4ᵗʰ attempt).

The mean result = (45 + 49 + 44 + 48) ÷ 4 = 46.5 s

The range of the repeats is 44 s to 49 s = 5 s

So, a reasonable estimate of the uncertainty in the mean value would be half of the range.

In this case, we could say the time taken was 46.5 s plus or minus ±2.5 s.

You can include a final column in your table of results to record the 'estimated uncertainty' in your mean measurements.

The level of uncertainty can also be shown when plotting your results on a graph (Figure 9).

As well as this, there will be some uncertainty associated with readings from any measuring instrument. You can usually take this as:

● half the smallest scale division. For example, 0.05 cm³ for each burette reading, or

● on a digital instrument, half the last figure shown on its display. For example, on a balance reading to 0.01 g the uncertainty would be ±0.005 g (Topic C1.2).

Anomalous results

Anomalies (or outliers) are results that are clearly out of line compared with others. They are not those that are due to the natural variation that you get from any measurement. Anomalous results should be looked at carefully. There might be a very interesting reason why they are so different.

If anomalies can be identified while you are doing an investigation, then it is best to repeat that part of the investigation. If you find that an anomaly is due to poor measurement, then it should be ignored.

Evaluation

If you are still uncertain about a conclusion, it might be down to the repeatability, reproducibility and uncertainty in your measurements. You could check reproducibility by: looking for other similar work on the Internet or from others in your class, getting somebody else, using different equipment, to redo your investigation (this occurs in peer review of data presented in articles in scientific journals), trying an alternative method to see if it results in you reaching the same conclusion.

When suggesting improvement that could be made in your investigation, always give your reasoning.

Study tip

The method chosen for an investigation can only be evaluated as being valid if it actually collects data that can answer your original question. The data should be repeatable and reproducible, and the control variables should have been kept constant (or taken into account if they couldn't be directly manipulated).

Synoptic link

See the Maths Skills section to learn how to use SI units, prefixes and powers of ten for orders of magnitude, significant figures, and scientific quantities.

Glossary

Accurate A measurement is considered accurate if it is judged to be close to the true value.

Acid When dissolved in water, its solution has a pH value less than 7. Acids are proton (H^+ ion) donors.

Activation energy The minimum energy needed for a reaction to take place.

Alkali metal Elements in Group 1 of the periodic table.

Alkali Its solution has a pH value more than 7.

Alkane Saturated hydrocarbon with the general formula C_nH_{2n+2}, for example, methane, ethane, and propane.

Alkene Unsaturated hydrocarbon which contains a carbon–carbon double bond. Its general formula is C_nH_{2n}, for example, ethene, C_2H_4.

Alloy A mixture of two or more elements, at least one of which is a metal.

Anhydrous Describes a substance that does not contain water.

Anode The positive electrode in electrolysis.

Anomalies Results that do not match the pattern seen in the other data collected or are well outside the range of other repeat readings (outliers).

Aqueous solution The mixture made by adding a soluble substance to water.

Atmosphere The relatively thin layer of gases that surround planet Earth.

Atom The smallest part of an element that can still be recognised as that element.

Atom economy A measure of the amount of starting materials that end up as useful products.

Atomic number The number of protons (which equals the number of electrons) in an atom. It is sometimes called the proton number.

Avagadro constant The number of atoms, molecules, or ions in a mole of any substance (i.e., 6.02×10^{23} per mol).

Balanced symbol equation A symbol equation in which there are equal numbers of each type of atom on either side of the equation.

Base The oxide, hydroxide, or carbonate of a metal that will react with an acid, forming a salt as one of the products. (If a base dissolves in water it is called an alkali). Bases are proton (H^+ ion) acceptors.

Biodegradable Materials that can be broken down by microorganisms.

Biofuel Fuel made from animal or plant products.

Blast furnace The huge reaction vessels used in industry to extract iron from its ore.

Bond energy The energy required to break a specific chemical bond.

Burette A long glass tube with a tap at one end and markings to show volumes of liquid; used to add precisely known volumes of liquids to a solution in a conical flask below it.

Carbon footprint The total amount of carbon dioxide and other greenhouse gases emitted over the full life cycle of a product, service or event.

Carbon steel Alloy of iron containing controlled, small amounts of carbon.

Categoric variable Categoric variables have values that are labels. For example types of material.

Catalyst A substance that speeds up a chemical reaction by providing a different pathway for the reaction that has a lower activation energy. The catalyst is chemically unchanged at the end of the reaction.

Catalytic converter Fitted to exhausts of vehicles to reduce pollutants released.

Cathode The negative electrode in electrolysis.

Ceramics Materials made by heating clay, or other compounds, to high temperatures (called firing) to make hard, but often brittle, materials, which make excellent electrical insulators.

Chromatography The process whereby small amounts of dissolved substances are separated by running a solvent along a material such as absorbent paper.

Climate change The change in global weather patterns that could be caused by excess levels of greenhouse gases in the atmosphere.

Closed system A system in which no matter enters or leaves.

Collision theory An explanation of chemical reactions in terms of reacting particles colliding with sufficient energy for a reaction to take place.

Composites Materials made of two or more different materials, containing a matrix or binder surrounding and binding together fibres or fragments of another material which acts as the reinforcement.

Compound A substance made when two or more elements are chemically bonded together.

Continuous variable Can have values (called a quantity) that can be given by measurement (for example, mass, volume, temperature, etc.).

Continuous data Data that can take any value.

Control group If an experiment is to determine the effect of changing a single variable, a control is often set up in which the independent variable is not changed, thus enabling a comparison to be made. If the investigation is of the survey type a control group is usually established to serve the same purpose.

Control variable A variable which may, in addition to the independent variable, affect the outcome of the investigation and therefore has to be kept constant or at least monitored.

Covalent bond The bond between two atoms that share one or more pairs of electrons.

Covalent bonding The attraction between two atoms that share one or more pairs of electrons.

Cracking The reaction used in the oil industry to break down large hydrocarbons into smaller, more useful ones.

Data Information, either qualitative or quantitative, that has been collected.

Delocalised electron Bonding electron that is no longer associated with any one particular atom.

Dependent variable The variable for which the value is measured for each and every change in the independent variable.

Diffusion The automatic mixing of liquids and gases as a result of the random motion of their particles.

Directly proportional A relationship that, when drawn on a line graph, shows a positive linear relationship that crosses through the origin.

Discrete data Data that can only take certain values.

Displacement reaction A reaction in which a more reactive element takes the place of a less reactive element in one of its compounds or in solution.

Distillation Separation of a liquid from a mixture by evaporation followed by condensation.

Dot and cross diagram A drawing to show only the arrangement of the outer shell electrons of the atoms or ions in a substance.

DNA (deoxyribonucleic acid) A large organic molecule that encodes genetic instructions for the development and functioning of living organisms and viruses.

Double bond A covalent bond made by the sharing of two pairs of electrons.

Electrical (chemical) cells Contain chemicals that react to produce electricity.

Electrolysis The breakdown of a substance containing ions by electricity.

Electrolyte A liquid, containing free-moving ions, which is broken down by electricity in the process of electrolysis.

Electron A tiny particle with a negative charge. Electrons orbit the nucleus of atoms or ions in shells.

Electronic structure A set of numbers to show the arrangement of electrons in their shells (or energy levels).

Element A substance made up of only one type of atom. An element cannot be broken down chemically into any simpler substance.

End point The point in a titration where the reaction is complete and titration should stop.

Endothermic A reaction that takes in energy from the surroundings.

Equilibrium The point in a reversible reaction at which the forward and backward rates of reaction are the same. Therefore, the amounts of substances present in the reacting mixture remain constant.

Exothermic A reaction that transfers energy to the surroundings.

Fair test A fair test is one in which only the independent variable has been allowed to affect the dependent variable.

Fermentation The reaction in which the enzymes in yeast turn glucose into ethanol and carbon dioxide.

Filtration The technique used to separate substances that are insoluble in a particular solvent from those that are soluble.

Flame emission spectroscopy A method of instrumental analysis in which the light given off when a sample is placed in a flame produces characteristic line spectra to identify and measure the concentration of metal ions in the sample.

Flammable Easily ignited and capable of burning rapidly.

Formulation a mixture that has been designed as a useful product.

Fraction Hydrocarbons with similar boiling points separated from crude oil.

Fractional distillation A way to separate liquids from a mixture of liquids by boiling off the substances at different temperatures, then condensing and collecting the liquids.

Fuel cells Sources of electricity that are supplied by an external source of fuel.

Fullerene Form of the element carbon that can exist as large cage-like structures, based on hexagonal rings of carbon atoms.

Functional group An atom or group of atoms that give organic compounds their characteristic reactions.

Galvanised Iron or steel objects that have been protected from rusting by a thin layer of zinc metal at their surface.

Giant covalent structure A huge 3D network of covalently bonded atoms.

Giant lattice A huge 3D network of atoms or ions.

Giant structure See Giant lattice.

Gradient A measure of the slope of a straight line on a graph.

Group All the elements in the columns (labelled 1 to 7 and 0) in the periodic table.

Half equation An equation that describes reduction (gain of electrons) or oxidation (loss of electrons).

Halogens The elements found in Group 7 of the periodic table.

Hazard A hazard is something (e.g., an object, a property of a substance or an activity) that can cause harm.

Homologous series A group of related organic compounds that have the same functional group.

LIEBIG CONDENSER

Hydrated Describes a substance that contains water in its crystals.

Hydrocarbon A compound containing only hydrogen and carbon.

Hypothesis A proposal intended to explain certain facts or observations.

Incomplete combustion When a fuel burns in insufficient oxygen, producing carbon monoxide as a toxic product.

Independent variable The variable for which values are changed or selected by the investigator.

Inert Unreactive.

Intermolecular forces The attraction between the individual molecules in a covalently bonded substance.

Ion A charged particle produced by the loss or gain of electrons.

Ionic bond The electrostatic force of attraction between positively and negatively charged ions.

Ionic equation An equation that shows only those ions or atoms that change in a chemical reaction.

Isotope Atoms that have the same number of protons but different number of neutrons, i.e., they have the same atomic number but different mass numbers.

Law of the conservation of energy The total mass of the products formed in a reaction is equal to the total mass of the reactants.

Le Châtelier's Principle When a change in conditions is introduced to a system at equilibrium, the position of equilibrium shifts so as to cancel out the change.

Life cycle assessment Carried out to assess the environmental impact of products, processes or services at different stages in their life cycle.

Line graph Used when both variables are continuous. The line should normally be a line of best fit, and may be straight or a smooth curve.

Line of best fit A straight line that represents the general trend of data. An equal number of data points should be above and below the line of best fit.

Mass number The number of protons plus neutrons in the nucleus of an atom.

Mean The arithmetical average of a series of numbers.

Mixture When some elements or compounds are mixed together and intermingle but do not react together (i.e. no new substance is made). A mixture is not a pure substance.

Mole The amount of substance in the relative atomic or formula mass of a substance in grams.

Molecular formula The chemical formula that shows the actual numbers of atoms in a particular molecule.

Monomers Small reactive molecules that react together in repeating sequences to form a very large molecule (a polymer).

Nanoscience The study of very tiny particles or structures between 1 and 100 nanometres in size – where 1 nanometre = 10^{-9} metres.

Neutral A solution with a pH value of 7 which is neither acidic nor alkaline. Alternatively, something that carries no overall electrical charge.

Neutralisation The chemical reaction of an acid with a base in which a salt and water are formed. If the base is a carbonate or hydrogen carbonate, carbon dioxide is also produced in the reaction.

Neutron A dense particle found in the nucleus of an atom. It is electrically neutral, carrying no charge.

Noble gases The very unreactive gases found in Group 0 of the periodic table. Their atoms have very stable electronic structures.

Non-renewable Something which cannot be replaced once it is used up.

Nucleus (of an atom) The very small and dense central part of an atom that contains protons and neutrons.

Order of magnitude A comparison of the size of values. Two values are the same order of magnitude if their difference in size is small in comparison to other values being compared.

Ore Ore is rock which contains enough metal to make it economically worthwhile to extract the metal.

Oxidation The reaction when oxygen is added to a substance / or when electrons are lost.

Oxidised A reaction where oxygen is added to a substance / or when electrons are lost from a substance.

Particulate Small solid particle given off from motor vehicles as a result of incomplete combustion of its fuel.

Percentage yield The actual mass of product collected in a reaction divided by the maximum mass that could have been formed in theory, multiplied by 100.

Periodic table An arrangement of elements in the order of their atomic numbers, forming groups and periods.

pH A number which shows how strongly acidic or alkaline a solution is.

Pipette A glass tube used to measure accurate volumes of liquids.

Polymer A substance made from very large molecules made up of many repeating units.

Precipitate An insoluble solid formed by a reaction taking place in solution.

Precise A precise measurement is one in which there is very little spread about the mean value. Precision depends only on the extent of random errors – it gives no indication of how close results are to the true (accurate) value.

Prediction A forecast or statement about the way something will happen in the future.

Product A substance made as a result of a chemical reaction.

Proton A tiny positive particle found inside the nucleus of an atom.

Qualitative data Data that is descriptive or categorical.

Quantitative data Data that is numerical or a measurement.

Range The maximum and minimum values of the independent or dependent variables.

Ratio A way of comparing two or more quantities, showing how many times one quantity is contained within the other.

Reactant A substance we start with before a chemical reaction takes place.

Reaction profile The relative difference in the energy of reactants and products.

Reactivity series A list of elements in order of their reactivity.

Recycle The process in which waste materials are processed to be used again.

Reduction A reaction in which oxygen is removed or electrons are gained.

Relationship The link between the variables that were investigated.

Relative atomic mass A_r The average mass of the atoms of an element compared with carbon-12 (which is given a mass of exactly 12). The average mass must take into account the proportions of the naturally occurring isotopes of the element.

Relative formula mass M_r The total of the relative atomic masses, added up in the ratio shown in the chemical formula, of a substance.

Repeatable A measurement is repeatable if the original experimenter repeats the investigation using the same method and equipment and obtains the same or precise results.

Reproducible A measurement is reproducible if the investigation is repeated by another person, using different equipment and the same results are obtained.

Respiration The process by which food molecules are broken down to release energy for the cells.

Reversible reaction A reaction in which the products can re-form the reactants.

R_f (retention factor) A measurement from chromatography: it is the distance a spot of substance has been carried above the baseline divided by the distance of the solvent front.

Risk The likelihood that a hazard will actually cause harm.

Rusting The corrosion of iron.

Sacrificial protection An effective way to prevent rusting whereby a metal more reactive than iron (such as zinc or magnesium) is attached to or coated on an object.

Salt A salt is a compound formed when some or all of the hydrogen in an acid is replaced by a metal.

Saturated hydrocarbon Describes a hydrocarbon with only single bonds between its carbon atoms. This means that it contains as many hydrogen atoms as possible in each molecule.

Shell An area in an atom, around its nucleus, where electrons are found.

SI system of units A system of units for physical quantities that are considered the standard units.

Significant figures (s.f.) The important digits within a number. All non-zero digits are significant. Zeros may be significant if followed by another non-zero digit.

Stainless steel A chromium-nickel alloy of steel which does not rust.

Standard form A way of displaying large and small numbers.

State symbol The abbreviations used in balanced symbol equations to show if reactants and products are solid (s), liquid (l), gas (g) or dissolved in water (aq).

Strong acids These acids completely ionise in solution and have a high concentration of $H^+(aq)$ ions in solution.

Thermal decomposition The breakdown of a compound by heating it.

Thermosetting polymer Polymer that can form extensive cross-linking between chains, resulting in rigid materials which are heat-resistant.

Thermosoftening polymer Polymer that forms plastics which can be softened by heating, then remoulded into different shapes as they cool down and set.

Titration A method for measuring the volumes of two solutions that react together.

Transition element Element from the central block of the periodic table.

Universal indicator A mixture of indicators that can change through a range of colours to show how strongly acidic or alkaline liquids and solutions are.

Unsaturated hydrocarbon A hydrocarbon whose molecules contains at least one carbon–carbon double bond.

Valid Suitability of the investigative procedure to answer the question being asked.

Variable Physical, chemical or biological quantity or characteristic.

Viscosity The resistance of a liquid to flowing or pouring; a liquid's 'thickness'.

Weak acids Acids that do not ionise completely in aqueous solutions.

Word equation A way of describing what happens in a chemical reaction by showing the names of all reactants and the products they form.

Index

Appendix 1: the periodic table

key

relative atomic mass
atomic symbol
name
atomic (proton) number

1	hydrogen
H	1

1	2											3	4	5	6	7	0
																	4 **He** helium 2
7 **Li** lithium 3	9 **Be** beryllium 4											11 **B** boron 5	12 **C** carbon 6	14 **N** nitrogen 7	16 **O** oxygen 8	19 **F** fluorine 9	20 **Ne** neon 10
23 **Na** sodium 11	24 **Mg** magnesium 12											27 **Al** aluminium 13	28 **Si** silicon 14	31 **P** phosphorus 15	32 **S** sulfur 16	35.5 **Cl** chlorine 17	40 **Ar** argon 18
39 **K** potassium 19	40 **Ca** calcium 20	45 **Sc** scandium 21	48 **Ti** titanium 22	51 **V** vanadium 23	52 **Cr** chromium 24	55 **Mn** manganese 25	56 **Fe** iron 26	59 **Co** cobalt 27	59 **Ni** nickel 28	63.5 **Cu** copper 29	65 **Zn** zinc 30	70 **Ga** gallium 31	73 **Ge** germanium 32	75 **As** arsenic 33	79 **Se** selenium 34	80 **Br** bromine 35	84 **Kr** krypton 36
85 **Rb** rubidium 37	88 **Sr** strontium 38	89 **Y** yttrium 39	91 **Zr** zirconium 40	93 **Nb** niobium 41	96 **Mo** molybdenum 42	[98] **Tc** technetium 43	101 **Ru** ruthenium 44	103 **Rh** rhodium 45	106 **Pa** palladium 46	108 **Ag** silver 47	112 **Cd** cadmium 48	115 **In** indium 49	119 **Sn** tin 50	122 **Sb** antimony 51	128 **Te** tellurium 52	127 **I** iodine 53	131 **Xe** xenon 54
133 **Cs** caesium 55	137 **Ba** barium 56	139 **La*** lanthanum 57	178 **Hf** hafnium 72	181 **Ta** tantalum 73	184 **W** tungsten 74	186 **Re** rhenium 75	190 **Os** osmium 76	192 **Ir** iridium 77	195 **Pt** platinum 78	197 **Au** gold 79	201 **Hg** mercury 80	204 **Tl** thallium 81	207 **Pb** lead 82	209 **Bi** bismuth 83	[209] **Po** polonium 84	[210] **At** astatine 85	[222] **Rn** radon 86
[223] **Fr** francium 87	[226] **Ra** radium 88	[227] **Ac*** actinium 89	[261] **Rf** rutherfordium 104	[262] **Db** dubnium 105	[266] **Sg** seaborgium 106	[264] **Bh** bohrium 107	[277] **Hs** hassium 108	[268] **Mt** meitnerium 109	[271] **Ds** darmstadtium 110	[272] **Rg** roentgenium 111	[285] **Cn** copernicium 112	[286] **Uut** ununtrium 113	[289] **Fl** flerovium 114	[289] **Uup** ununpentium 115	[293] **Lv** livermorium 116	[294] **Uus** ununseptium 117	[294] **Uuo** ununoctium 118

*The lanthanides (atomic numbers 58–71) and the actinides (atomic numbers 90–103) have been omitted.

Relative atomic masses for **Cu** an **Cl** have not been rounded to the nearest whole number.